INTRODUCTION TO
BIOTECHNOLOGY

INTRODUCTION TO BIOTECHNOLOGY

Ravi Pathak

ATLANTIC
PUBLISHERS & DISTRIBUTORS (P) LTD

Published by

ATLANTIC

PUBLISHERS & DISTRIBUTORS (P) LTD

B-2, Vishal Enclave, Opp. Rajouri Garden,
New Delhi-110027
Phones : 25413460, 25429987, 25466842

Sales Office
7/22, Ansari Road, Darya Ganj,
New Delhi-110002
Phones : 23273880, 23275880, 23280451
Fax : 91-11-23285873
web : www.atlanticbooks.com
e-mail : info@atlanticbooks.com

ISBN 81-269-0598-0

Printed in India
at Syndicate Binder, Noida

Preface

Due to enormous progress in the field of Biology, Biotechnology has shown spectacular development and widens its webs all around. It is gaining importance in the modern world and is often quoted as the next big thing after information technology, owing to its benefits to man. It has enabled the organisms to become more resistant to disease, influenced the rate of fruit ripening and has increased productivity of crops, thereby solving the world food shortages.

The present book is an attempt to furnish precise, illustrative, to the point information about techniques in Biotechnology. With the balanced coverage of basic molecular biology, historical developments and contemporary applications, the book describes in detail the processes and methods used to manipulate living organisms or the substances and products from these organisms for medical, agricultural and industrial purposes. It acquaints the readers with genetic engineering, bioinformatics, animal and plant biotechnology, environmental biotechnology, bioethics and biosafety. As an aid to the smooth reading of the text, a glossary of difficult and obscure words and terms has been appended to the book. In addition, a bibliography has also been provided which will facilitate further reference.

Introduction to Biotechnology stands apart from other entries in the market for its ability to make the fundamentals of Biotechnology easily accessible and relevant to both the beginners as well as advanced learners. It is hoped that it will prove equally useful to the teachers of the subject and the general readers interested in biotechnology and keen to know the latest developments, methods and applications in this arena.

Suggestions and comments from readers are welcome, which will in fact strengthen my knowledge and help me to gear up for the further improvement of the contents of this book.

Ravi Pathak

Table of Contents

1 Biotechnology: The Genesis of Revolution

- Introduction
- Historical Background
- Business of Biotechnology
- Boom or Bust

Introduction

Man has always been intrigued by nature. It is this intrigue that has driven him to investigate and discover life. Biology is one area of science that has traditionally been the source of inspiration for individuals from diverse areas, be it Leonardo da Vinci who sketched anatomical details of humans with amazing precision or the Wright brothers, who tried to mimic the flight of a bird and designed the first flying machine. If you need a quick insight into this fascinating area, imagine yourself as being built and managed by molecules. It is due to the dynamics of these particles that we grew from microscopic fertilized eggs into functioning human beings.

Molecules are nothing special. They are combinations of six elements that occur very naturally in nature, namely carbon, hydrogen, nitrogen, oxygen, phosphorous and sulphur. All living beings are made up of combinations of these molecules. In a sense, every living being functions fundamentally in the same way. When it comes to molecules, there is really no major difference between a whale and a bacterium. These molecules perform various intricate tasks by means of various chemical reactions under rigidly defined conditions. Biology, as a subject, seeks to understand how this is performed and controlled. All along, this study has been a challenge for mankind.

The Industrial Revolution demonstrated the power of technology to human beings. It made seemingly impossible tasks possible and today, man is on the threshold of conquering the stars, thanks to technology. It is thus imperative that technology plays a major role in the challenge that biology has had encountered all along. That is truly the genesis of biotechnology. Hence, biotechnology can be defined as the fusion of technology with the fundamentals of biology. It can be defined as the use of such biological forms as the living organisms to make or process different products for better living, an example of which is genetic engineering.

At the outset, biotechnology is a difficult subject to get into one's head. It is vast, technical subject and changes at an amazing pace. While there is an ever-increasing curiosity in the minds of people about this subject, lack of a comprehensible language makes it even more difficult to pursue.

Biotechnology operates at the molecular level of life, where most boundaries between species disappear. What makes it so fascinating and powerful is its ability to synchronise the complex waltzes among molecules that make every living being what it is. At a more general level, biotechnology is an umbrella term that covers various techniques that use the properties of living things to make products and provide services. What has intrigued most people is the rapid pace with which biotechnology has taken over biology and allied subjects. It is worth mentioning that scientists in nondescript laboratories are doing things that are changing the face of life. Gene cloning is a case in point.

The fact remains that, like everything else, the skills of biotechnology have been developed from whatever knowledge we possessed before. The feeling of missing out on what it is all about is similar to reading a book from halfway through. In order to truly make sense of it, one has to start from the very beginning and move one step at a time.

As is typical of new technologies affecting the society at large, biotechnology has its own share of allies and enemies. The battle lines are clearly drawn between these warring factions. Such issues are vital with regard to the technology. Another important aspect is that this science is largely in its state of infancy in our country. Then there is the issue of intellectual property rights, or plainly speaking, who owns the information that we generate using biotechnology.

The following chapters will attempt at covering some of these fundamental questions that biotechnology has raised. To know the future, one has to understand history. That is precisely what the subsequent sections will try to unravel.

Historical Background

Most of the time, making discoveries in biology has been nothing but a refashioning of pre-existing concepts and trends. In most cases, the impact of such innovations or new discoveries has not been immediate. It has been an arduous task in compiling subject-specific information and putting it in conceivable text.

This is true for all the branches of science, and more so, in case of biotechnology, since it requires an amalgamation of two diverse fields. To talk about biology's contribution to the same, one has to look back to 1665 when, for the very first time, Dr. Robert Hooke published a review of a few of his observations he made while peering down a microscope. He coined the word "cell". Ten years after this, a skilful lens grinder named Anton van Leeuwenhoek made history by designing microscopes that could magnify objects to more than 200 times their size.

The Dutch lens grinder, Anton van Leeuwenhoek, looked at extremely small life forms and called them "animalcules".

During the 1700s, many other scientists used the microscope to study cells of various organisms. However, it was left to two German biologists, Matthias Schlieden and Theodor Schwann, to come up with a concrete theory on cells, which for the first time proved, beyond doubt that all organisms are made of cells. The next question that bothered scientists was how cells came into existence and what is it that makes different organisms so diverse. This challenge was taken up by Charles Darwin and Gregor Mendel. Their epoch-making contributions on the concepts of origin of life and the theory of inheritance laid the foundations for modern biology. Darwin made evolution comprehensible on the basis of his theory of Natural Selection in 1859. At around the same time, Mendel, an Austrian monk who was fascinated by biology, meticulously studied the mode of transmission of heritable traits in peas. His work, unlike Darwin's, was recognised after his death by the rediscovery of his work by two scientists, Correns and de Vries, thirty years later.

Morgan was yet another scientist who revolutionised the area of genetics by establishing *Drosophila melanogaster* (fruit fly) as a model system to study various concepts of genetic traits and chromosomal abnormalities.

The seminal works that contributed to the growth of biotechnology as a subject are enlisted in Table 1.1.

TABLE 1.1 History of Biotechnology

Year	Summary of Discovery
1665	Robert Hooke observes tiny microscopic structures under the first microscope and coins the word "cell".
1830	Scottish physician Robert Brown identifies an opaque area in plant cells and calls it "nucleus".
1857	Biologists first observe small rod-like bodies in dividing cells. These are called chromosomes (coloured bodies) based on their properties of absorbing certain coloured dyes.
1869	Swiss biologist Johann Freidrich Miescher isolates nuclein, now called DNA. Its structure is unravelled more than a decade later by Watson and Crick.

(Continued)

TABLE 1.1 *(Continued)*

Year	Summary of Discovery
1871	Deoxyribonucleic acid (DNA) was first identified in the sperm of the Rhine River trout. However, its exact role could not be determined.
1879	German biochemist Albrecht Kossel discovers the composition of nucleic acid: adenine, thymine, cytosine, guanine, and uracil.
1882	Belgian biologist Eduard van Beneden discovers that every species characteristically has a specific number of chromosomes.
1900	Gregor Mendel's classic experiments in plant genetics and heredity, published 35 years earlier, are rediscovered by Correns and de Vries.
1902	Walter S. Sutton reports, after studying grasshopper sperm, that each chromosome pairs with another physically similar one and that they separate during meiosis. The individual members of each pair are divided among the cells.
1907	Tissue culture techniques developed.
1909	Danish botanist Wilhelm Ludwig Johannsen coins the word "genes".
1919	Thomas Hunt Morgan identifies XY male and XX female chromosomes and suggests that some traits are sex-linked. Later he demonstrated the mechanism of mutation of genes.
1930	Electron microscope invented, enabling researchers to see viruses. With better contrast and staining agents, macromolecules like DNA are eventually exposed.
1937	Invention of electrophoresis (use of an electric field to move charged particles dispersed in liquid) by Arne Tiselius. Emerges as the most vital tool in biochemistry for studying proteins and nucleic acids.
1944	Oswald T. Avery and colleagues demonstrate that DNA carries genetic information.
1949	Sequence of DNA molecules found to control cell growth via some sort of code.
1950	Alfred Hershey and Martha Chase use radioactively labeled E. coli to prove that DNA, and not protein, are the carriers of heredity.
1951	Maurice Wilkins and Rosalind Franklin determine DNA "structure" for the first time using crystallography and X-ray diffraction.
1953	James Watson and Francis Crick propose a double helix based on Rosalind Franklin's model as the molecular structure of DNA.
1958	Matthew Meselson and Franklin Stahl concluded that an old strand of DNA acts as a template for new strand formation and synthesis of new DNA occurs while old chromosome is in process of uncoiling.

(Continued)

TABLE 1.1 *(Continued)*

Year	Summary of Discovery
1961	French molecular biologists Francois Jacob and Jacques Monod propose that differences in the structure and function of cells result from differential gene expression.
1964	Marshall Nirenburg cracks genetic code.
1970	First restriction enzyme, a specialised protein used to cut DNA strands precisely at specific locations, is isolated.
1972	Paul Berg and Stanford colleagues synthesize the first recombinant DNA molecule by linking different DNA fragments in a test tube.
1973	UCSF's Herbert Boyer and Stanford's Stanley Cohen perform first genetic engineering experiment by linking toad genes with E. coli bacteria.
1976	UCSF's J. Michael Bishop and Harold Varmus discover genes, called oncogenes, which can lead to cancer, winning the Nobel Prize for the discovery. Yuet Wai Kan pioneers molecular techniques that allow the first fetal test to identify sickle cell anaemia, a hereditary blood disease.
1977	K. Itakura and colleagues clone the first human gene using the recombinant DNA techniques.
	William Rutter and UCSF colleagues isolate gene for rat insulin and transplant it into bacteria. Five years later, genetically engineered human insulin is introduced in the market. DNA-sequencing techniques developed.
1979	The biotechnology company Genentech Inc., in collaboration with UCSF, develops synthetic human growth hormone.
	Using gene mapping techniques, UCSF's Y.W. Kan and Judy Chang discover the single genetic mutation responsible for beta thalassemia, the most common form of life-shortening blood disease.
1980	Virus-fighting interferon cloned.
1981	Transgenic mice and transgenic fruit flies are produced. They soon become model systems for studying mutations, gene expression, and human disease. Seven years later, a cancer prone transgenic mouse becomes the first patented life form.
1983	Barbara McClintock wins the Nobel Prize in physiology for her discovery that some genes can "jump around" from one chromosome to another. These are called transposons.
1987	Biochemist William Rutter and his Emeryville Chiron Corp. produce the first commercial genetically engineered vaccine against hepatitis B.

(Continued)

TABLE 1.1 *(Continued)*

Year	Summary of Discovery
1990	NIH approves use of gene therapy on four-year-old girl with rare immune system disorder.
	America's Human Genome Project begins. Its goal is to map and clone every gene in the human body. UCSF's Rick Myers, David Cox, and colleagues begin mapping chromosomes 4 and 21.
1990	Chymosin, an enzyme used in cheese making, becomes the first product of genetic engineering to be introduced into the food supply.
1991	The gene implicated in the inherited form of breast cancer is discovered. This paves the way for the treatment of other similar forms of cancer.
1992	Techniques for testing embryos for inherited diseases such as cystic fibrosis are developed.
1994	First commercial approval by the U.S. FDA of a transgenic plant, the Flavr-Savr tomato that has an increased shelf life than its wild counterparts.
1996	The first trial of gene therapy for a neurological disorder is conducted in New Zealand.
	Dolly the sheep is cloned from a cell from an adult sheep making it the first cloned organism.
1998	New Zealand researchers produce Elsie, a clone of Lady, the last of the rare Enderby cattle line, saving it from certain extinction. Embryonic stem cells are grown successfully, opening new doors to cell-or tissue-based therapies.
1999	New Zealand researchers develop a new vaccine for bovine and human tuberculosis.
	Chinese scientists clone a giant panda embryo.
2003	The human genome is sequenced.

Business of Biotechnology

Much of the attention that biotechnology has attracted is due to the promise it holds for the industry and the economy. In fact, the euphoria is such that biotechnology is quoted as the next big thing after information technology. However, the definition of what exactly comes under the ambit of biotechnology is still blurred, at least as far as the Indian scenario is concerned. People often confuse pharmaceutical companies with biotechnology firms. While it is true that pharmaceutical companies are using biotechnology for research and development, it is certainly not a *de novo* biotechnology industry. It is against this backdrop that one can attempt to define this seemingly ambiguous term of "biotech firms".

Biotechnology is in its state of infancy in India, both at the general as well as the corporate level. The American corporate and the government set most of the initial benchmarks. If one follows the American definition, biotechnology is defined as "the use of cellular and molecular processes to solve problems or make products" by the Biotechnology Industry Organization (BIO). This definition includes all firms that use cells and biological molecules for applications in medicine, agriculture, and environmental management. However, it explicitly excludes pharmaceutical companies (except for distinct biotechnology divisions), contract research organizations, and equipment manufacturers. India is trying to adopt a similar mandate to earmark the sector as an independent sector. Unlike other industries, which are defined by the products manufactured or services offered, the biotechnology industry is defined uniquely by the nature of technologies it uses. Figure 1.1 shows the distribution of American firms under the biotechnology umbrella.

Indian Scenario

Biotechnology in India is mostly related and limited to conventional biotechnology, concerned with fermentation processes using microorganisms

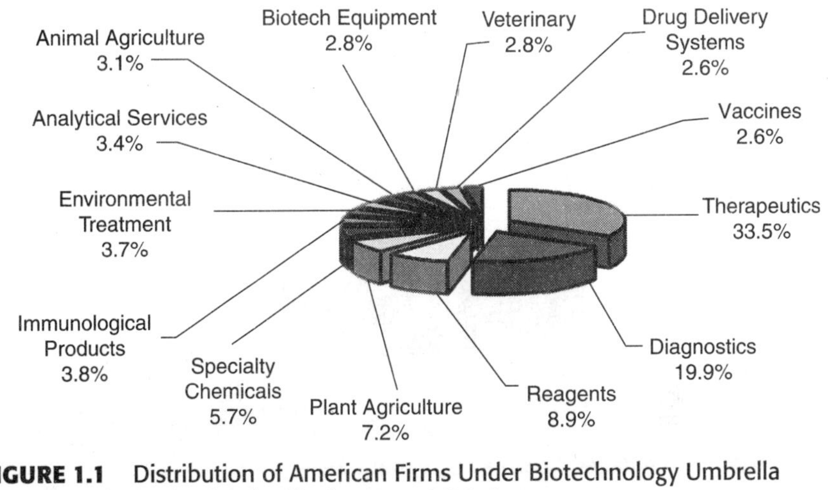

FIGURE 1.1 Distribution of American Firms Under Biotechnology Umbrella
Source: Standard Industrial Classification Manual, U.S. Government

for manufacturing products like alcohol, cheese, and bread. The microorganisms can also be used to produce medicinal and industrial products such as antibiotics and enzymes, plant tissue culture, bio-fertilizers, and bio-pesticides. Other applications are extraction of active therapeutic ingredients from plant or animal sources, conventional plant breeding techniques, artificial insemination of animals, etc. In the last few years, the focus has shifted to the growing awareness and demand for *modern* biotechnology products such as recombinant vaccines, diagnostics, biopharmaceuticals, and hybrid seeds.

As compared to conventional biotechnology, modern biotechnology encompasses recombinant DNA technology, hybridoma technology, gene transfer, cell fusion techniques, structural and functional genomics, protein engineering, molecular design, monoclonal antibodies, genetic modification of crops or foods, etc. Currently, the modern biotechnology sector in India comprises three main segments:

- Healthcare biotechnology
- Agricultural biotechnology
- Veterinary biotechnology

Figure 1.2 depicts the percentages occupied by the various sectors in India.

India's biotechnology industry is much smaller compared to its pharmaceuticals industry, which comprises about 20,000 small and large drug manufacturers (This data is as per the survey conducted by the Confederation of Indian Industry.) However, most of the major biotech activities are

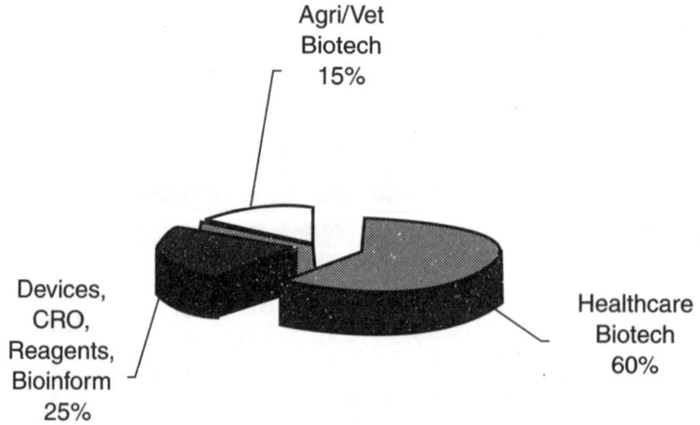

FIGURE 1.2 Percentages Occupied by Various Sectors

concentrated within the large pharmaceuticals or agricultural companies. Unlike in developed countries, in India, there are not many small or medium-sized biotechnology enterprises, established by academic researchers or supported by venture capitalists.

Currently, the Indian biotechnology sector is in its nascent stage and forms a mere 1 per cent of the global biotechnology market. However, it is slated for an exponential growth over the next five years with an expected global market share of 10%. The Department of Biotechnology (DBT) estimates that the overall biotechnology market (including conventional and modern biotechnology) in India is about USD 2.5 billion for 2001, and is expected to touch USD 4 billion by 2005, growing at a Compound Annual Growth Rate (CAGR) of 11%. The market for modern biotechnology products and services in India is estimated to be about USD 150 million (about INR 7,500 million) for 2002, of which about 40% (USD 60 million) are estimated to be export revenues. Figure 1.3 shows the growth in the Indian biotechnological sector.

The consumption of biotechnology products in India is estimated to grow ten-fold to USD 1.5 billion by 2007 and to USD 4.5 billion by 2010. Based on DBT estimates, the biotechnology equipment market in India is about USD 30 million and growing at 2–3% with focus gradually shifting from government institutes and laboratories to private enterprises.

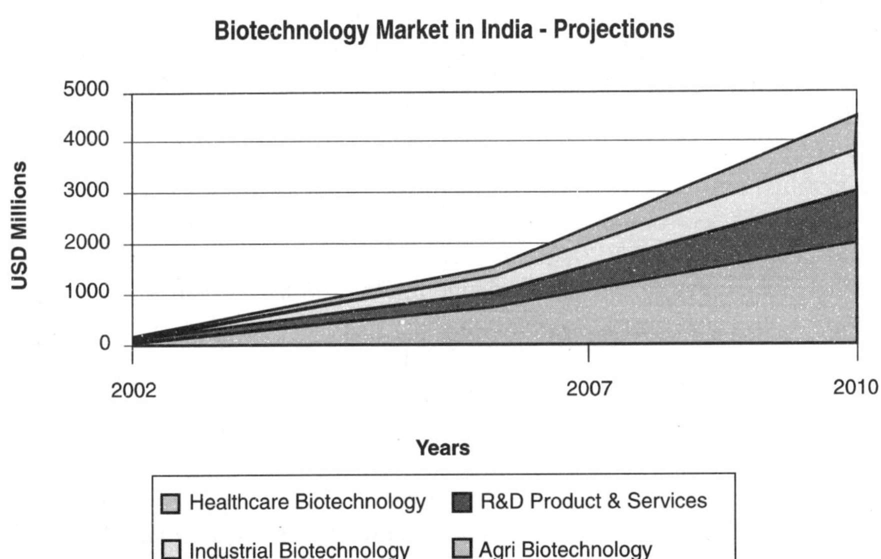

FIGURE 1.3 Graphical Representation of Growth in Indian Biotechnological Sector

Boom or Bust

India today holds a small share of the global biotech market, but has the capability to become a dominant player in the field. It has a rich human capital, which is the strongest asset for this knowledge-based industry. India has a large English-speaking base and, according to the Confederation of Indian Industry, produces roughly 2.5 million graduates in Information Technology, engineering and life sciences, about 6,50,000 postgraduates and nearly 1,500 PhDs qualified in biosciences and engineering every year. India has proved its competence and skills in such selected areas of biotechnology as the bioprocess engineering, gene manipulation of microbes and animal cells, downstream processing and isolation methods, and recombinant DNA technology of plants and animals. India has also allowed and assisted stem cell research that permits researchers to use embryos from fertility clinics upon informed consent of the donors, thus giving it a clear headstart in this new and promising field in biotech. Clearly, India has the strength and capabilities in this industry, and a definite advantage to forge ahead and become the chosen location for many biotech companies looking for large markets and low cost, qualified manpower to work in their research and development division.

While much of the hype around biotechnology has revolved around the scientific breakthroughs at the academic level, biotechnology is still perceived as a far too complicated sector to invest in. Unlike other sectors that are well understood and predictable to some extent, biotechnology is still in its nascent stage. In Europe, opposition is already mounting against biotechnological products with activists boycotting the same. While this has not happened in India, investors are wary of investing in the same.

Drawing parallels with Information Technology (IT) at this stage cannot be justified since the premises of these technologies are entirely different. Unlike the IT sector, biotechnology is capital intensive and revolves around highly skilled manpower. Sustaining a high level of investment and providing quality manpower is a major challenge for the industry. In spite of the best efforts of the government, private funding in the industry is negligible and venture capitalists are reluctant since there is a high level of skepticism regarding the growth and development of biotechnology in India. In addition to this, difficult EXIM policies are forcing them to invest at various other viable sectors.

The biggest stumbling block is the government policies on the intellectual property and patent regulations. Being a signatory of TRIPS (Trade-Related aspects of Intellectual Property Rights), India is committed to setting up patent regime in place and regulating the same. However, owing to various factors, the precise policy guidelines are still to be in place. This has directly affected foreign investment in the sector, compounding the already poor growth rate of the industry.

While it may be too early to write the biotechnology industry off, it is not yet the time to rejoice either. What is needed is a calculated approach to the overall scenario. The next decade will witness a boom in the industry as is projected by various corporates across the globe. Diversification and increased state funding is certainly the need of the hour. Boom or bust is a question that can be addressed at a stage when the pros and cons are more evenly balanced.

2 Fundamentals: The Groundwork

- Introduction
- Molecules of Life
- Cell Architecture
- Life Cycle of a Cell

Introduction

Since the beginning of time, man has been confronted by a question: What distinguishes living organisms from inanimate life forms? After all, living beings themselves are composed of lifeless molecules that conform to the existing chemical and physical laws. How is it that living beings are "alive" and not like the inanimate entities that are composed of more or less similar components?

The most simplistic logical answer to the query lies in the difference in the degree of complexity and organization of these molecules. The other difference between living organisms and inanimate life forms is in the living organisms' ability to extract, transform and use the energy in the environment to their growth and benefit.

This energy, present in our environment, is also responsible for sustaining molecular complexity and functionality. However, the most crucial difference is in the ability of living organisms to self-replicate precisely. Each component of a living being has a preordained well-defined function. The interplay among various moieties is highly fluid in nature. This fluidity or flexibility makes these systems dynamic in nature where changes in one component can affect the other.

The system can be, to some extent, compared to a computer program algorithm where every single equation is responsible in directing actions of the machine and gives it dynamic properties. In spite of this high level of coordination, there is a marked level of dissimilarity among various life forms. This variation is a result of the dynamism of molecules and the processes they regulate based on environmental conditions. Yet they are very much alike at the cellular and chemical levels. It is this complex symphony of similarities and differences that one has to appreciate in order to comprehend the true nature of life and the efforts to manipulate the same. To put things into perspective, the living world has many shades of grey to it unlike the non-living one. The trick is to look through the grey and make sense of the cryptic nature of life.

The following sections will help you to know more about the cell and its architecture and the dynamism involved.

MOLECULES OF LIFE

We are more or less familiar with the concept of the Big Bang theory, according to which, the Earth and other planets were created due to a massive cosmic explosion. Although this theory is widely accepted, it probably will never be proved; consequentially, leaving a number of tough, unanswered questions. Till date, there is no standard theory of the origin of life but there are some accepted models around which theories of existence are formed. They are:

- Prebiotic conditions resulting in creation of basic small molecules of life as per the Urey-Miller experiment in 1953.
- Phospholipids form the lipid bilayers, which is the basic structure of a cell membrane.

Life originated on Earth about four billion years ago as the simple unicellular bacteria that, for the first time, used inanimate chemical molecules to extract energy from the environment for its survival. The bacteria were created due to the atmospheric conditions prevailing at the time as explained by scientist Stanley L. Miller in his classic experiment. Figure 2.1 shows Miller's experimental set-up.

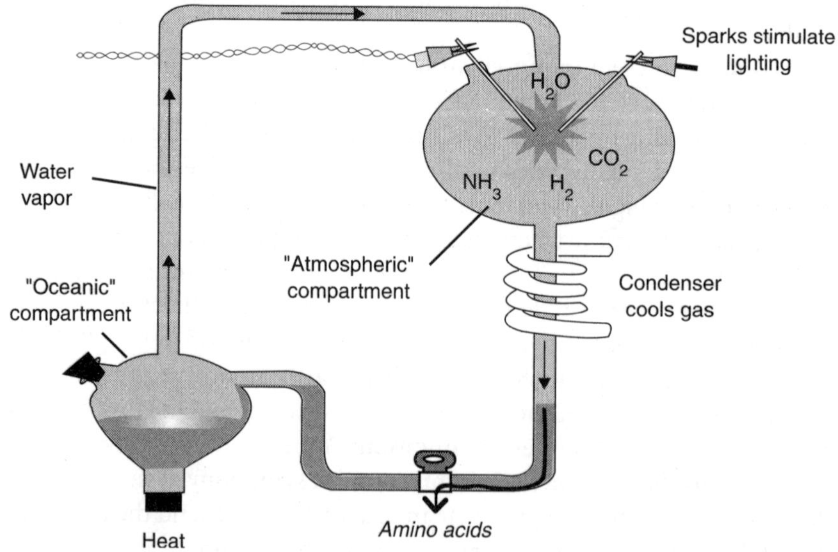

FIGURE 2.1 Experimental Set-up of Miller's Experiment

At the University of Chicago, under the guidance of his advisor, Nobel laureate Harold C. Urey, Miller carried out a series of laboratory experiments to determine if complex organic compounds could be synthesized under conditions that were chemically similar to those of the early Earth. Miller used compounds such as methane and ammonia as well as water in his experiment. An electric discharge (simulating lighting) passed through an atmosphere of CH_4, NH_3, H_2O, and H_2 lead to the generation of key organic compounds such as amino acids. He discovered that many carbon-bearing or organic compounds were formed under these circumstances, thus providing considerable support for the concept that life (or at least the chemical building blocks of life) originated on Earth through naturalistic chemical processes.

Life on Earth revolves around carbon, which accounts for more than half the dry weight of the cell. It is capable of forming either single bonds with hydrogen or double bonds in case of oxygen and nitrogen. In addition to this, carbon can form bonds with itself. This is also the reason as to why nature would have selected carbon over the other molecules to form the crux of life.

It is from this that the basic building blocks of life, proteins, carbohydrates and nucleic acids are derived. Some small molecules function as precursors for synthesis of macromolecules, and the cell is careful to provide the appropriate mix of small molecules needed. Small molecules also store and distribute the energy for all cellular processes; they are broken down to extract this chemical energy as when sugar is degraded to carbon dioxide and water, it releases energy bound up in the molecule.

Macromolecules are the most interesting and characteristic molecules of living systems. In a true sense, the evolution of life, as we know it, is the evolution of macromolecular structures. As is evident from Miller's experiment, life did evolve in water, thus making it one of the most important macromolecules. The next section will describe some of the macromolecules.

Water

Water is the most abundant substance in the living systems, comprising 70–90% of the total weight. In humans, about two-thirds of the body is made up of water. The physical and chemical properties of water are central to biological structure and function. The evolution of life on Earth has been influenced to a great extent by both the solvent and reactant properties of water. It is a tiny bipolar molecule that has a high freezing and boiling point. In a water molecule, the oxygen and hydrogen atoms are linked by a hydrogen bond. There are basically four types of interactions that occur within and between biomolecules in water acting as a solvent. They are:

- Ionic
- Hydrophobic

- Van der Waals interactions
- Hydrogen bonding

Water maintains the molecular confirmations and stabilizes the structural organization of living organisms. It also provides a stable pH-controlled environment for the various biochemical reactions that are vital to life.

Proteins

These are the most abundant and one of the most vital biomolecules in the body. To appreciate the abundance of protein within a cell, we can estimate the number of protein molecules in a typical eukaryotic cell, such as a hepatocyte in the liver. This cell, roughly a cube 15 m (0.0015 cm) on a side, has a volume of 3.4×10^{-9} cm^3 (or millilitres).

Assuming a cell density of 1.03 g/ml, the cell would weigh 3.5×10^{-9} g. Since protein accounts for approximately 20% of a cell's weight, the total weight of cellular protein is 7×10^{-10} g. The average yeast protein has a molecular weight of 52,700 (g/mol); this value is typical of eukaryotic proteins. One can calculate the total number of protein molecules per liver cell as about 7.9×109 from the total protein weight and the number of molecules per mole, which is a constant (Avogadro's number). To carry this calculation one step further, consider that a liver cell contains about 10,000 different proteins; thus, a cell contains close to a million molecules of each protein on average.

Its importance stems from the fact that it is the molecular instrument through which the genetic information is expressed, as we shall see in the subsequent chapters. In fact the word protein is derived from the Greek word *protos*, which means, "first or foremost".

All proteins are derived from relatively simple monomeric units called amino acids. All proteins—from the simple bacteria to the gigantic whale—are constructed from the same ubiquitous set of 20 amino acids (list) that are covalently linked in linear sequences.

Figure 2.2 shows the formation of a peptide bond.

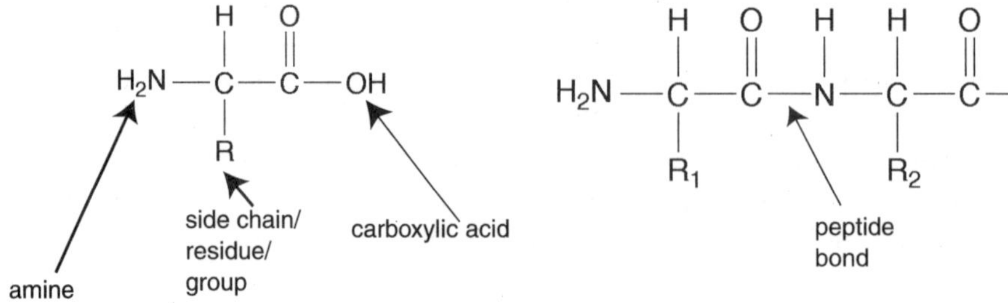

FIGURE 2.2 Formation of Peptide Bond

Figure 2.3 shows the structure of all twenty amino acids.

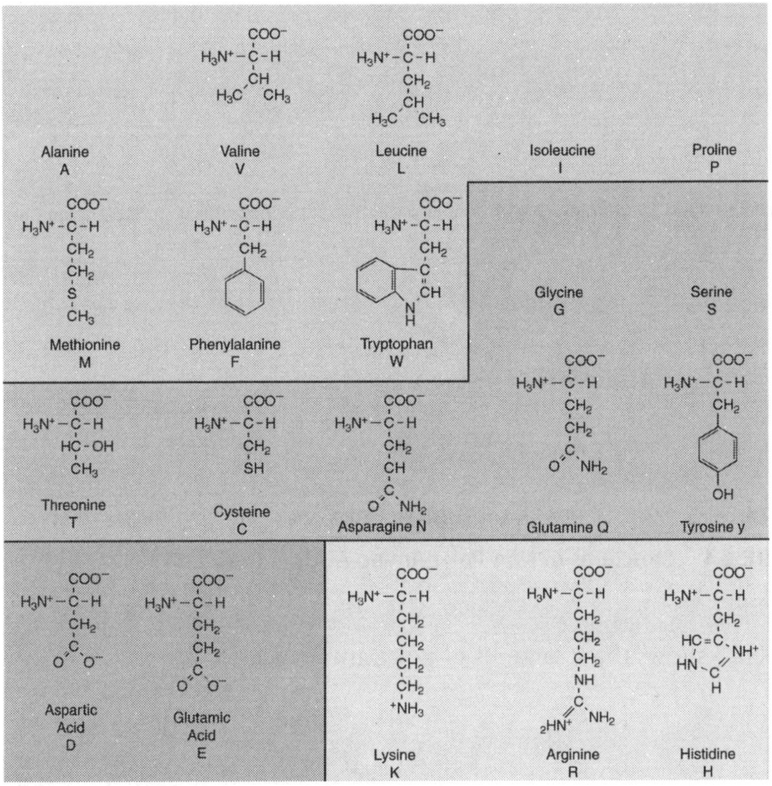

FIGURE 2.3 Structures of Amino Acids

Each of these amino acids has a side chain with distinct chemical properties. The most intriguing fact is that cells can produce different proteins with varying properties from the same 20 amino acids in various combinations and sequences. Some examples of these variations can be seen in case of diverse products such as enzymes, hormones, antibodies, antibiotics and even poisons. One of the most important proteins is the enzyme.

Life depends on the existence of these powerful and extremely specific catalysts that catalyze virtually every single biochemical reaction in the body. The amino acids are basically classified into various groups on the basis of the side chains attached to them. They are divided as follows:

- Nonpolar aliphatic R groups
- Aromatic R groups
- Polar charged R groups
- Positively and negatively charged R groups

Figure 2.4 shows the structures of non-polar amino acids.

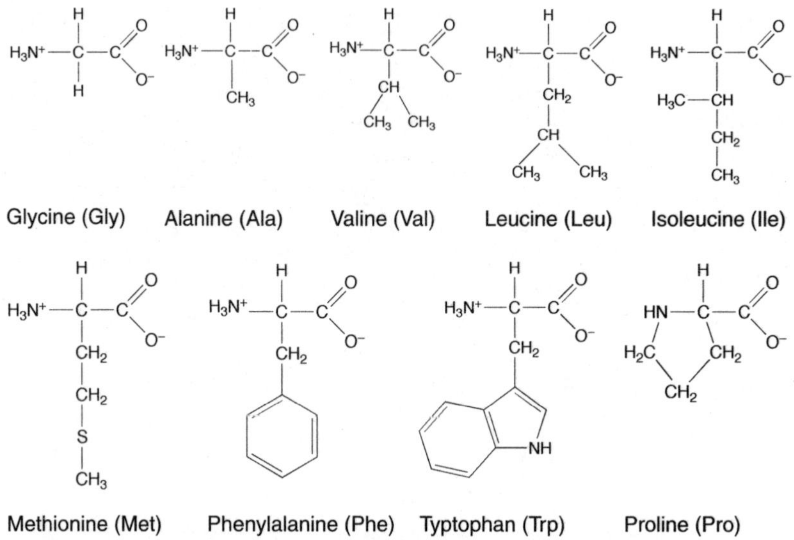

FIGURE 2.4 Structures of Non-Polar Amino Acids

Figure 2.5 shows the structures of polar amino acids.

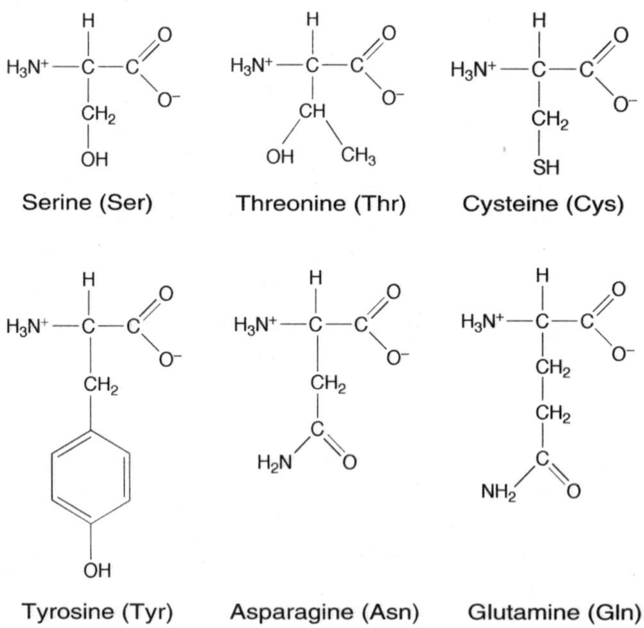

FIGURE 2.5 Structures of Polar Amino Acids

The structure of the proteins is crucial in the role that they play, and this is dictated by the arrangement of the amino acids. The differences in the function of the proteins result from the differences in the amino acid composition and sequence. There are four levels of protein structures that are commonly defined as shown in Figure 2.6.

a Primary
b Secondary
c Tertiary
d Quaternary structure

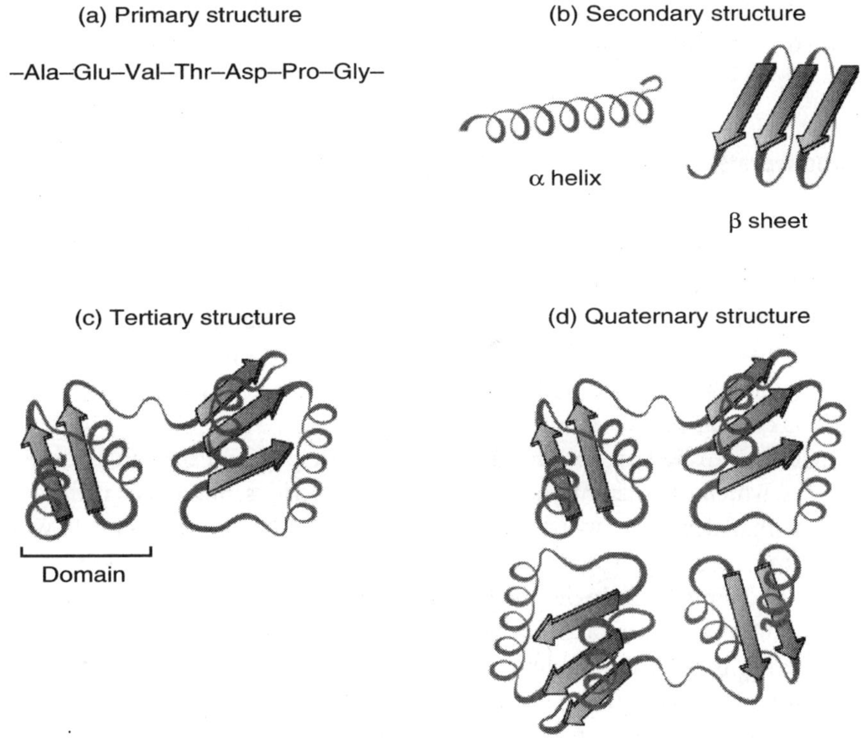

FIGURE 2.6 Types of Protein Structures

The primary structure consists of a sequence of amino acids linked together by peptide bonds and includes many disulfide bonds (bonds between sulfur molecules). The resulting polypeptide can then fold into coiled units called the secondary structure, such as a α helix. The tertiary structure defines all the three-dimensional folding of the polypeptide. When a protein has two or more polypeptide subunits, their arrangement in space is referred to as the quaternary structure.

Many of the proteins within cells are enzymes, which accelerate (catalyze) reactions involving small molecules. Other proteins allow cells to move and do work, maintain internal cell rigidity, and transport molecules across membranes. Proteins even direct their own synthesis and that of other macromolecules. The individual polypeptide chains in the proteins are called the subunits and the active protein is called the multimer. The most commonly occurring multimers are the dimers, trimers, pentamers, and decamers. Once a protein has been synthesized, it has to be folded in its correct structure. Only then can a protein interact correctly with molecules that complement the same. One example of this is the antibodies that are complex glycoproteins with specific regions that interact with bacteria and other pathogens. The functional three-dimensional state of a protein is termed as the native state.

Carbohydrates

These are the most abundant biomolecules on Earth. They are used as a dietary supplement (sugar and starch) in most parts of the world, and it is the breakdown of these carbohydrates that provides energy necessary to sustain life. Carbohydrates are also part of the structural components of various organisms (cell wall and connective tissues). Certain complex forms of carbohydrates conjugate with proteins and lipids to act as signals regulating various cellular processes in the body.

Carbohydrates are predominantly cyclized polyhydroxy aldehydes or ketones, which can occur in nature as monosaccharides, oligosaccharides, and polysachharides. The word saccharide is derived from the Greek word *sakcharon* meaning "sugar". The monosaccharides are the simplest of the sugars and contain a single polyhydroxy aldehyde or ketone unit. There are two families of monoasaccharides—the aldoses and the ketoses—depending on whether they contain aldehyde or ketone residues. Figure 2.7 shows some common monosaccharides occurring in nature.

Oligosaccharides consist of short chains of monosaccharide units or residues joined together by characteristic linkages called the glycosidic bonds. The most abundant are the disaccharides that contain two monosaccharide units. Figure 2.8 shows the structure of a disaccharide molecule.

Sugar polymers containing more than 20 monosaccharide units are generally called polysachharides. These may contain anywhere between hundreds to thousands of monosaccharide units. Some of them are linear in nature while a few others are in the form of branched chains. These are also called glycans and mainly differ from each other in the identity of the recurring monosaccharide units, in the length of the chains, in the type of the bonds

Structures of Common Monosaccharides

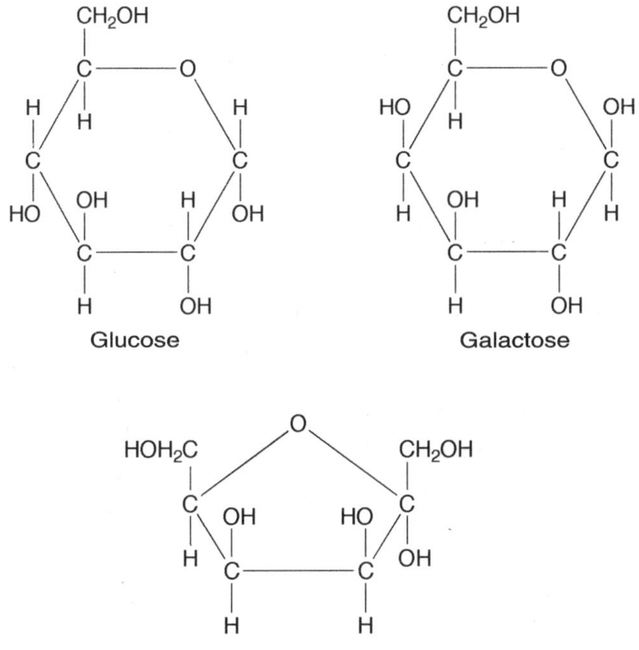

FIGURE 2.7 Common Monosaccharides in Nature

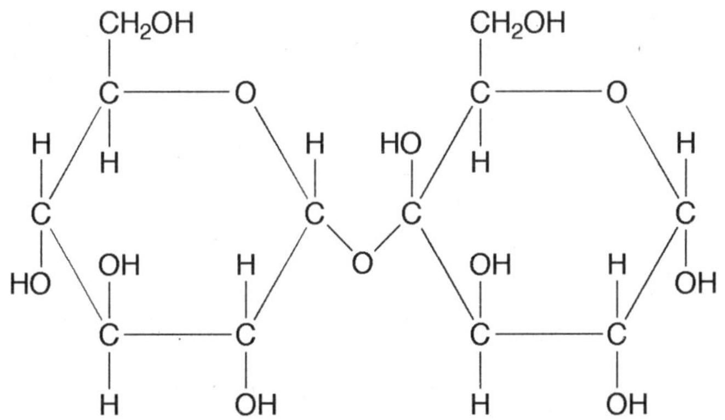

Maltose
(A Compound of 2 glucose molecules)

FIGURE 2.8 Structure of Disaccharide Molecule

linking the units and in the degree of branching. Figure 2.9 shows the structure of Xanthan Gum, a polysaccharide.

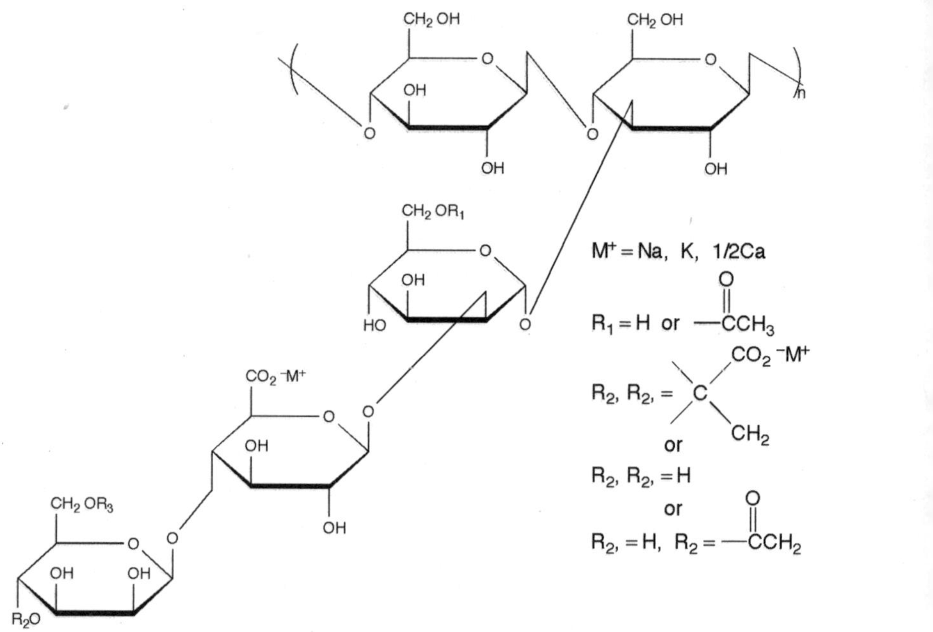

FIGURE 2.9 Structure of Xanthan Gum

Nucleotides and Nucleic Acids

Nucleotides have a range of functions in the body. They are the driving force behind the various chemical reactions in the body, essential links in the cellular responses to various signals, integral structural components of the various enzyme cofactors and metabolic intermediates.

Their importance is, however, elevated due to the fact that they are the constituents of nucleic acids: Deoxyribonucleic Acids (DNA) and Ribonucleic Acids (RNA), which are the molecular repositories of genetic information. These molecules perform the important role of storing and transmitting genetic information from one generation to the other, which is a fundamental condition for life.

Nucleotides have certain characteristic components: a nitrogenous base (containing nitrogen), a pentose and a phosphate group. Figure 2.10 shows nucleotides and their derivatives.

The molecules that lack phosphate groups are called nucleosides. The nitrogenous bases are derivates of two compounds: pyrimidine and purines.

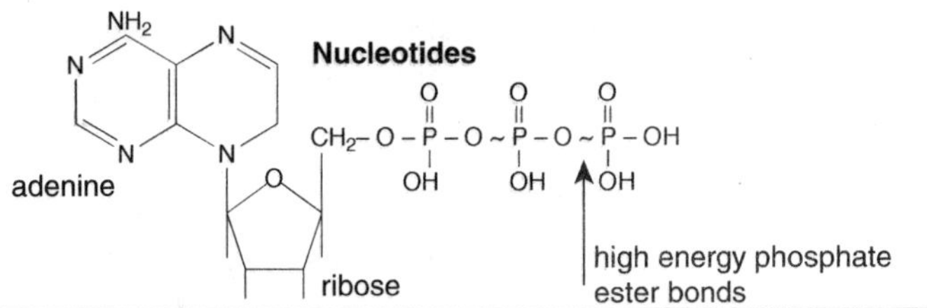

Adenosine triphosphate (ATP)

Adenosine diphosphate (ADP)
adenine - ribose - P P

Adenosine monophosphate (AMP)
adenine - ribose - P

Other Possibilities:

GTP	CTP	UTP	TTP
GDP	CDP	UDP	TDP
GMP	CMP	UMP	TMP

FIGURE 2.10 Nucleotides and Derivatives

Both DNA and RNA contain two major purine bases, Adenine (A), and Guanine (G), and two major pyrimidines. In DNA and RNA, one of the major pyrimidines is cytosine. However, the second pyrimidine is Thymine (T) in case of DNA and Uracil (U) in case of RNA. Figure 2.11 displays purines and pyrimidines, the building blocks of genetic material.

The nitrogen base is covalently bound to the sugar ribose in RNA and to deoxyribose (it has one oxygen less than ribose) in DNA. Phosphate is a derivative of phosphoric acid, providing the final covalent bond that connects

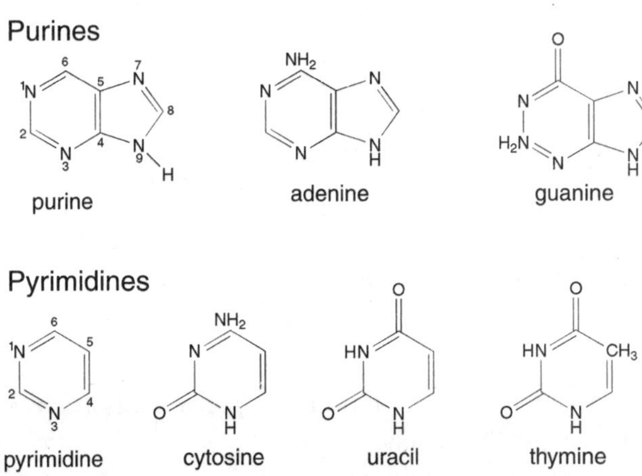

FIGURE 2.11 Purines and Pyrimidines

sugar molecules. Thus, in a sense, the backbone of the nucleic acid is a chain of alternating sugar-phosphate-sugar molecules.

Deoxyribose Nucleic Acid: The Double Helix

The structure of this enigmatic molecule was deciphered for the first time in 1953 by two young researchers at Cambridge University. They built a model as big as themselves from pieces of bent wire and brass and tried various combinations till they got a fairly stable structure, the now famous double helix. Basically, DNA is a long molecule that is formed by two long polynucleotide strands held together by hydrogen bonds. These bonds occur between the complementary pairs of nitrogen. Figure 2.12 shows the single stranded view of DNA.

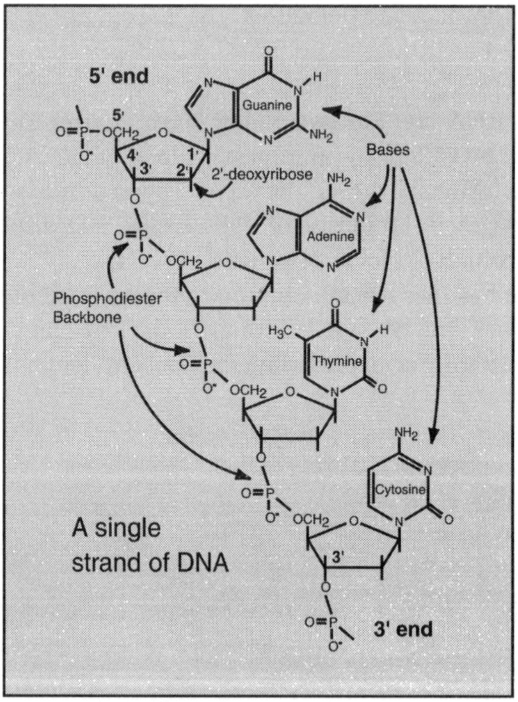

FIGURE 2.12 DNA: Single Stranded View

DNA is a nucleic acid that is supposed to contain the genetic instruction for the biological development of the whole life of living organisms including humans.

The pairing occurs according to a preordained pattern: Adenine pairs with Thymine by two hydrogen bonds (A=T) and Cytosine with Guanine by three hydrogen bonds (C=G). Figure 2.13 shows the complementary base pairing of nucleotides.

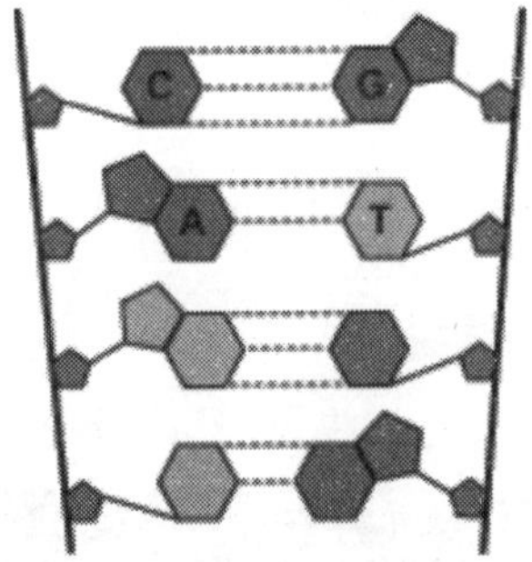

FIGURE 2.13 Complementary Base Pairing of Nucleotides

The DNA strands can be compared to a twisted ladder with the sugar-phosphate backbone representing the rails and the paired nitrogen bases the steps. The actual configuration of the molecule looks like a double helix.

RNA

As opposed to DNA, RNA has a single strand containing ribose sugar and uracil instead of thymine. Figure 2.14 shows the RNA architecture.

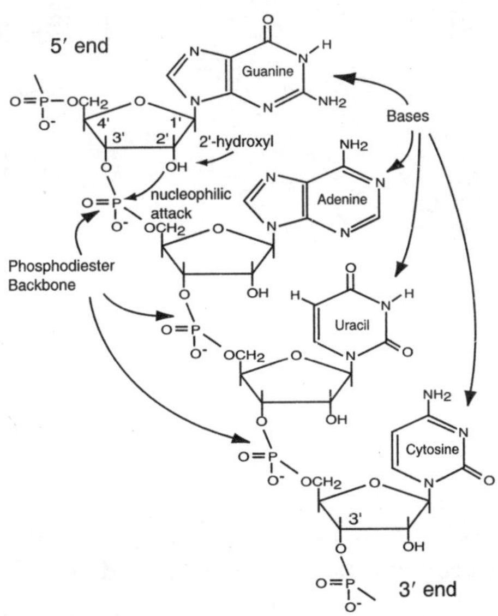

FIGURE 2.14 RNA Architecture

There are three main types of RNA molecules that are derived from DNA templates:

- Messenger RNA (mRNA) is a copy of the gene (DNA) giving the sequence of different amino acids to be incorporated into a protein.
- Transfer RNA (tRNA) is a carrier that delivers the correct amino acid for protein assembly.
- Ribosomal RNA (rRNA) is a major component of ribosomes.

Figure 2.15 shows the types of RNA moieties.

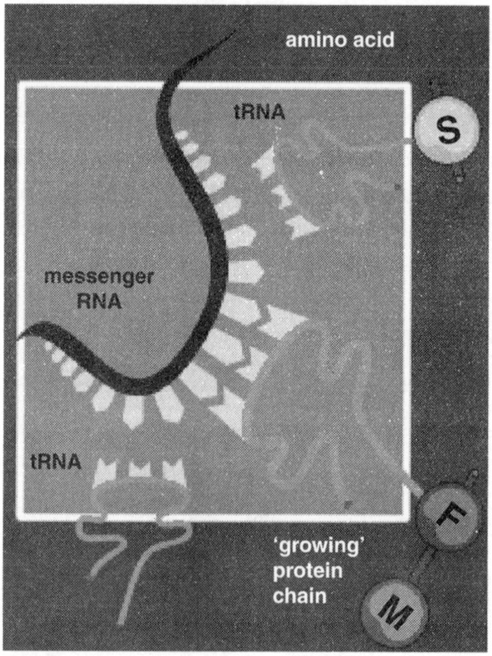

FIGURE 2.15 Types of RNA Moieties

Adenosine Triphosphate

ATP (Adenosine Triphosphate) is a nucleotide containing adenine, ribose and three phosphates, and is the energy currency of the cell. It releases energy when the bond between the second and third phosphate is broken. The hydrolytic cleavage of the molecule generates energy and Adenosine Diphosphate (ADP). This ADP can be reconverted into ATP by addition of a phosphate group. Figure 2.16 shows the structure of ATP.

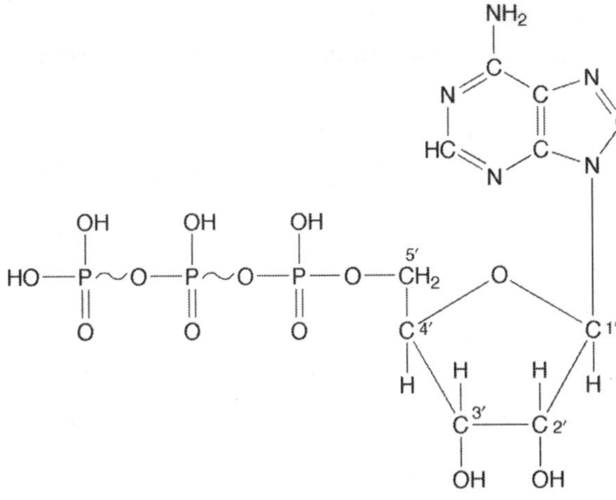

FIGURE 2.16 Structure of ATP

Lipids

The term lipid is derived from the Greek word "lipos", meaning fat. It is used to define biomolecules that are not soluble in polar solvents such as water but dissolve in non-polar solvents like benzene, ether and chloroform. Lipids are mainly composed of long hydrocarbon chains that are both nonpolar and hydrophobic. The main groups of compounds under lipids are triglycerides, phospholipids and steroids and waxes. Figure 2.17 shows some commonly occurring lipids.

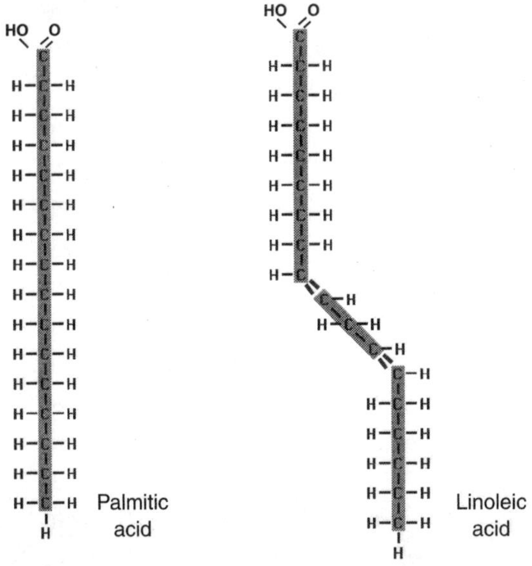

FIGURE 2.17 Commonly Occurring Lipids

The way in which the lipids naturally assume single and double-layered configuration makes them ideal candidates to make the structural framework of the cell. Steroids are complex compounds that reinforce the structure of the cell membrane in animal cells and form a vital component of animal hormones, e.g., cholesterol. Figure 2.18 shows the structures of cholesterol and phospholipid.

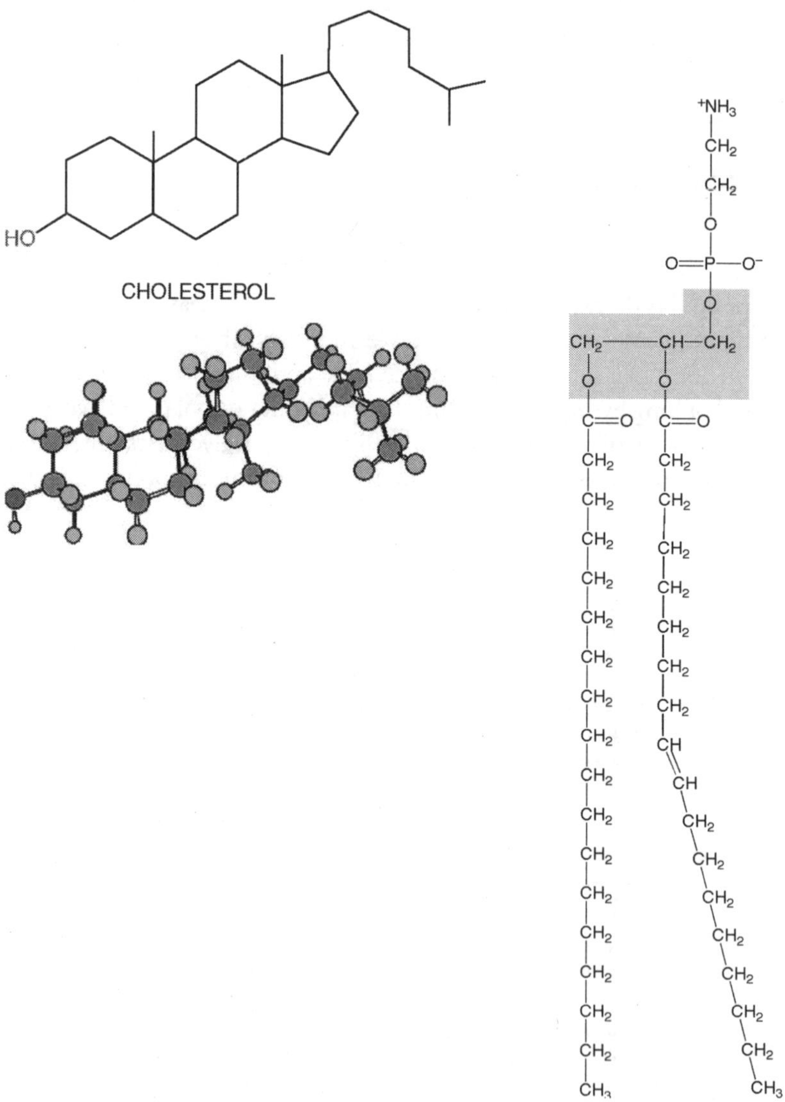

FIGURE 2.18 Structures of Cholesterol and Phospholipid

Classes of lipids that serve as a major structural component are the phospholipids. They are similar to the triglycerides in containing glycerol and fatty acids. They, however, differ in the fact that they have only fatty acids attached to the glycerol, while the third glycerol-binding site holds a phosphate group.

We have, in the above sections, closely studied the different macromolecules. Let us now see how they all make up the "cell", the structural and functional unit of all living organisms.

TIP

Cell is also sometimes called the "building blocks of life". Some living beings have a single cell only, but others like the human beings are multicellular.

Cell Architecture

The word cell, coined by one of the greatest experimental scientists of the seventeenth century, Robert Hooke, is derived from the Latin word "cella", which means a small room. Some of the properties of a cell are:

- It reproduces by cell division.
- It has the capability of taking in raw material, building cell components, creating energy molecules and releasing by-products.

All organisms are made up of cells, which are membrane-bound compartments filled with an aqueous solution of chemicals. The simplest life forms are solitary cells that propagate by dividing into two. The cellular architecture is one of the most fascinating and important aspects of the life forms. In general, life forms can be divided into two main types:

- **Prokaryotes:** They are the simplest life forms, which are unicellular in nature and do not have nucleus as compared to other organisms. Most prokaryotes are bacteria.
- **Eukaryotes:** The organisms falling under this category have complex cells. The eukaryotic cells, by definition, have the following components:
  - Cell wall (absent in case of animals)
  - Plasma membrane
  - Cytoplasm
  - Cell organelles

Figure 2.19 shows the structures of animal and plant cells.

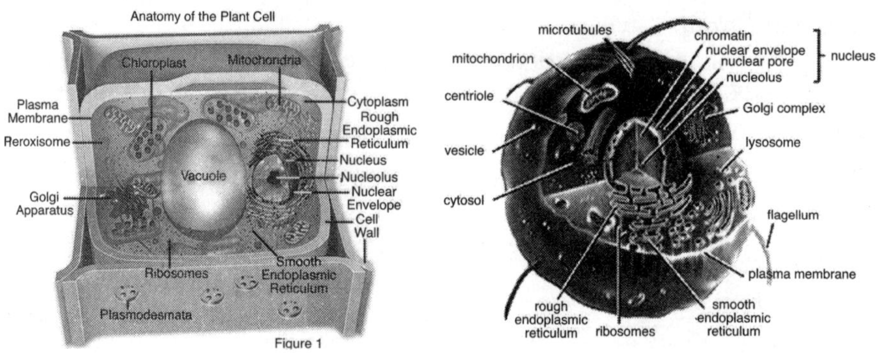

FIGURE 2.19 Structure of Animal and Plant Cell

Cell Wall

The composition and appearance of the cell wall differs according to the cell type and the function it performs. The plant cells have a fixed shape and size due to the presence of a cell wall. In higher plants, the cell wall is made up of the polysaccharide cellulose and embedded in a highly cross-linked matrix of polysaccharides such as pectin, lignin and hemicellulose. In plants, the cell wall is divided into three layers—the primary cell wall, the secondary cell wall and the tertiary cell wall. The fungal cell wall has a thick layer of polysaccharide fibres composed of chitin or cellulose and a thin layer of glycans.

Plasma Membrane

This is an absolute requirement for the survival of all living organisms. In prokaryotes, it forms the boundary of the cytoplasm. It is a typical bilayer of lipids in which the protein molecules are anchored and also contains sterols. The sterols being more rigid in nature confer structural stability to the cell. These molecules are selectively permeable and work as barriers for certain molecules while allowing the entry of others. The membrane, therefore, controls the flow of specific molecules into and out of the various compartments. The chemical composition of the membrane is highly variable: lipids (20–79%), proteins (20–70%), oligosaccharides

(1.5%) and water (20%). Figure 2.20 shows the plasma membrane showing lipid bilayer.

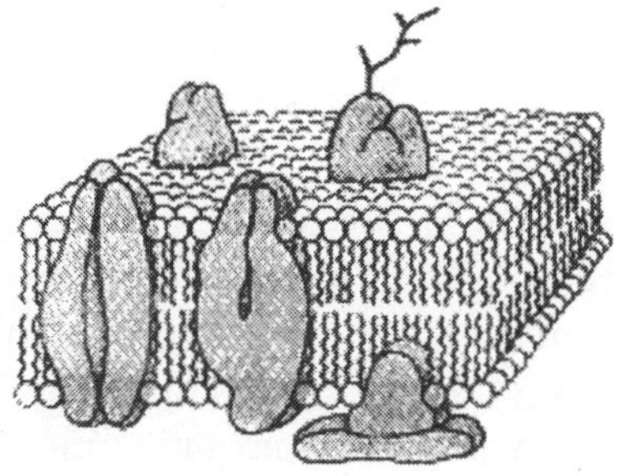

FIGURE 2.20 The Fluid Mosaic Model

Cytoskeleton

The cytoskeleton confers the ability of adopting various shapes and carrying out directed movements to the cell. The principal components are microfilaments, microtubules and intermediate filaments that constitute the cytoskeleton. The microfilaments are about 8nm in diameter and scattered or organised into a network or matrix. The microfilaments are made up of actin-like proteins. The microtubules are hollow tubes of about 25 nm formed of tubulin protein. They help in maintaining the cell shape and participate in intracellular transport. The intermediate filaments are tough and durable protein fibres or filaments of diameter ranging between 8–10nm. Figure 2.21 shows the components of cytoskeleton.

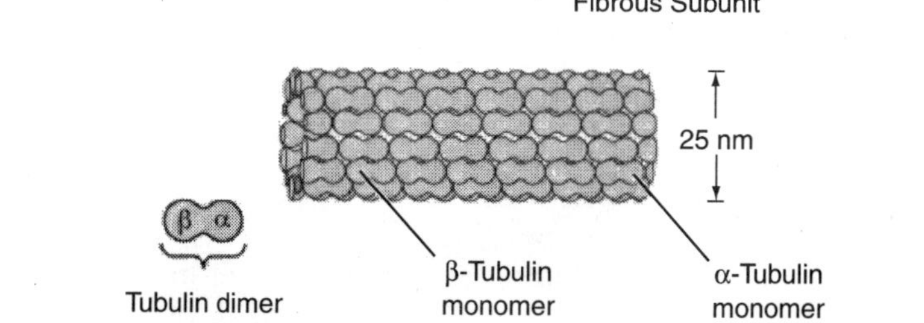

FIGURE 2.21 Components of Cytoskeleton

Cell Organelles

They are specialised structures inside the eukaryotic cells performing specialized functions. Some of the organelles are:

Endoplasmic Reticulum

The term was coined by Keith Porter, in 1953, to identify a fine network in the endoplasm of the cell. It is an irregular network of membranous tubules that form a continuous sheet enclosing a unitary space. In the cells that are involved in the secretion of various body fluids, the Endoplasmic Reticulum (ER) is embedded with ribosomes and termed Rough Endoplasmic Reticulum (RER). Figure 2.22 shows the endoplasmic reticulum and its components.

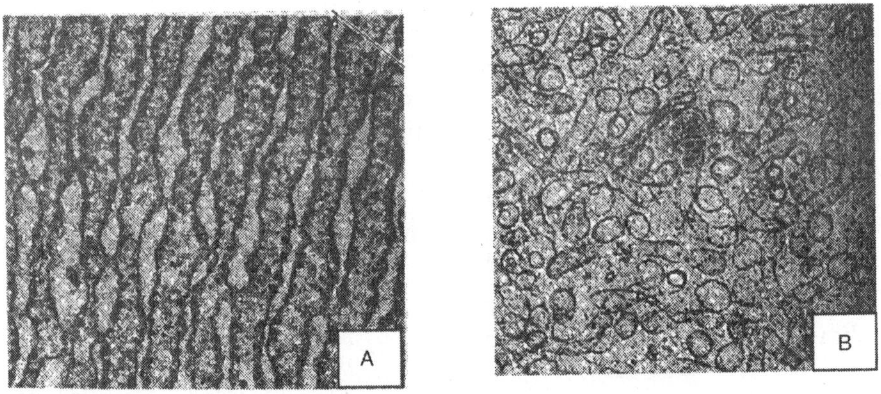

FIGURE 2.22 Electron Micrographs of Rough ER and Smooth ER

The cells that produce large quantities of lipids have ER that is not studded with ribosomes, called the Smooth Endoplasmic Reticulum (SER). The ER provides structural support to the cytoplasmic matrix and aids the synthesis of secretory proteins, lipids, in the detoxification of drugs and is also associated with muscular contraction by triggering the release of calcium ions.

Golgi Apparatus

It is a membranous organelle composed of flattened sac-like cisternae stacked on one another. These resemble the SER. On an average, there are around 4–8 cisternae in a stack. There exits a complex network of tubules and vesicles on the outer edges of the cisternae. The two ends of the cisternae are different in nature. The forming face (cis face) is associated with ER and is different from the sacs present on the maturing face (trans face) in thickness and contents. Figure 2.23 shows the structure of Golgi.

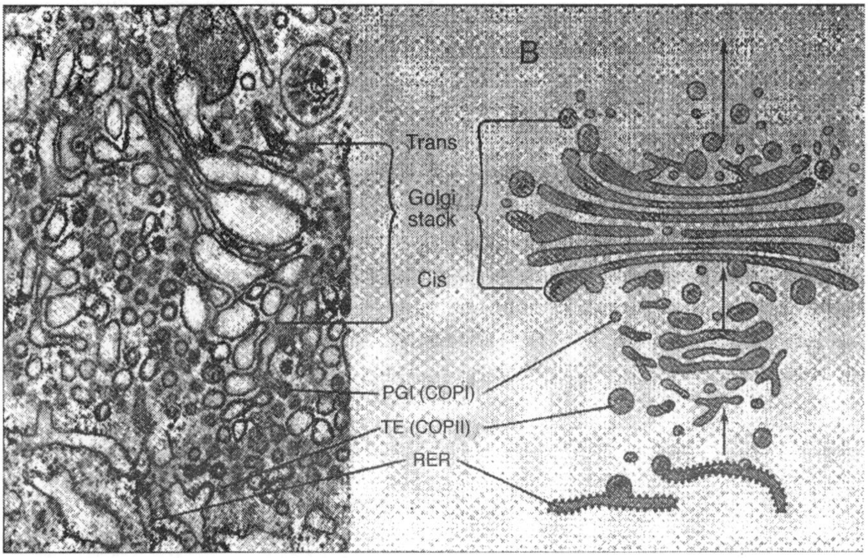

FIGURE 2.23 Golgi Apparatus

The materials appear to be transported form the cis face to the trans face by vesicles that keep budding off from the cisternal edge. The most important function of the Golgi apparatus is to package the material and prepare for secretions. The material to be secreted moves from ER to the Golgi during which the vesicles are pinched off from the ER. These travel to the Golgi and fuse with the cis face. Thus there is a close association between the two.

Ribosomes

These are super molecular complexes between the rRNAs and proteins, approximately 18–22nm in diameter. They are either associated with the endoplasmic reticulum or may be free in the cytoplasmic matrix. The ribosomes in case of eukaryotes are of 80S type (S stands for sedimentation coefficient) consisting of the 60S and 40S subunits. The smaller subunit is oblate ellipsoid and cap like in structure with a platform, a cleft, a head and a base. The larger subunit, on the other hand, has a protuberance, a ridge and a stalk. The ribosomes form a complex with the mRNA and are the sites for protein synthesis. Figure 2.24 displays the assembly of the ribosomal apparatus.

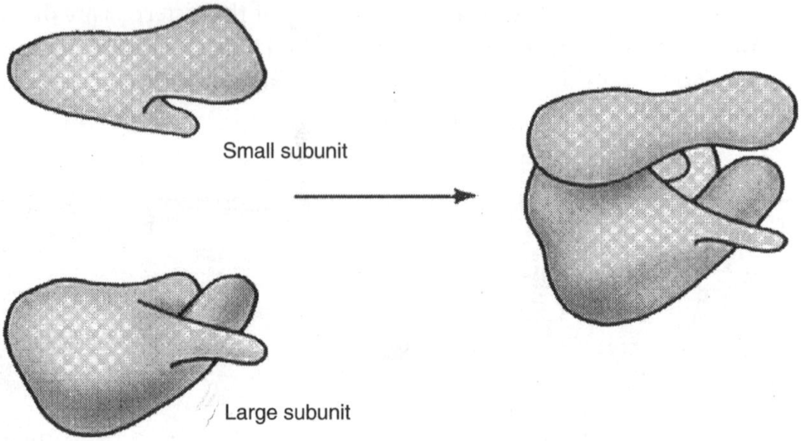

FIGURE 2.24 Assembly of Ribosomal Apparatus

Mitochondria

These are known as the powerhouse of the cell and are associated with the generation of ATP. They measure approximately 0.6–2.0nm in diameter and 5–10nm in length. There are more than a thousand or more mitochondria in the cell. They are bound by a double membrane—outer and inner one separated by 6–8nm intermembrane space. Figure 2.25 shows the electron micrograph structure of mitochondria.

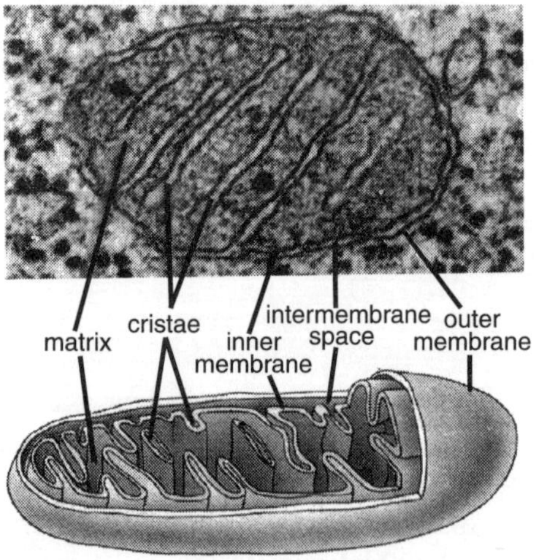

FIGURE 2.25 Electron Micrograph Structure of Mitochondria

The inner membrane forms many folds called crista, which increases the surface area available for absorption. The inner membrane encloses the mitochondrial matrix, which is quite dense and is loaded with ribosomes and

DNA. The mitochondrial DNA is a closed circular molecule similar to the bacterial DNA. The inner membrane has small spheres called the F1 particles and is associated with the ATP production. Presence of DNA and ribosomes in the mitochondria makes them independent and can synthesize their own proteins. The mitochondria multiply by means of binary fission, which is a bacterial character proving evolutionary links.

Nucleus

It is a relatively large organelle controlling all the activities of the eukaryotic cells. Some cells have more than one nucleus (multinucleate) while others lack nucleus (anucleate). The nucleus is enclosed by two membranes, which form a nuclear envelope. The outer and inner membranes are separated by a perinuclear space. The outer membrane is in continuation with the ER and the inner one surrounds the nuclear components. The outer membrane is interrupted in certain spaces by gaps called pores. These enable the exchange of various cellular materials between the nucleoplasm and cytoplasm. The nucleoplasm contains nucleolus and chromatin. The nucleolus helps in RNA synthesis and the ribosome subunit synthesis. The cell organelles composed of the various macromolecules play a vital role in the survival and proliferation of cells. Figure 2.26 shows the nucleus and components.

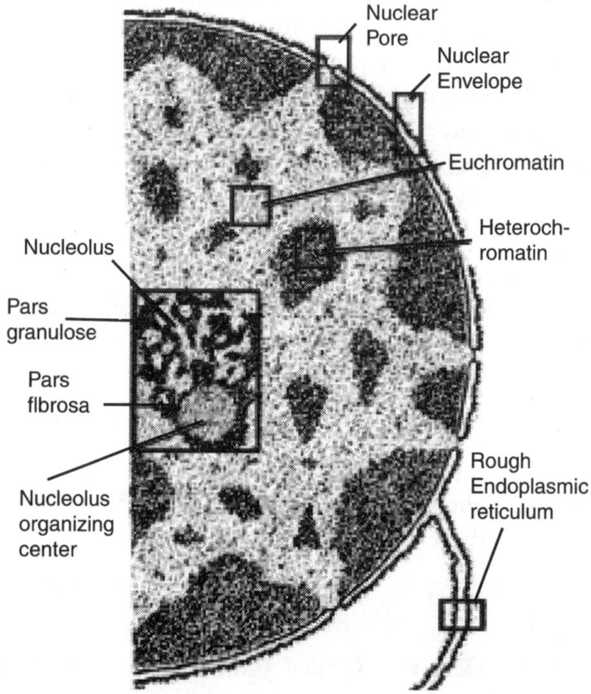

FIGURE 2.26 Electron Micrograph of Nucleous Showing Nucleolus

The mechanism of this proliferation and the ultimate fate of cells is an area that will be explored in the subsequent chapter.

LIFE CYCLE OF A CELL

Life in case of higher organisms originates with the formation of a zygote (fertilized egg). From this single cell, billions of cells develop by successive divisions. The cells then proceed further developing into various tissues and organs, performing a variety of functions. The entire process of cell division is highly regulated and there are sequences of events that occur in close synchrony. This orderly sequence of events by which the cell duplicates its contents and divides into two is termed as the cell cycle. The two main phases of a cell cycle are:

- Interphase
- Mitosis

Interphase

The cell spends most of its lifetime in this stage. The interphase is a period of intense biosynthetic and chemical activity in which the cell doubles in size and duplicates with a high level of precision. The nerve cells, however, are an exception since they never divide at all after birth, thus remaining in a state of perpetual interphase. Though this stage appears to be a "resting stage", it is in fact a period of intense activity. It is during this phase that the replication of DNA, nuclear proteins like histones, duplication of the centrioles and synthesis of energy-rich proteins that power mitosis take place. Figure 2.27 shows the phases of cell cycle.

The interphase is divided into three periods: G1 or the post-mitotic phase, the S phase or the synthesis phase and the pre-mitotic phase or the G2 phase.

G1 Phase

The post-mitotic gap phase. During this stage, the RNA and the proteins are synthesized, however, there is no synthesis of DNA.

S Phase

During this stage, the DNA is replicated and the DNA content of the nucleus is doubled.

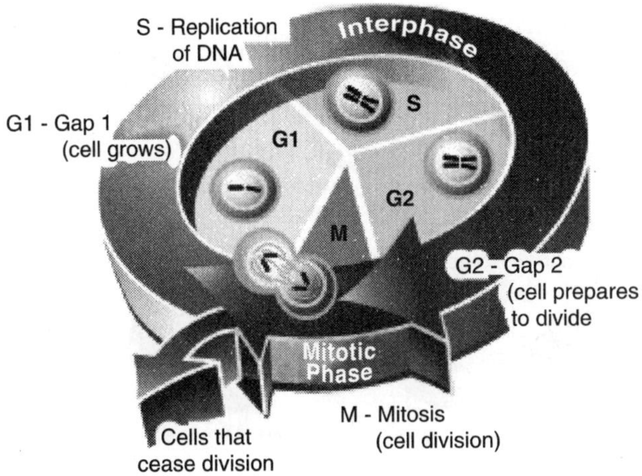

FIGURE 2.27 Phases of Cell Cycle

G2 Phase

During this stage, also called the pre-mitotic gap phase, the synthesis of RNA and protein continues, but the synthesis of DNA stops.

The lengths of the phases vary according to the level of activity of the cell. In cells that do not divide often, the G2 phase is longer. The cell has various options during this stage, one of them being to permanently stop cell division and enter the G0 phase. Such a cell is considered to be withdrawn from the cell division. The decision as to whether the cell will divide further or enter the G0 phase is taken during the G1 phase. Once the cell passes this checkpoint, it will continue with the cycle.

A G_0 cell is often called "quiescent". They are busy carrying out their functions in the organism, for example, secretion, attacking pathogens. Often G_0 cells are terminally differentiated. G_0 does not simply represent the absence of signals for mitosis but an active repression of the genes needed for mitosis. Cancer cells cannot enter G_0 and are destined to repeat the cell cycle indefinitely.

Mitosis

Also called the reduction division, this is the process in which the chromosomes are duplicated and distributed equally to the daughter cells. It is divided into five main stages as shown in Figure 2.28:

1. Prophase
2. Prometaphase

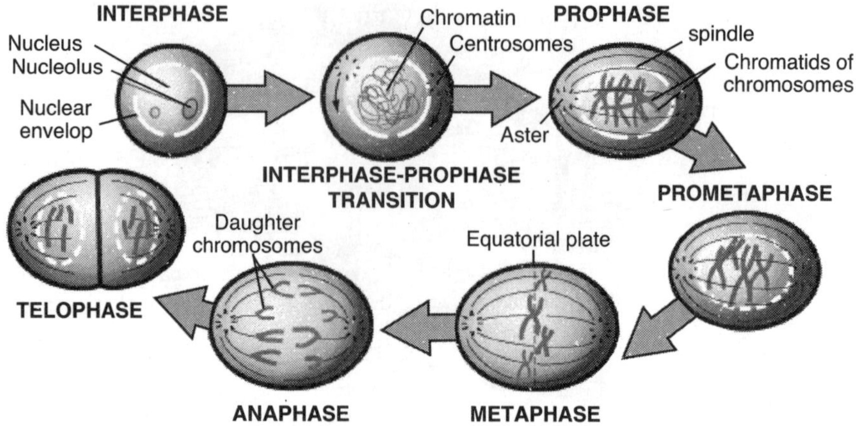

FIGURE 2.28 Phases of Meiosis

3. Metaphase
4. Anaphase
5. Telophase

Various important functions are performed during mitosis. The most essential feature is that the chromosomes are distributed equally between the two daughter cells. Another important feature is that the mitosis restores the surface/volume ratio of the cell. The process of repair of the body takes place because of the addition of cells by mitosis.

Meiosis

This is a process that occurs in the germ cells that are destined to produce the gametes (sperm and ova) in sexually reproducing organisms. Most of the stages of meiosis are similar to mitosis. It is basically divided into two parts: Meiosis I and Meiosis II. Figure 2.29 shows meiotic cell division.

Each one includes prophase, metaphase, anaphase and telophase. During the first phase, the members of homologous pair of chromosomes separate and are distributed into separate cells. In the second division, the sister chromatids of each chromosome separate and are distributed among the daughter cells. In this way, the number of daughter cells is ultimately reduced by half. The process of meiosis helps to maintain a definite constant number of chromosomes in sexually reproducing organisms. Due to the process of crossing over (exchange of genes), the process serves as the raw material for the evolutionary process. Throughout its life, the

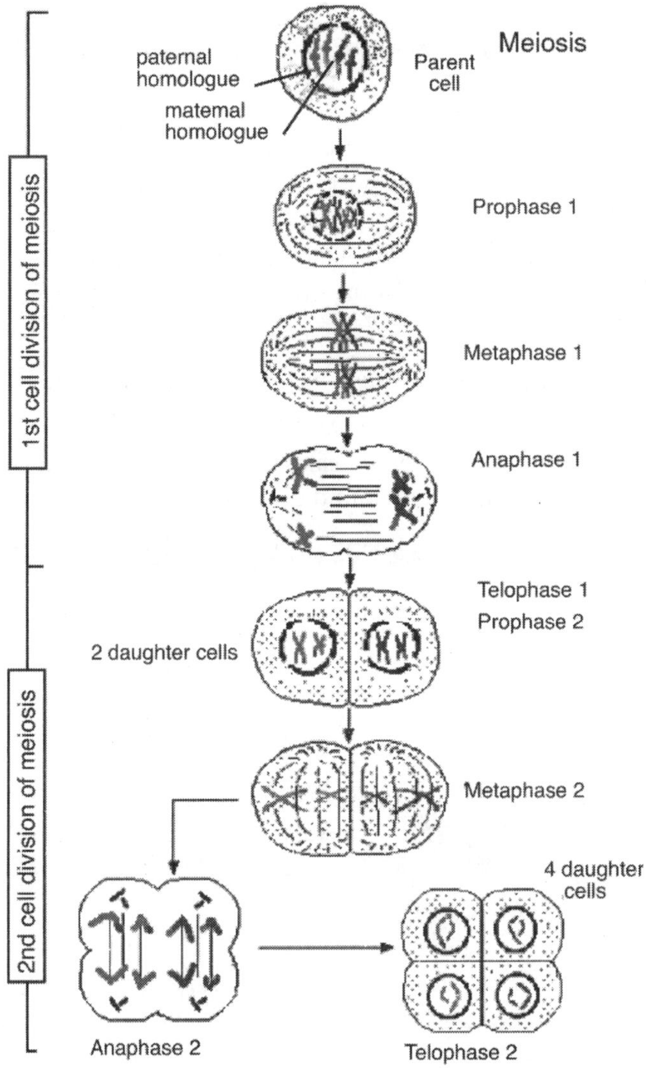

FIGURE 2.29 Meiotic Cell Division

cell proliferates and plays a major role in transfer, multiplication and regulation of various life processes. However, like every other living entity, death is inevitable and the cell meets its ultimate fate in a classically orchestrated waltz.

Regulation of the Cell Cycle

Cell cycle, as we have seen, is a highly regulated and complex process. Some of the features in regulation are critical for the survival of the cell where errors, if any, can lead to cancer. Cancer is a disease where regulation of the cell cycle goes awry and normal cell growth and behaviour are lost.

MPF (Maturation Promoting Factor)

This is a factor that triggers progression through the cell cycle. It's a protein complex of *CDK* (cyclin dependent kinases—a class of protein) and cyclin. MPF is a kinase enzyme; once active it switches on/off the target proteins by phosphorylating (adding a phosphate molecule) them. MPF basically promotes mitosis by phosphorylating other proteins including one that leads to destruction of cyclin itself.

CDK (Cyclin Dependent Kinase)

These add phosphate to a protein and along with cyclins, are the major control switches for the cell cycle, causing the cell to move from G_1 to S or G_2 to M. They are active only when bound to cyclins.

G_1 Cdk (Cdk4), **S-phase Cdk** (Cdk2), **M-phase Cdk** (Cdk1)

Figure 2.30 shows the regulation of cell cycle by regulators.

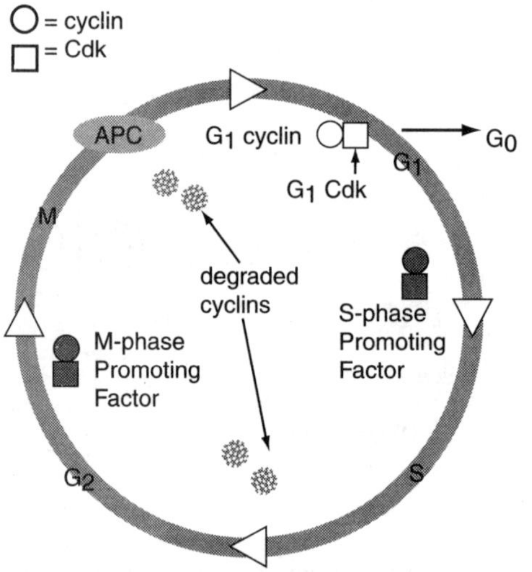

FIGURE 2.30 Regulation of Cell Cycle by Regulators

Cyclin

This is a protein whose amount varies cyclically; when in high concentrations it binds to cdk and makes MPF ... [cyclin + cdk = MPF] ... favours mitosis

p53

It is a protein that functions to block the cell cycle if the DNA is damaged. If the damage is severe, this protein can cause apoptosis (cell death). p53 levels are increased in damaged cells. This allows the cell enough time to repair DNA by blocking the cell cycle. A p53 mutation is the most frequent mutation leading to cancer. An extreme case of this is Li Fraumeni syndrome, where a genetic defect in p53 leads to a high frequency of cancer in affected individuals.

p27

It is a protein that binds to cyclin and CdK blocking entry into S phase.

Anaphase-Promoting Complex (APC)

The APC is also called the **cyclosome** and the complex is often designated as the **APC/C**. The APC/C triggers the events leading to destruction of the cohesions thus allowing the sister chromatids to separate. It also degrades the mitotic cyclin B.

Apoptosis: Cell Death

If a cell has an error in its DNA that cannot be repaired, it undergoes a process called programmed cell death (apoptosis). This process is shown in Figure 2.31.

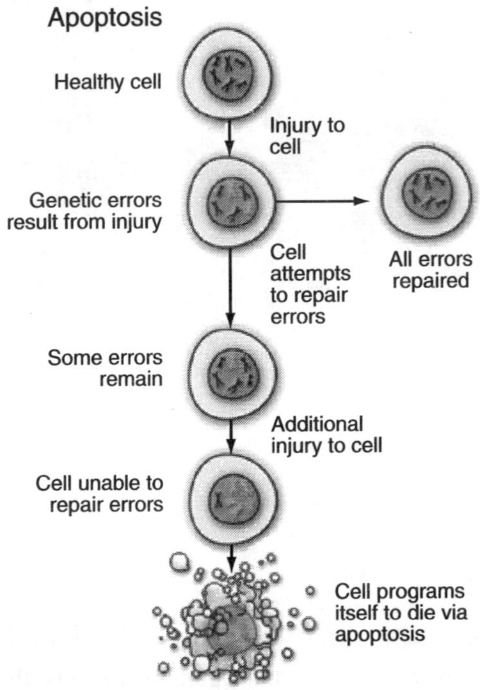

FIGURE 2.31 Death of a Cell
Source: U.S. National Library of Medicine

Apoptosis is a common process throughout life that helps the body get rid of cells that have outlived its purpose or are defective in nature. Cells that undergo apoptosis break apart and are recycled by a type of white blood cell called a macrophage. The macrophages are involved in the vital process of immune defence in the body. Apoptotic death can be triggered by a wide variety of stimuli, and not all cells necessarily will die in response to the same stimulus. Among the more studied death stimuli is DNA damage (by irradiation or drugs used for cancer chemotherapy), which in many cells leads to apoptotic death via a pathway dependent on p53. Figure 2.32 shows the final stages of a cell.

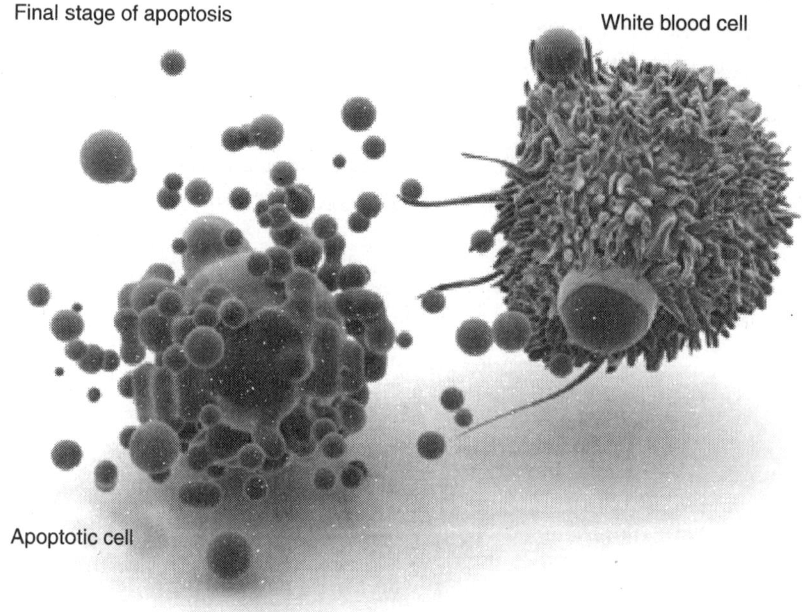

FIGURE 2.32 Cell Death
Source: U.S. National Library of Medicine

Apoptosis protects the body by removing genetically damaged cells that could lead to cancer, which results from a disruption of the normal regulation of the cell cycle. When the cycle proceeds without control, cells can divide without order and accumulate genetic defects that can lead to a cancerous tumour. Figure 2.33 shows the mutation of genes and progression of cancer.

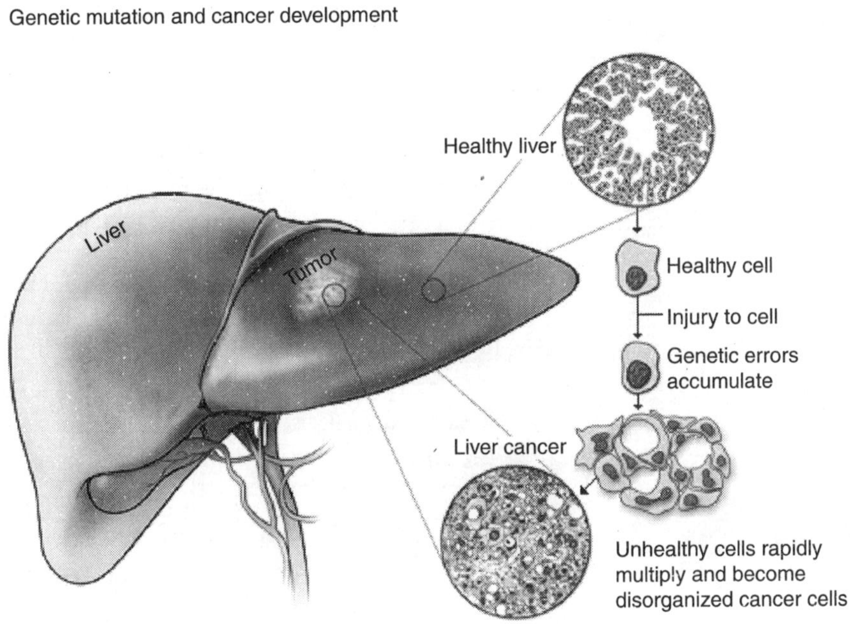

FIGURE 2.33 Gene Mutation and Progression of Cancer
Source: U.S. National Library of Medicine

In this chapter, we have explored the elements of a cell. We have seen how the different cells and their evolution have shaped the present state of the human beings and other living organisms. The evolution of the plant and animal lives has involved considerable level of development and diversification at the cellular level.

TERMS TO KNOW

Term	Explanation
cell	Fundamental structural unit of life, consisting primarily of outer plasma membrane, which separates it from the environment; the genetic material (DNA), which encodes heritable information for the maintenance of life; and the cytoplasm, a heterogeneous assemblage of ions, molecules, and fluid.
cell cycle	Complete sequence of steps performed by a cell to replicate itself. Most of the cycle consists of a growth period in which the cell takes on mass and replicates its DNA.
cytoskeleton	Integrated system of molecules within eukaryotic cells, which provides them with shape, internal spatial organization, motility, and may assist in communication with other cells and the environment. Red blood cells, for instance, would be spherical instead of flat if it were not for their cytoskeleton.
Endoplasmic Reticulum (ER)	Network of membranes in eukaryotic cells, which helps in the control of protein synthesis and cellular organization.
extracellular matrix (ECM)	Region outside of metazoan cells, which includes compounds, attached to the plasma membrane, as well as dissolved substances attracted to the surface charge of the cells. The ECM functions both to keep animal cells adhered together as well as buffering them from their environment.
flagellum	Hair-like structure attached to a cell, used for locomotion in many protists and prokaryotes. The prokaryotic flagellum differs from the eukaryotic flagellum in that the prokaryotic flagellum is a solid unit composed primarily of the protein flagellin, while the eukaryotic flagellum is composed of several protein strands bound by a membrane, and does not contain flagellin. The eukaryotic flagellum is sometimes referred to as an undulipodium.

Golgi apparatus	Eukaryotic organelle, which packages cell products, such as enzymes and hormones, and coordinates their transport to the outside of the cell.
lysosome	Eukaryotic organelle, which carries digestive enzymes. The lysosome fuses with a vacuolar membrane containing ingested particles, which are then acted upon by the enzymes.
microtubules	Type of filament in eukaryotic cells composed of units of the protein tubulin.
mitochondrion	Complex organelle found in most eukaryotes; believed to be descended from free-living bacteria that established a symbiotic relationship with a primitive eukaryote. Mitochondria are the sites of most of the energy production in most eukaryotes; they require oxygen to function the eukaryotic flagellum.
nuclear membrane	The double membrane, which surrounds the eukaryotic nucleus. It has many pores in its surface, which regulate the flow of large compounds into and out of the nucleus.
nucleoid	Region in prokaryotes where the DNA is concentrated. Unlike a nucleus, it is not bound by a membrane.
nucleus	Membrane-bound organelle, which contains the DNA in the form of chromosomes. It is the site of DNA replication, and the site of RNA synthesis.
plasma membrane	Outer membrane of a cell, sometimes called the cell membrane. The term plasma membrane is used more frequently when discussing prokaryotes.
chloroplast	A chlorophyll-containing plastid found in algal and green plant cells.
chromosome	Linear piece of eukaryotic DNA, often bound by specialised proteins known as histones.
contractile vacuole	In many protists, a specialised vacuole with associated channels is designed to collect excess water in the cell. Microtubules periodically contract to force this excess water out of the cell, regulating the cell's osmotic balance.

plastid	Any of several pigmented cytoplasmic organelles found in plant cells and other organisms, having various physiological functions, such as the synthesis and storage of food.
nucleotide	Unit from which nucleic acids are constructed by polymerization. It contains a sugar, a phosphate group, and an organic base. ATP is a nucleotide.
chlorophyll	The green-coloured pigment that absorbs light during photosynthesis, often found in plants, algae, and some bacteria; it includes a porphyrin ring, and often has a long hydrophobic tail.
cellulose	Carbohydrate polymer of the simple sugar, glucose. It is found in the cell walls of plants and green algae, as well as dinoflagellates. Cellulose is the most abundant compound on Earth that is manufactured by living things.
glycoprotein	Membrane-bound protein that has attached branching carbohydrates. These may function in cell-cell recognition, such as in human blood groups and immune system response, as well as in resisting compression of cells.
hydrophobic	"Water fearing". Hydrophobic compounds do not dissolve easily in water, and are usually non-polar. Oils and other long hydrocarbons are hydrophobic.
peptidoglycan	Carbohydrate polymer cross-linked by proteins. It is found in the cell wall of Gram positive bacteria, where it stains with the dye crystal-violet.
photosynthesis	Biochemical process in which light energy is absorbed by chlorophyll, and is used to fuel the building of sugar molecules.

3 Molecular Cycles: 'Let There be Life'

- Introduction
- Genes and Chromosomes
- Proteins: The Holy Grail

Introduction

Each and every living organism, leave alone human beings, has genes and chromosomes that determine their progress in life right from birth to death. For this reason, they are aptly called the "building blocks of life". These molecules of life help to lay down rules for the growth, development and various functions of the organisms' physical and biochemical systems. This chapter will explore these molecules of life and their cycles. This chapter will also study the mechanisms that regulate the synthesis of DNA (Deoxyribonucleic acid), RNA (Ribonucleic acid) and protein molecules.

Genes and Chromosomes

For a layman, the word "gene" refers to a hereditary cause of a trait, conditions or disease. This word is commonly used in different branches of study and research such as classical genetics, molecular genetics and evolutionary biology. Put simply, or rather in molecular biological terms, genes refer to segments of DNA that are transcribed by the cell into RNA that are further translated into proteins.

The study of genes began in the early nineteenth century when biologists concentrated on the phenomenon of heredity. The true chemical nature of genes was not being studied during this time period. However, Frederick Griffith's discovery of virulent and non-virulent strains of pneumonia changed the course of research and study.

Griffith conducted a series of experiments on a family of mice. He injected them with various mixtures of bacteria. He observed that the mice injected with virulent S. pneumoniae bacteria developed pneumonia and died, while those injected with an avirulent type of bacteria remained healthy.

On the other hand, when mice were injected with a mixture of live avirulent and virulent cells, the rodents died. The inferences Griffith derived were phenomenal.

A detailed report showcasing the biopsy reports of the dead mice showed that their respiratory tracts contained virulent bacteria. This clearly showed that the living harmless bacteria had acquired the ability of the dead bacteria. This phenomenon was called transformation and the unknown component responsible for this was the transforming principle. This inference formed the foundation of the idea of an independent genetic material that was hereditary in nature.

The challenge to unravel the true nature of genes and to determine the transforming principle was taken up by Oswald Avery. He showed, after a series of exhaustive experiments, that this transforming principle had properties that were similar to DNA.

DNA was discovered in 1869 by Johann Friedrich Miescher, a Swiss biochemist working in Tübingen, Germany. Most people do not know this little-known fact and credit Watson and Crick for the discovery. However, at the time when Miescher discovered the DNA, its precise biological function was not known.

The experiment conducted by Griffith and several other biologists' experiments influenced Watson and Crick to study DNA and discover the double helix structure of the DNA molecule. The double helix structure of DNA opened new vistas in the study of how genes replicate. Further, in 1903, W.S. Sutton proved that genes are located in chromosomes and their inheritance patterns are similar to that of chromosomes during cell division.

A chromosome is a long piece of DNA, which has embedded many genes and other elements. Scientists who study the discipline of cytochemistry have proved that the chromosomes comprise DNA and protein dispersed in equal quantities.

TIP

The word "chromosome" comes from the Greek words "chroma" and "soma". Chroma means colour while soma means body.

It was assumed that DNA, as a molecule, was not versatile enough to carry out the thousands of biological processes. The scientists concluded that proteins are the ultimate regulators of the hereditary function in cells. This assumption about DNA molecules was widely accepted because proteins were known, correctly, to be highly variable polymeric molecules, each one made up of a different combination of 20 chemically distinct amino-acid monomers. After extensive research, the scientific community has accepted that the key to all questions relating to heredity lies in DNA.

Organization of Eukaryotic Genome

In order to understand the complex nature of molecular processes, one has to first look at the complexity of the genomes. A genome contains the entire genetic information about the hereditary material possessed by an organism, simple or complex. Organisms with complex cells are known as Eukaryotes. Such organisms have many chromosomes in the cells' nuclei. Eukaryotic genomes are split into two major components:

1. Nuclear genome
2. Mitochondrial genome

The bulk of the genome is contained in the chromosomes and the mitochondria and, in the case of photosynthetic organisms, in the chloroplasts. In case of the nuclear genome, it is found in the chromosomes that act as the vehicle of inheritance. Figure 3.1 depicts the high level of complexity that is involved in the organization of the genomic material within the cells.

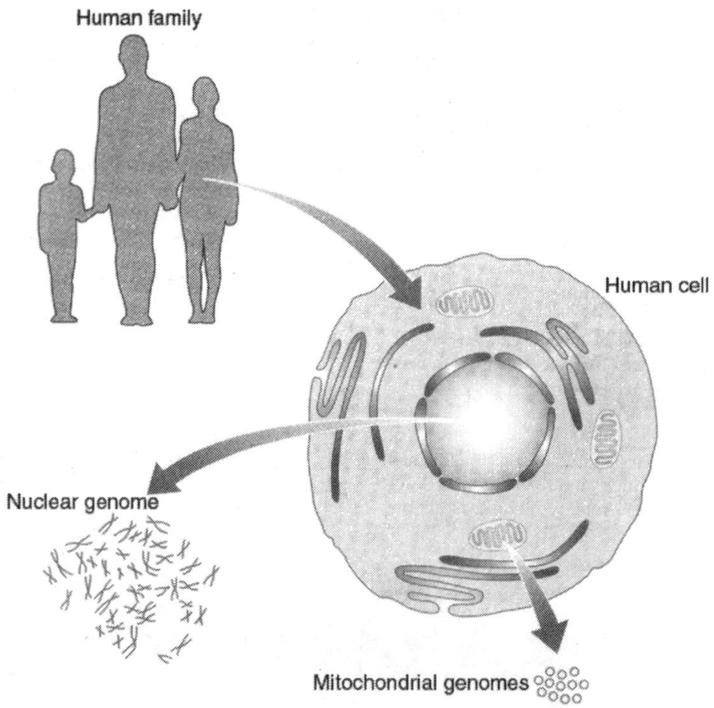

FIGURE 3.1 The Nuclear and Mitochondrial Components of the Human Genome

Packaging of DNA

Nature is an efficient packer. This has already been seen when we discussed about the organization of the cell. DNA molecules are bound together with the help of special proteins called histones. Further biochemical analysis proved that there are four classes of histones, namely H2A, H2B, H3 and H4. Two molecules of each class of histone together form a barrel-shaped core octamer with the DNA wound around the octamer, making 1¾ turns around the protein complex. The amount of DNA associated with the histone octamer is around 140–150 bp. The octamer plus the DNA comprise what is called the nucleosome core. A small stretch of DNA (50–70 bp) runs between adjacent nucleosome cores, and is known as the **linker**. A single nucleosome consists of one core plus a linker. This formation and the structure vary with the different organisms. This model is depicted in Figure 3.2.

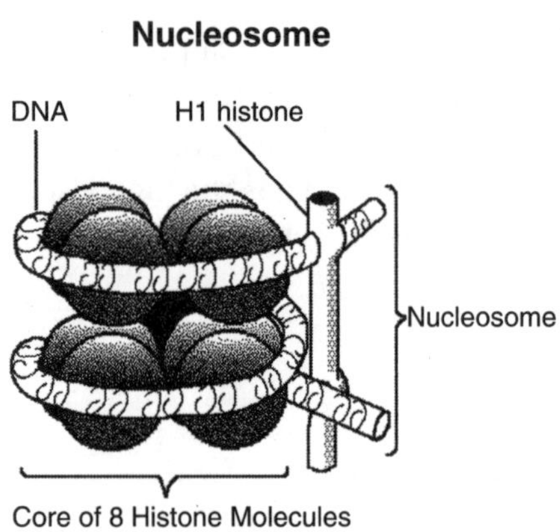

FIGURE 3.2 Structure of a Nucleosome

Figure 3.3 shows the unpacked form of the nucleosome. It is popularly termed as the beads on string model.

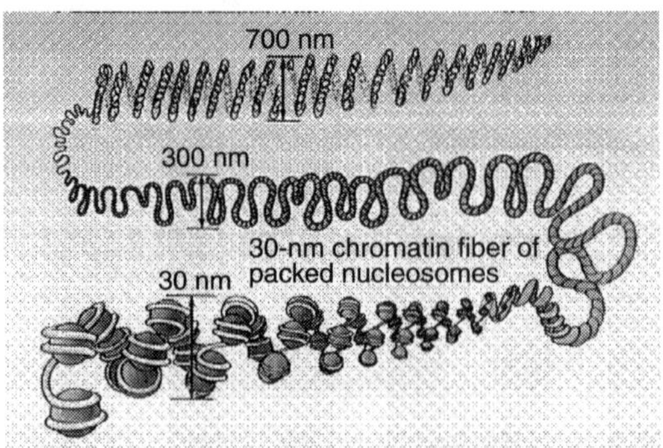

FIGURE 3.3 The Unpacked Form of the Nucleosome

A highly condensed version of this model is called the 30nm fibre. However, the exact mechanism of the nucleosome's association to form the 30nm fibre is not clear. Out of the several models that have been proposed, the most popular is the solenoid structure.

Classification of Genes

Genes can be classified using different parameters. However, we will discuss the classification based on the functional properties. Such a classification suggests that all eukaryotes possess the same basic set of genes. This classification also reveals that the number of genes present corresponds to the complexity of the organism. Figure 3.4 depicts the classification of genes on the basis of functional properties.

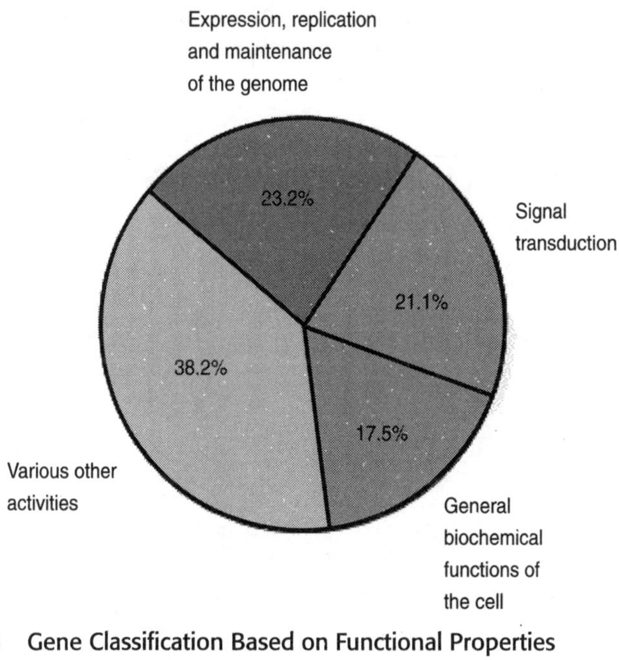

FIGURE 3.4 Gene Classification Based on Functional Properties

Functional classification is more clear and advantageous because it enables us to break down the complex functional genes into further subsets. These subsets can be even further classified on the basis of their subfunctions. However, this approach also has limitations due to the lack of functional assignments to genes on the basis of sequencing data. The functional classification approach can be modified by classifying the structures of proteins rather than their functions.

Central Dogma

"The Central Dogma is a paradigm of molecular biology": This hypothesis was described by Crick in 1958. As per this hypothesis, the overall flow of genetic information is depicted as:

DNA → RNA → Protein

An organism replicates both DNA and RNA in its body. RNA can be made or transcribed from DNA. This process of deriving is called transcription since nucleic acids are the templates that are used in the process. In some cases, RNA is used to make DNA (i.e. "reverse transcription" which is discussed later) using a particular enzyme called reverse transcriptase.

The central dogma suggests that DNA directs the synthesis of RNA, and RNA makes proteins, which ultimately makes the cell. In other words, we can say that life can be traced back to DNA. Despite the universality of the central dogma, there are important variations in the way information flow from DNA to protein. However, the dogma, depicted in Figure 3.5, is one of the most

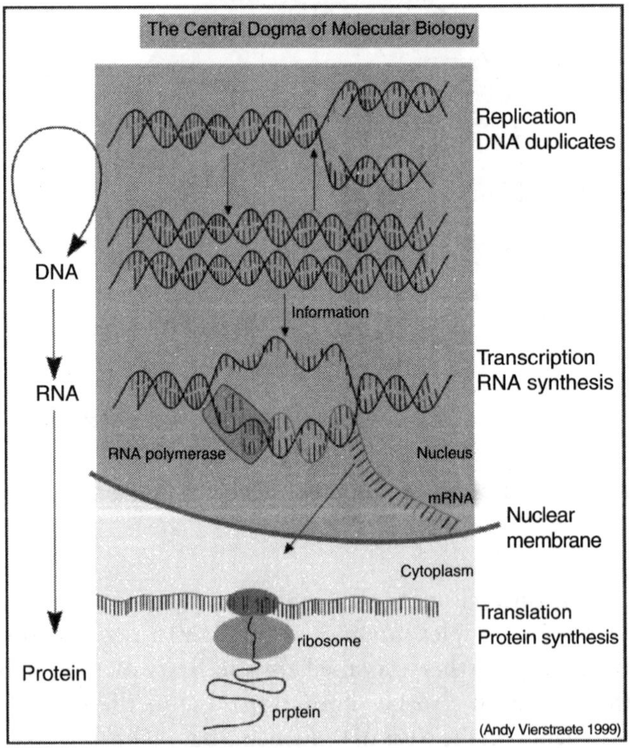

FIGURE 3.5 The Central Dogma

basic tenets of molecular biology that has to be studied in detail in order to appreciate the complexity of the genetic organization.

Commencement of Replication

DNA duplication is regarded as a precursor to cell division. In eukaryotes, this occurs during S phase of the cell cycle. The replication process always begins at the points called the origins of replication on a DNA molecule. Once the process is initiated, two replication fork-like structures emerge from the origin and progress in opposite directions along the DNA. This kind of replication methodology is called the bidirectional mode of replication.

The replication forks then progress along the DNA and participate in the process of genome replication. This process also helps in directing the synthesis of new complementary strands of DNA. At the chemical level, the synthesis of DNA seems similar to the synthesis of RNA that occurs during transcription, even though the actual mechanics of these two processes are very different.

The two strands of the double helix are copied prior to the process. This is important considering DNA polymerase enzymes (one of the several classes of enzymes that polymerize DNA nucleotides) are only able to synthesize DNA in the 5′→3′ direction. While one strand of the parent double helix, called the leading strand, can be copied in a continuous manner, the replication of the lagging strand has to be carried out in a discontinuous fashion, resulting in a series of short segments called Okazaki fragments that must be connected together to produce the intact daughter strand. This process is shown in Figure 3.6.

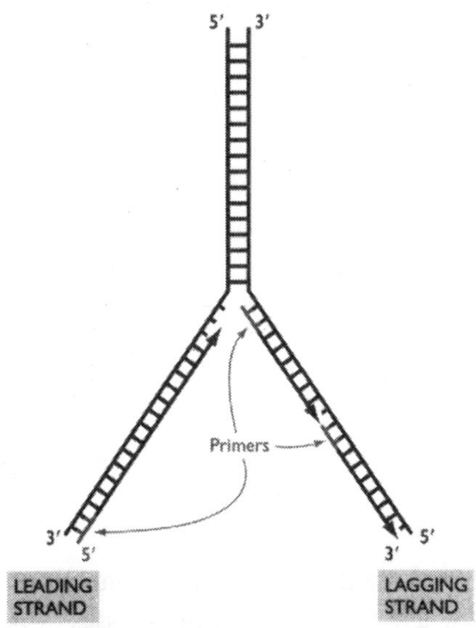

FIGURE 3.6 Complications with DNA Replication

One important constraint in this process is that the DNA polymerases cannot initiate DNA synthesis on a molecule that is entirely single stranded. This makes it necessary for the presence of a short double-stranded region to provide a 3′ end onto which the enzyme can add new nucleotides called a primer. These primers are required to initiate complementary strand synthesis on the leading polynucleotide, and for every segment of discontinuous DNA synthesized on the lagging strand. The mechanism of ligation (joining) of DNA fragments by DNA ploymerase is shown in Figure 3.7.

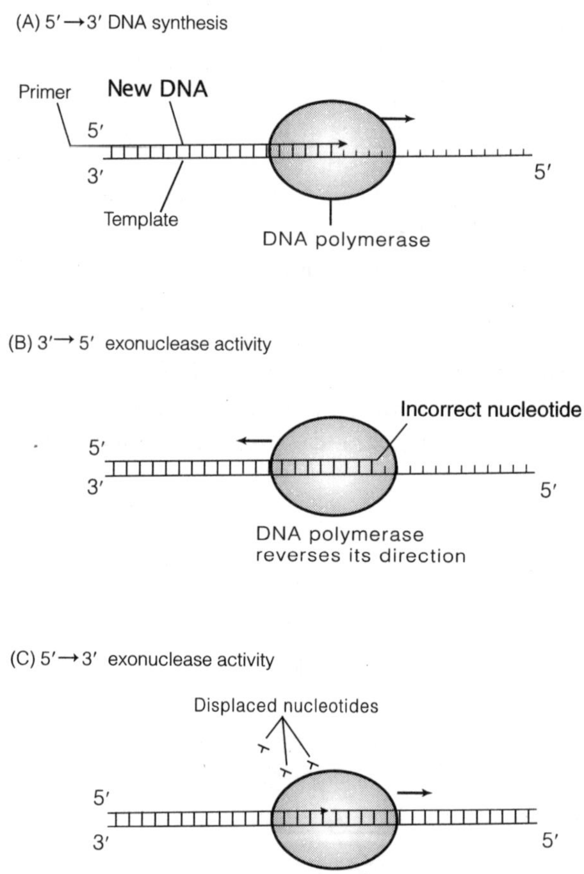

FIGURE 3.7 The DNA Synthesis and Exonuclease Activities of DNA Polymerases

It was in the mid 1950s that scientists realized that DNA synthesis was the key to replication of genes. There are three main classes of polymerases that were attributed to this process and were first identified in *E. coli*.

1. DNA polymerases I
2. DNA polymerases II
3. DNA polymerases III

DNA polymerases I and II are made up of single polypeptides while DNA polymerase III is a multi-subunit enzyme, with a molecular mass of approximately 900kDa. The three main subunits that form the core enzyme are: α, ε and θ, with the polymerase activity specified by the α subunit and the $3' \rightarrow 5'$ exonuclease by ε. The function of θ is not clear though it is supposed to play a crucial role in the binding of the subunits.

Eukaryotic replication system consists of nine DNA polymerases, which in mammals are distinguished by Greek suffices (α, β, γ, δ, etc.). DNA polymerase δ, the main replicating enzyme has two subunits and works in conjunction with an accessory protein called the proliferating cell nuclear antigen (PCNA). PCNA is the functional equivalent of the β subunit of *E. coli* DNA polymerase III and performs the function of holding the enzyme tightly to the template. DNA polymerase α also has an important function in DNA synthesis and initiates eukaryotic replication. DNA polymerase α participates in the replication of the mitochondrial genome. Figure 3.8 depicts Initiation of DNA synthesis in bacteria and eukaryotes.

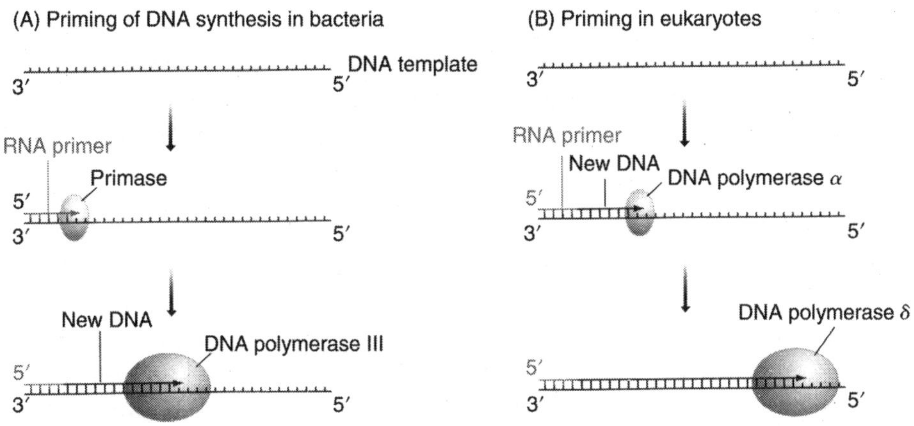

FIGURE 3.8 Priming of DNA Synthesis in Bacteria and Eukaryotes

Polynucleotides are synthesized by DNA polymerases in the $5' \rightarrow 3'$ direction. This results in synthesis of a lagging strand on one of the parent templates in a broken fashion. This was demonstrated in 1969 when Okazaki fragments were first isolated from *E. coli.* These fragments are 1,000 to 2,000 nucleotides in length in case of the bacteria, but in higher eukaryotes, Okazaki fragments are much shorter. Also remember that a primer is needed for the initiation of the reaction.

The reason for this dependency is due to the proofreading activity. This primer is synthesized on the DNA template by an enzyme called RNA primase. Any nucleotide that has been inserted incorrectly at the extreme 3′ end of a growing DNA strand, and which is non complementary is excised by the 3′→5′ exonuclease activity of the DNA polymerase. Complementary base pairing is one of the crucial factors for the reaction to occur.

This is the reason why an entirely single-stranded template, lacking a base-paired 3′ nucleotide, cannot be used as a template by a DNA polymerase.

Since RNA polymerases can use single stranded templates for replication, the primer is derived from RNA. In case of *E. coli,* primers are synthesized by an RNA polymerase enzyme called primase that is unrelated to the transcribing enzyme, with each primer 4 to 15 nucleotides in length. The strand synthesis is resumed by DNA polymerase III after the completion of primer synthesis. The RNA primer of 8 to 12 nucleotides synthesized by primase is passed over to DNA polymerase α, which extends the same by adding about 20 nucleotides of DNA. After completion of the synthesis of the RNA-DNA primer, DNA replication is continued by the main replicative enzyme, DNA polymerase δ.

Once primed, the leading-strand copy is synthesized continuously until replication is completed. However, synthesis of a new Okazaki fragment is initiated every time on the lagging strand. In *E. coli,* which makes Okazaki fragments of 1,000 to 2,000 nucleotides in length, approximately 4,000 priming actions are required every time the genome is replicated.

Events at Bacterial Replication Fork

There are combinations of events that occur at the replication fork during the elongation phase of genome replication. Within the elongation phase, there are two processes—initiation and elongation overlap. During the replication process, the helicase structure binds itself to its origin to form a pre-priming complex. This results in the formation of primosome. Primosome is a protein complex that is involved in genome replication. It is responsible for initiating replication of the leading strand.

The processes such as transcription, recombination and DNA repair require unwinding of the DNA; all these processes are carried out by the helicases. These enzymes bind to a single-stranded DNA, and migrate along the polynucleotide in either the 5′→3′ or 3′→5′ direction. The energy required for the breakage of base pairs is generated by hydrolysis of ATP.

There are some mechanisms and process in order to prevent the complementary base pairing of the single stranded DNA and digestion of the same by nucleases. For the prevention, proteins called the Single-Strand Binding proteins (SSBs) are recruited to polynucleotides, which attach to them and prevent them from reassociating and from degradation. Figure 3.9 depicts the role of the SSBs in replication.

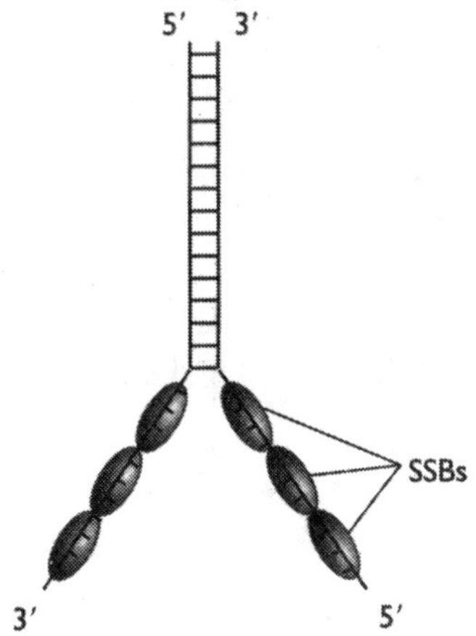

FIGURE 3.9 SSBs Attach to the Unpaired Polynucleotides

Bacterial SSBs are made up of four identical subunits. The bacterial SSBs are hypothesized to work in a similar way to the eukaryotic SSBs. The process in this case is by enclosing the polynucleotide in a channel formed of a series of SSBs attached side by side on the strand. The replication mediator proteins bring about the detachment of the SSBs, which is essential when the replication complex starts replicating the single strands.

The discontinuous strand synthesis on the lagging strand generally begins after more than a thousand nucleotides are synthesized on the leading strand. The primase, which is associated with the DnaB helicase in the primosome, makes an RNA primer. This is then extended by DNA polymerase III. Figure 3.10 shows the initiation and synthesis of lagging strand.

The complex formation of the DNA polymerase III with the primosome is called the replisome. The replication process is terminated by the unison of the individual Okazaki fragments. One member of each pair of adjacent Okazaki fragments retains the RNA primer attached at the point where ligation should occur. At this point the DNA polymerase III is replaced by the DNA polymerase I which removes the primer along with a small fragment of the Okazaki fragment.

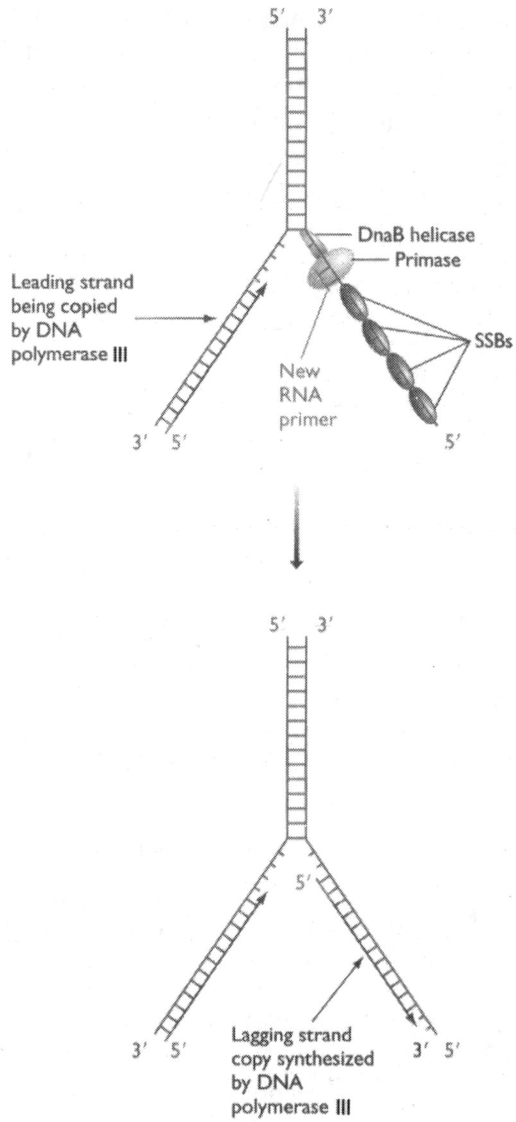

FIGURE 3.10 Priming and Synthesis of the Lagging Strand Copy During DNA Replication of *E. coli*.

The two Okazaki fragments now lie next to each other, with the terminal regions of both composed only of DNA. A phosphodiester bond is added in place by a DNA ligase (an enzyme that synthesizes phosphodiester bonds as part of DNA replication, repair and recombination processes) that completes the replication process.

As compared to the RNA synthesis, the rate of DNA synthesis is approximately five times more. Hence it is imperative that the replication complex easily leaves behind an RNA polymerase; to explain this phenomenon a model proposed by Deshpande and Newlon in 1996 states that the replication fork pauses behind the RNA polymerase, it proceeds only when the transcript has been completed. In the end, the replication fork reaches the end of the molecule or meets a second replication fork moving in the opposite direction.

Shortening of Chromosomes

It is observed that in some cases the extreme 3′ end of the lagging strand is not copied because the final Okazaki fragment is not primed. The absence of this Okazaki fragment leads to the shortening of the lagging-strand copy. This means that during the next round of replication the resulting daughter molecule will be shorter than its grandparent.

Telomeres are the unusual DNA sequences at the ends of eukaryotic chromosomes. The solution to the end-shortening problem is related with the way in which this telomeric DNA is synthesized. In most cases, the telomere is replicated rather faithfully by the replication machinery. However, as has been exemplified above, the shortening of the telomere is not inevitable. This has to be countered by the action of an enzyme called telomerase, which brings about the extension of the telomere. Telomerase is composed of a protein and RNA subunit. In the human telomerase, the RNA component is 450 nucleotides long, containing a 5′-CUAACCCUAAC-3′ sequence near its 5′ end, whose central region is the reverse complement of the human telomere repeat sequence 5′-TTAGGG-3′. This enables telomerase to extend the telomeric DNA by a mechanism in which the telomerase RNA is used as a template for each extension step and the DNA synthesis being carried out by the protein component of the enzyme, which is a reverse transcriptase. Due to the prevalence of the guanine nucleotides in the strand synthesized by telomerase, it is called the G-rich strand.

Once the G-rich strand is long enough, the primase-DNA polymerase α complex attaches at its end initiating synthesis of a complementary DNA. A new RNA primer is recruited for the purpose, which brings about the shortening of the C-rich strand, nevertheless ensuring that the overall length of the chromosomal DNA is maintained.

Transcription

Transcription is the process through which a DNA sequence is enzymatically copied by an RNA polymerase to produce a complementary RNA.

RNA Polymerase I is located in the nucleolus and transcribes ribosomal RNA (rRNA).

RNA Polymerase II is localized to the nucleus, and transcribes messenger RNA (mRNA) and most small nuclear RNAs (snRNAs).

RNA Polymerase III transcribes transfer RNA (tRNA) and other small RNAs.

Transcription is the beginning of the process that culminates into the translation of the genetic code into a functional peptide or protein. The actual process by which a gene is transcribed is divided into three main phases: initiation, elongation and termination.

Initiation

One of the key events in the initiation of transcription is the transcription of genes by RNA polymerases rather than just fragments of DNA. This means that the initial binding of the RNA polymerase enzyme to the DNA must occur at a specific predetermined location. This is crucial considering the fact that in case of higher eukaryotes most of the genome is made up of introns (non-coding regions) and the presence of such locations would ensure the fidelity of transcription.

The start site of transcription is signalled by means of the attachment points called the promoters. A promoter is a short strand of nucleotide sequence that is recognized by the RNA polymerase enzyme as a point at which it binds to the DNA. These are always located upstream of the DNA sequence. The promoters are more or less similar in sequence, enabling the RNA polymerase to recognize it. In *E. coli*, it is made up of two distinct components, the 35 box and the 10 box, as shown in Figure 3.11. The latter is called the pribnow box.

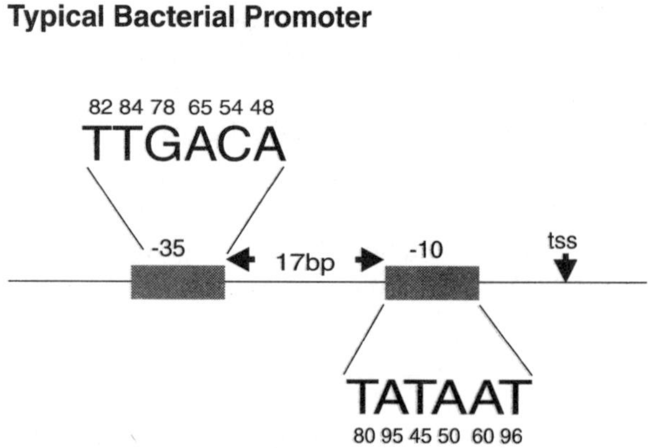

FIGURE 3.11 A Typical Bacterial Prometer

In case of the eukaryotes, the promoters are slightly different in construct. The genes trancribed by RNA polymerase are preceeded by -75 box (5′-GGN-NCAATCT-3′) and -25 box also called the Hogness box (5′-TATAAAT-3′). These are consensus sequences with N indicating a position occupied by any of the four nucleotides.

RNA polymerase (RNAP), upon recognizing the promoter, winds along the double stranded DNA and the resulting structure is known as the *closed complex*. In this complex, it covers around 60bp of the helix. Upon unwinding, the DNA becomes single-stranded in the vicinity of the initiation site. This is denoted as +1. This unwound form bound to the enzyme is called the *open complex*.

Elongation

The basics behind the process of elongation are the same in bacteria and eukaryotes, with the length of the transcript to be synthesized being the only difference. The RNA polymerase core enzyme migrates along the DNA molecule adding ribonucleotides to the 3′ end of the growing RNA molecule unwinding a small region of the double helix at any one time. The time taken for the synthesis of the transcript may vary between a few seconds in case of prokaryotes, to hours in case of the eukaryotes. This is due to the occurrence of multiple introns that increase the length of the transcript.

Due to its instability, RNA polymerase II requires the action of other factors to mediate elongation of lengthy transcripts. Elongation factors are proteins that associate with the polymerase after it has cleared the promoter and left behind the transcription factors involved in initiation. Thirteen elongation factors are currently known in mammalian cells, displaying a variety of functions. Table 3.1 shows some instances of elongation factors for mammalian RNA polymerase II.

TABLE 3.1 Elongation Factors and their Functions

Elongation Factor	Function
CSB, ELL, Elongin	Suppress 'pausing' of RNA polymerase II, which can occur when the enzyme transcribes through a region where intrastrand base pairs (e.g., a hairpin loop) can form.
SII	Prevents arrest (complete cessation of elongation).
FACT, HMG14	Modify chromatin in order to assist elongation.

Termination

There are two well known termination mechanisms:

Intrinsic termination

This mechanism is also known as the *Rho-independent termination.* This involves terminator sequences within the RNA that signal the RNA polymerase to stop. The terminator sequence is usually a palindromic sequence (reading the same in either direction) that forms a stem-loop hairpin structure that leads to the dissociation of the RNA polymerase from the DNA template.

Rho-dependent termination

Some termination signals are recognized by the *E. coli* RNA polymerase core enzyme alone, while some depend on the presence of an additional polypeptide termination factor called ρ-factor that stops RNA synthesis at specific sites. This is bound to the mRNA and moves towards the RNA polymerase. When ρ-factor reaches the RNAP, it causes RNAP to dissociate from the DNA, terminating transcription. Termination can also occur where RNAP comes across a region with repetitious thymidine residues in the DNA template.

The final event in the process of transcription is dissociation of the RNAP enzyme from the DNA and the release of the RNA transcript. The next sets of events in case of the eukaryotes are called the post-transcriptional modifications. These are essential due to the higher level of compartmentalization in the eukaryotic systems and the mRNA has to be transported to the cytoplasm before translation can occur. The following modifications occur in case of eukaryotic mRNA. A newly made mRNA copy of a gene (pre-mRNA), must undergo three major modifications before it can be translated into a protein:

1. Capping: A special nucleotide is attached at the 5′ end of the mRNA. This modification is necessary for efficient initiation of protein synthesis and serves as stabilization.
2. Poly(A)-tail: A special enzyme attaches a chain of 150–200 adenine nucleotides to the 3′ end of the pre-mRNA directly after transcription. This increases the stability and prolongs the lifetime of an mRNA molecule.
3. Splicing: Non-coding sequences, introns, are removed from the pre-mRNA to create mRNA, containing only the coding sequences for a protein. This process is carried out by the spliceosome. The spliceosome is a special complex made up by proteins and a catalytic kind of small nuclear RNA molecules, the snRNAs.

PROTEINS: THE HOLY GRAIL

The molecular cycles that govern the transfer of genetic material form one generation to the other are both complex and highly efficient. However, the end point of this transfer culminates with the formation of a functional protein. For all the fidelity that is maintained in the process of transcription and translation, a non-functional protein can lay waste to the entire process. Hence the focus has shifted form the study of molecular aspects of genomes (information about the genetic make-up of organisms) back to proteins. As far as the biological fraternity is concerned, the search for novel proteins is analogous to the quest for the elusive and legendary Holy Grail.

Proteomics: The Next Revolution

Rapid advances have been made in obtaining genomic information, however, very little progress has been seen in the ability to relate this information to the gene products, i.e., proteins. In the last few years, it has become widely recognized that the genome only represents the first layer of complexity. The biological function is not carried out by the static genome but mainly by the dynamic population of proteins determined by the interplay of gene and protein regulation with extracellular influences. Even the list of mature proteins, including splice variants and post-translational modifications, cannot be predicted from the genome sequence. There are numerous instances of differential splicing and many post-translational protein modifications (e.g., phosphorylation, glycosylation, ubiquitination and methylation) that can govern the behaviour of proteins more than differing rates of synthesis. Proteins are the main carriers of biological activity that embody the next level of complexity. Protein function depends on the precise amino acid sequence, the modifications (especially regulatory ones like phosphorylation), the three-dimensional (3D) structure, the protein concentration, the association with other proteins and the extracellular environment.

With the realization creeping in that the genome sequence alone is insufficient to provide insights into complex cellular, organ and system level phenomena, an additional enterprise has emerged that encompasses the study of the structure, function and interaction of proteins, and this enterprise forms the realm of "proteomics".

The term proteome refers to the protein equivalent of the genome. However, this definition does not take into account that the proteome is a highly dynamic entity that will change based on cellular state and the extracellular milieu. A precise definition of a proteome is the one that specifies that it is the protein complement of a given cell at a specified time, including the set of all protein isoforms and protein modifications. It comprises the complete set of gene products that are synthesized by the genome, and the analysis of these gene products has been termed proteomics. There has been great progress in the area owing to novel instrumentation, experimental strategies and bioinformatics methods. Figure 3.12 depicts the shift from genomics to proteomics.

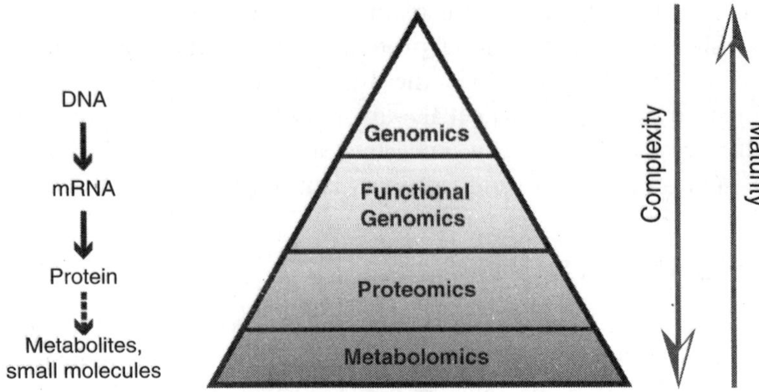

FIGURE 3.12 Shift from Genomics to Proteomics

Whether proteomics becomes an intellectually coherent discipline rather than a sprawling agglomeration of separate technologies, is difficult to predict. However, it is a useful organizing principle that encourages researchers to think in an interdisciplinary manner and to venture beyond their individual universes. The post-genomic era certainly belongs to the study and discovery of novel proteins.

Terms to Know

Aminoacyl or A site	The site in the ribosome occupied by the aminoacyl-tRNA during translation.
Anticodon	The triplet of nucleotides, at positions 34–36 in a tRNA molecule, that base-pairs with a codon in an mRNA molecule.
Anticodon arm	Part of the structure of a tRNA molecule.
Base pair	The hydrogen-bonded structure formed by two complementary nucleotides. When abbreviated to 'bp', the shortest unit of length for a double-stranded DNA molecule.
Base-pairing	The attachment of one polynucleotide to another, or one part of a polynucleotide to another part of the same polynucleotide, by base pairs.
Beads-on-a-string	An unpacked form of chromatin consisting of nucleosome beads on a string of DNA.
Cap binding complex	The complex that makes the initial attachment to the cap structure at the beginning of the scanning phase of eukaryotic translation.
Concatamer	A DNA molecule made up of linear genomes or other DNA units linked head to tail.
Degeneracy	Refers to the fact that the genetic code has more than one codon for most amino acids.
DNA-binding motif	The part of a DNA-binding protein that makes contact with the double helix.
DNA ligase	An enzyme that synthesizes phosphodiester bonds as part of DNA replication, repair and recombination processes.
Elongation factor	A protein that plays an ancillary role in the elongation step of transcription or translation.
Enhanceosome	A structure formed by DNA bending that comprises a collection of proteins involved in activation of the RNA polymerase II transcription initiation complex.
Enhancer	A regulatory sequence that increases the rate of transcription of a gene or genes located some distance away in either direction.
fMet	*N*-formylmethionine, the modified amino acid carried by the tRNA that is used during the initiation of translation in bacteria.

Gene expression	The series of events by which the biological information carried by a gene is released and made available to the cell.
Heterochromatin	Chromatin that is relatively condensed and is thought to contain DNA that is not being transcribed.
Histone	One of the basic proteins found in nucleosomes.
Klenow polymerase	A DNA polymerase enzyme, obtained by chemical modification of *Escherichia coli* DNA polymerase I, used primarily in chain termination DNA sequencing.
Major groove	The larger of the two grooves that spiral around the surface of the B-form of DNA.
Single-strand binding protein (SSB)	One of the proteins that attaches to single-stranded DNA in the region of the replication fork, preventing base pairs forming between the two parent strands before they have been copied.
Transcript	An RNA copy of a gene.
Transcription	The synthesis of an RNA copy of a gene.
Upstream	Towards the 5′ end of a polynucleotide.
Zinc finger	A common structural motif for attachment of a protein to a DNA molecule.

4 Genetic Engineering

- Introduction
- Molecular Scissors
- Vectors: Nature's Trojan Horse
- Cloning a Gene
- Polymerase Chain Reaction (PCR): DNA Copiers
- DNA Fingerprinting: Molecular IDs
- Gene Engineering Product Optimization
- Biotechnology: The Ultimate Panacea

Introduction

The technique of genetic engineering involves several tools that enable one to manipulate and alter the genomic content. One of the most important and fundamental tools in this regard is the restriction enzyme that is used in "splicing" through the genetic material of organisms. The discovery and subsequent use of this enzyme is pivotal in determining the success of biotechnology and gene manipulation.

Molecular Scissors

There is a wide array of tools used in the field of genetic engineering and one of the basic tools is molecular scissors. These are groups of enzymes made of bacteria. These enzymes are used to cut DNA strands at a specific position.

The basic objective of this recombinant DNA technology is to create recombinant DNA (rec-DNA). One of the first requirements is to create small DNA fragments. In the early days of rec-DNA technology, this was a difficult proposition. Since very crude mechanical methods were used to shear DNA it resulted in random fragments that were of no significance.

The breakthrough came in 1970 with the discovery of a class of enzymes in the bacteria *Haemophilus influenzae* that are called restriction endonucleases or restriction enzymes. These are the most ideal DNA snippers that are routinely employed by the bacteria as a defence against viruses. The viruses hijack the bacterial cell by injecting their DNA or RNA into it. The bacteria in turn counterattack by chopping up the same with their restriction endonucleases, thus restricting the viral infection.

Mode of Action

The most crucial aspect of this mechanism is that each restriction enzyme cuts a DNA molecule at very specific points only identified by a particular stretch of nucleotides. This means that by using different restriction enzymes one can generate specific fragments of DNA of varying lengths.

Restriction Enzymes as Tools

Recognition sequences typically are only four to twelve nucleotides long. Because there are only so many ways to arrange the four nucleotides—A, C, G and T–into a four or eight or twelve nucleotide sequence, recognition sequences tend to "crop up" by chance in any long sequence. Furthermore, restriction enzymes specific to hundreds of distinct sequences have been identified. As a result, potential "restriction sites" appear in almost any gene or chromosome. Meanwhile, the sequences of some artificial plasmids include a "linker" that contains dozens of restriction enzyme recognition sequences within a very short segment of DNA. So no matter the context in which a gene naturally appears, there is probably a pair of restriction enzymes that can snip it out, and which will produce ends that enable the gene to be spliced into a plasmid (i.e. which will enable what molecular biologists call "cloning" of the gene).

Currently, there are more than a thousand restriction enzymes that have been discovered and the list is increasing. Table 4.1 shows some commonly used restriction enzymes.

TABLE 4.1 Commonly Used Restriction Enzymes

Bacterial Source	Enzyme Abbreviation	Sequence 5-3 3-5
Haemophilus aegyptius	HaeIII	GC---CC CC---GG
Staphylococcus aureus 3A	Sau3AI	GATC---CTAG
Bacillus amyloliquefaciens H	BamHI	G--GATCC CCTAG--G
Escherichia coli RY13	EcoRI	G--AATTC CTTAA--G

TABLE 4.1 *(Continued)*

Bacterial Source	Enzyme Abbreviation	Sequence 5-3 3-5
Haemophilus influenzae Rd	HindIII	GTPy---PuAC CTPu---PyTG
Providencia stuartii	PstI	CTGCA--G G--ACGTC
Serratia marcescens	SmaI	CCC--GGG GGG--CCC
Xanthomonas malvacearum	XmaI	C--CCGGG GGGCC--C
Moraxella bovis	MboII	GAAGAN* CTTCTN*

*This is a type II enzyme that does not cut within the restriction sequence, but at whatever sequence that lies to the right. N is any nucleotide.

Fragment Complementarity and Splicing

Once the DNA is cut using the restriction enzyme, it leaves both the fragments (remember DNA is double-stranded) dangling, with single-stranded 'tails' of unpaired bases. Complementary fragments of DNA bond together when bases on these ends pair up. Figure 4.1 shows the action of EcoRI restriction enzyme resulting in sticky ends. The complementary base pairing of the ends results in a recombinant DNA.

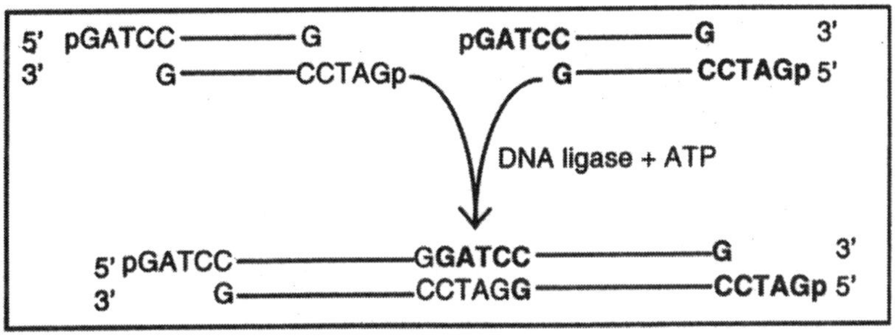

FIGURE 4.1 Ligation of Molecules with Cohesive Ends

These bonds are formed naturally but are very weak and hence they are made more secure by using a sealing enzyme called a 'ligase' in a process called ligation. These occur naturally to help synthesize DNA and repair minor damages to the molecule.

These ligase enzymes catalyze the formation of phosphodiester bonds between the 3' OH groups at the terminus of one of the strands, with the 5'PO_4 terminus of another strand. The DNA ligase encoded by the bacteriophage T4 is used routinely for the purpose of ligation. T4 DNA ligase activity requires ATP as a cofactor to form an enzyme-AMP intermediary complex. It then binds to the exposed ends of the interacting DNA molecules to create a covalent phosphodiester bond. Figure 4.2 shows the basic chemistry of ligation.

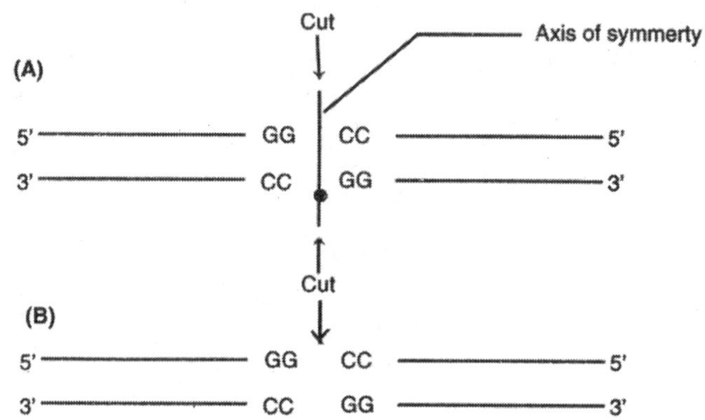

FIGURE 4.2 A - The Target Site for RE Hae III. Arrows Indicate the Sites where the Cuts will be Effected. B - The Target Site After the Cut

Figure 4.3 shows the ligation of fragments generated by EcoRI restriction enzyme.

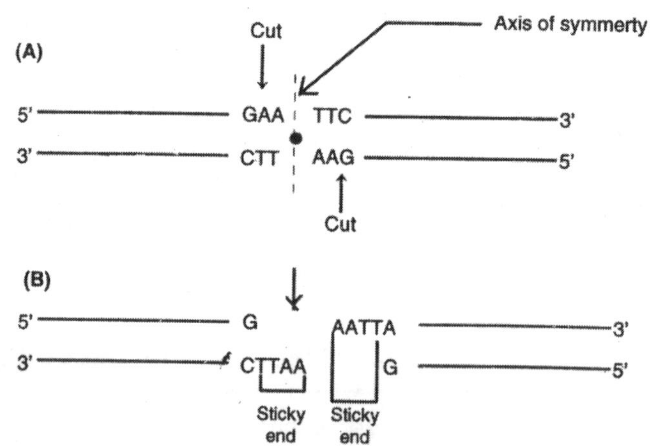

FIGURE 4.3 Recognition Site of the Restriction Endonuclease EcoRI

Cleavage or cutting with restriction enzymes generally results in two kinds of fragments: those that are terminal sequences with tails and those that have blunt heads. Based on the nature of the fragment, different methods are used to ligate them.

Many Recognition Sequences are Palindromic

While recognition sequences vary widely, many of them palindromic; that is, the sequence on one strand reads the same in the opposite direction on the complementary strand. The meaning of "palindromic" in this context is different from what one might expect from its linguistic usage: GTAATG is not a palindromic DNA sequence, but GTATAC is.

Types of Restriction Enzymes

There are three main types of restriction enzymes:

- Type I
- Type II
- Type III

Each type of enzyme is used depending on the type of fragments that needs to be generated. Type I enzymes are not used much due to their random nature of cleavage whereas Type III are extremely rare in nature and are used for very specific applications. Type II enzymes are the ones that are routinely used for genetic engineering purposes. The target sequences for the same are usually short, typically 4–6 bases in length. Table 4.2 shows different types of restriction enzymes and their characteristics.

TABLE 4.2 Restriction Enzymes and Characteristics

Characteristics	Type I	Type II	Type III
Restriction and modification activities	Single multifunctional enzyme	Separate endonuclease and methylation	Separate enzymes with a common subunit
Protein structure of restriction endonucleases	Three different subunits	Simple	Two different subunits
Requirements for restriction	ATP, Mg + 2, SAM*	Mg + 2	ATP, Mg + 2, SAM*
Cleavage sites	Possibly random. Up to 1000 bp from the host specificity site	At or near host specificity site	24–26 bp to 3 end of host specificity site
Enzyme turnover	No	Yes	Yes

*SAM (S-adenosylmethionine)

Source: Old and Primrose

The basic determining factor in the frequency of the action of these enzymes is the length of the restriction sites and their nucleotide composition in relation to the rest of the DNA. In a case, where, in a DNA all the four nucleotide bases occur randomly and with equal frequency, any given four bases sequence will occur on an average of once in every 256 base pairs (4^4), while six base pair recognition sequence will occur once in every 4096 base pairs (4^6).

Naming

The discovery of a large number of enzymes gradually made nomenclature a necessity in order to differentiate the different categories. The widely accepted norms for the nomenclature of enzymes are:

- Species name of the host organism is identified by the first letter of the genus name and the first two letters of the specific epithet to form a three-letter abbreviation in italics (as in the table).
- When multiple restriction systems are derived from the same host they are designated by numbers.
- All restriction enzymes have the general name endonuclease R, but in addition also carry the system name.

Having acquainted ourselves with the mechanism of generating DNA fragments the next logical question is what does one do with it. As mentioned earlier, the purpose of rec-DNA technology is to introduce foreign DNA into another species to create a recombinant.

This is easier said than done since integrating a foreign DNA into a host is not an easy task. A number of approaches have been adopted for the same. The most common techniques for direct transfer of DNA (or DNA fragments) include electroporation, use of spheroplasts and transfection. However, the most extensively used protocols involve the use of 'vectors'. Let us discuss vectors in detail.

VECTORS: NATURE'S TROJAN HORSE

A vector is nothing but a carrier on which the DNA fragment piggybacks on its way into the cell. One can compare it to the Trojan horse—it smuggles the DNA into the cell. The vector molecules exist as independent entities and do not become part of the chromosomal system of the organism.

There are certain prerequisites that have to be satisfied before a vector is employed. One of them is that the vector should be small in size and capable of replicating within the host. The other criterion is that it should have at least one site where the foreign DNA can be inserted. In addition to this, many vectors also have certain additional features like antibiotic resistance and promoters to control the expression of a gene.

One has to bear in mind that the same restriction enzyme has to be used to cut the vector as well as the foreign DNA.

Ligation of the insert and the vector is a crucial element for the success of any application. The procedure depends largely on the type of fragments that are generated by the restriction enzymes (sticky or blunt). This is the primary reason why both the vector and the DNA are cut with the same restriction enzyme.

Vectors are broadly divided into two main categories, namely:

- Plasmids—the small units of DNA, distinct from chromosomes.
- Bacteriophages—the bacterial viruses.

Plasmids and bacteriophages are shown in Figure 4.4.

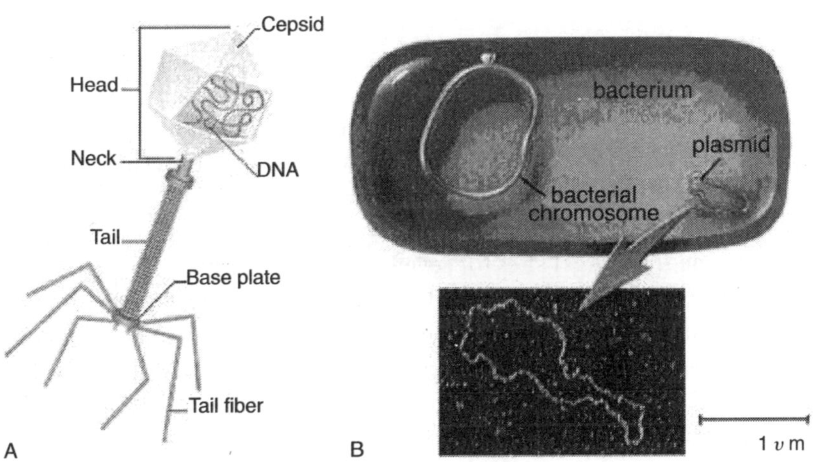

FIGURE 4.4 Bacterium Showing Chromosome and Plasmids

The different types of vectors used are as described below.

Pbr 322 and Related Plasmids

As mentioned earlier, naturally occurring plasmids were used for various applications during the early days of genetic engineering. This was, however, proving to be cumbersome and the success rate was extremely poor. In order to circumvent these shortcomings the first artificial plasmid was designed (Sutcliffe 1979). Figure 4.5 shows the restriction map of pBR322 vector denoting various restriction sites.

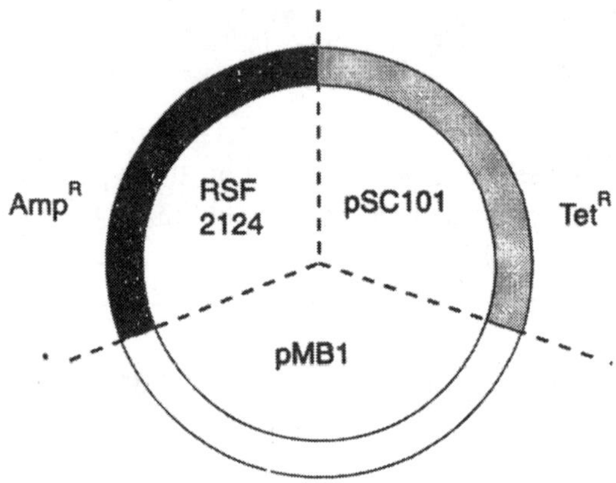

FIGURE 4.5 The Plasmid pBR322 Amp and Tet are Genes for Resistance to Ampicillin and Tetracycline

The plasmid has certain advantages over their natural counterparts:

- It contains the ampicillin and tetracycline resistance genes that are useful in detecting the efficiency of the transformation (insertion of the foreign DNA in the plasmid).
- There are over 20 enzymes with unique cleavage sites on the pBR322 genome. The target sites for seven of these enzymes lie within the tetracycline resistance gene, while there are three sites within the ampicillin resistance region. Thus, cloning in pBR322 with the aid of any one of these enzymes results in insertional inactivation of either the Ap or the Tc genes (the plasmid becomes susceptible to these antibiotics indicating successful integration of the gene of interest).

Over the years, numerous vectors have been derived from the pBR322 class of plasmids. These have been improved upon on various counts in order to allow larger inserts and better selection and screening strategies. Some of these are the pUC vector, pAT153, and the pBluescript series of plasmids.

Bacteriophage λ Vectors

Bacteriophage λ are a class of complex viruses infecting *E. coli*. The coliophage, as it is known, is used widely for a range of applications. One of the most interesting characteristics of these is the packaging system. The extreme ends of the λ DNA are known as COS sites, each is single-stranded, 12-nucleotides long. Because their sequences are complementary to each other, one end of λ DNA may base-pair with the other end of a different λ DNA, forming concatemers. The two ends of a λ DNA may also bind together, forming a circular DNA. In the host cell, the λ DNA circularizes because ligase may seal the join of the COS sites. In the assembly process of λ phages, two proteins Nu1 and A can recognize the COS site, directing the insertion of the λ DNA between them into an empty head. The filled head is then attached to the tail, forming a complete λ virion. The whole process normally takes place in the host cell. Figure 4.6 shows the mechanism of packaging of the phage head.

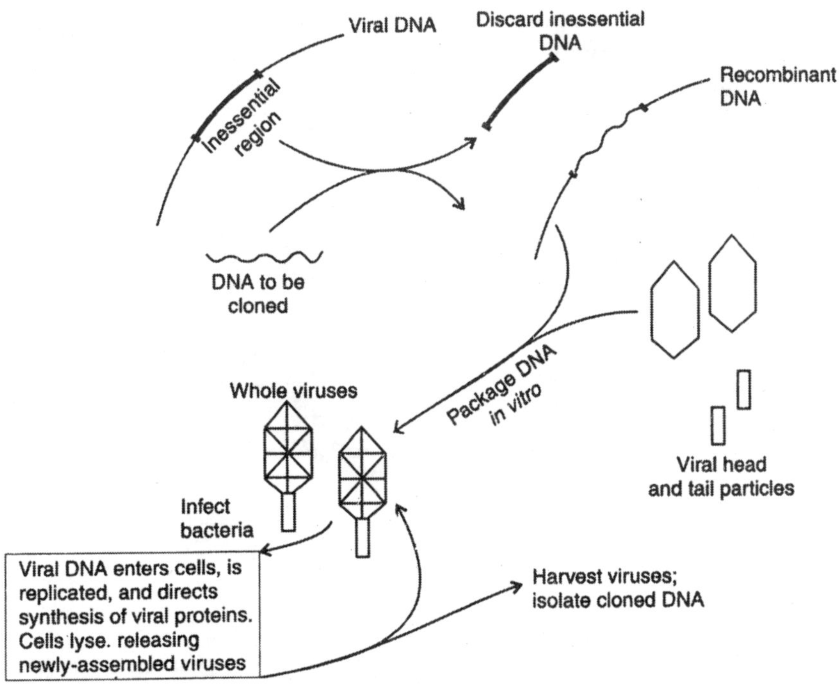

FIGURE 4.6 Cloning DNA in Bacteriophage

Further Proteins Nu1 and A are encoded by the genes in the λ genome. If the two genes are mutated, λ DNA cannot be packaged into the pre-assembled head. Because tails attach only to filled heads, the cell will accumulate separate empty heads and tails, which can then be extracted.

When the extract is mixed with recombinant λ DNA and proteins Nu1 and A, the complete λ virion carrying recombinant λ DNA is assembled. The major advantage of the λ phage vector is its high transformation efficiency, about 1000 times more efficient than the plasmid vector. This is made possible due to the fact that much of the phage genome is dispensable. By replacing a major part of the genome one can design what is called as a replacement vector. Some commonly designed vectors are λ gt 10 and 11.

Cosmids

The λ genome that indicates the genetic makeup of the bactriophage, is further exploited to develop, what are called the cosmid vectors. This allows the packaging of larger fragments of DNA. In this the *cos* sites of the lambda phage along with the other essential regions are ligated at either end of the DNA fragment that is to be packaged (cloned).

This allows the phage to package the same and infect the host. Once the DNA fragment enters the host it circularizes due to the cos sites and replicates within the host sans any viral activity. Figure 4.7 shows the mechanism of cosmid vectors.

Other different types of vectors are as follows:

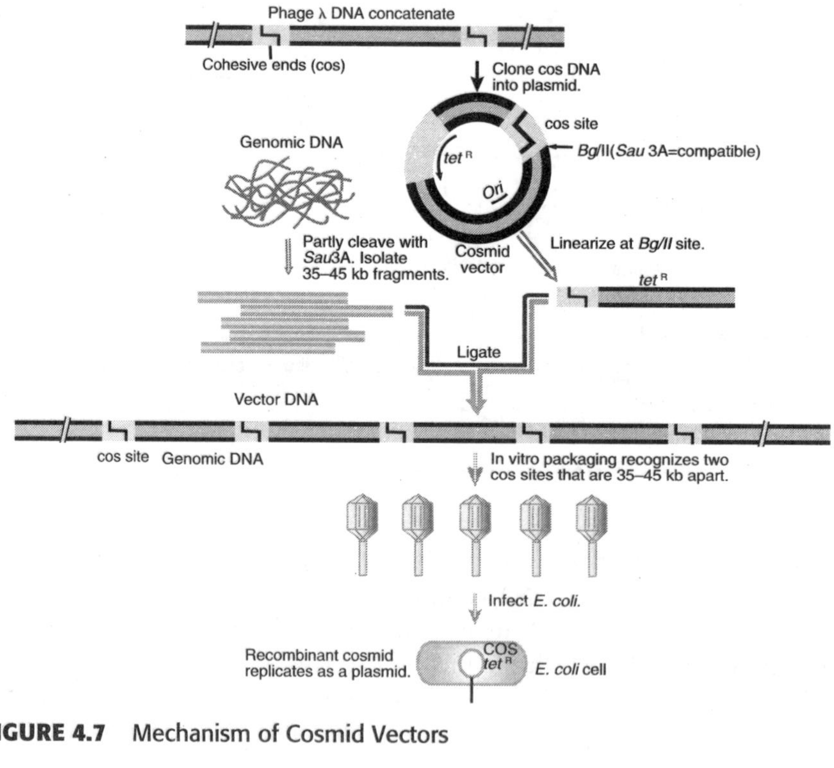

FIGURE 4.7 Mechanism of Cosmid Vectors

M13 Vectors

M13 is a filamentous, male-specific coliphage (*E. coli* infecting) composed of a single-stranded circular DNA genome and a protein coat. Because of its filamentous structure, additional DNA can be inserted into its genome and packaged into an extended phage capsid. During infection, the single-stranded DNA is converted to the double-stranded Replicative Form (RF) from which the mature single-stranded form is synthesized (Figure 3.7). One valuable feature of these vectors is that large sizes of DNA can be easily introduced in them without any packaging constraints like the lambda phages.

The M13 phage has been developed into a cloning vector by the introduction of the lactose operon region of *E. coli* into the phage, generating the M13 lac phage referred to as M13mp1. Gronenborn and Messing are the first scientists to design the M13 phage and further modified the phage M13mp1 by creating an EcoRI restriction site within the gene encoding b-galactosidase.

This phage (M13mp2) is easily screened for the introduction of a foreign DNA fragment by assaying for B-galactosidase activity in appropriate *E. coli* hosts. The properties described above make M13 an ideal vector for obtaining single-stranded templates. Figure 4.8 shows the structure of an M13 vector.

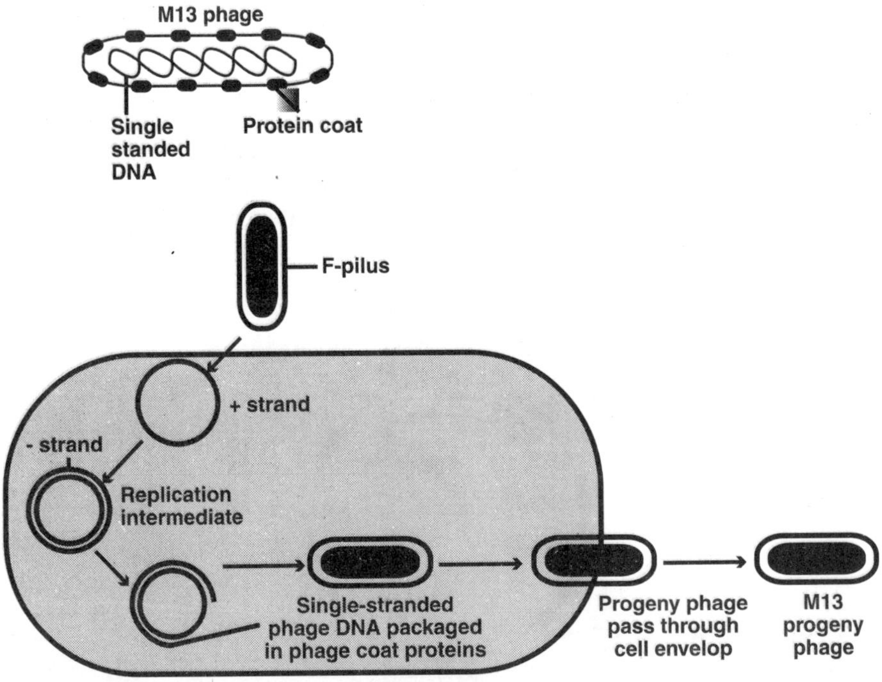

FIGURE 4.8 Life Cycle of Bacteriophage M13
Source: Hfr or *F. E. coli* Bacterium

Ti Plasmids

The virulent strains of bacteria *Agrobacterium tumefaciens*, known to cause crown gall disease in plants, contain large plasmids called the Ti-plasmids. A part of the Ti plasmid called the T-DNA was found to integrate with the plant nuclear DNA. It is this area that has been exploited for use as a vector in plant systems. When the foreign DNA is introduced along with the T-DNA it integrates into the host genome. Figure 4.9 shows the mechanism of transfer of Ti plasmid to the host.

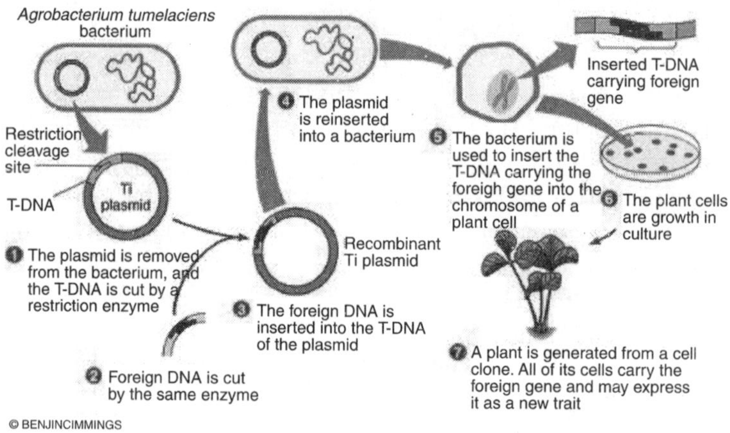

FIGURE 4.9 Transfer of Ti Plasmid to Host

The plant does not suffer from the disease because the regions responsible for the same are deleted prior to ligation of the foreign DNA. This is possible due to the fact that the region responsible for virulence is distinct form the T-DNA region. The vector is routinely employed in creating transgenic plants. Figure 4.10 shows the structure of Ti plasmid.

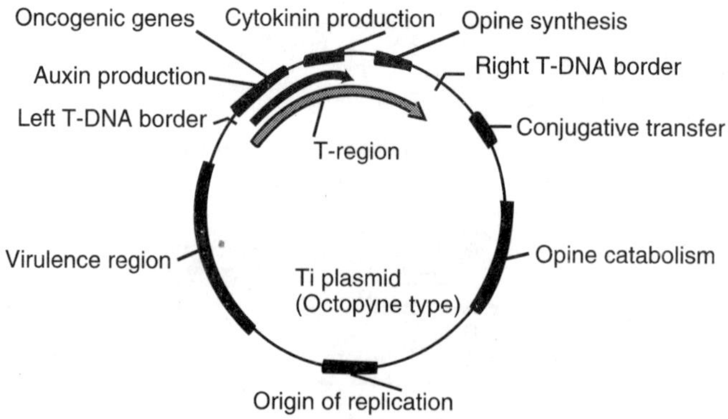

FIGURE 4.10 Gene Map of pSV1GT5 Vector

Yeast Derived Vectors

Genetic engineering, that is, transformation of yeast cells with recombinant DNA, became feasible for the first time in 1978. Since then, recombinant DNA technology in yeast has established itself, and a multitude of different vector constructs are available.

These are quite different to the plasmid vectors as they are linear in nature, unlike the circular plasmids. The viability of the yeast chromosome depends on the presence of centromeric sequences (CEN), telomers (TEL), and Autonomous Replicative Sequences (ARS), which ensure stability and replicative capacity. Some examples of vectors derived from yeast are:

- Integrative plasmids (**YIp**), which by homologous recombination are integrated into the host genome at the locus of the marker, when this is opened by restriction and linearized DNA is used for transformation. This (normally) results in the presence of one copy of the foreign DNA inserted at this particular site.
- Episomal plasmids (**YEp**), which carry part of the 2 u plasmid DNA sequence necessary for autonomous replication. Multiple copies of the transformed plasmid are propagated in the yeast cell and maintained as episomes.
- Autonomously replicating plasmids (**YRp**), which carry a yeast origin of replication (ARS sequence) that allows the transformed plasmids to be propagated several hundred-fold.
- Cen plasmids (**YCp**). In addition to an ARS sequence, these vectors carry a centromeric sequence (derived from one of the nuclear chromosomes), which normally guarantees stable mitotic segregation and reduces the copy number of self-replicated plasmids to just one.

Figure 4.11 shows the different types of yeast shuttle vectors.

YAC (*Yeast Artificial Chromo*somes): These are extremely large segments of DNA from another species spliced into DNA of yeast. YACs are used to clone up to one million bases of foreign DNA into a host cell, where the DNA is propagated along with the yeast cell's other chromosomes. The most notable features of YACs are that large DNA (>100 kb) can be ligated between two arms with each arm ending with a yeast telomere so that the product is stabilized in the yeast cell. Interestingly, larger YACs are more stable than shorter ones, which favours cloning of large stretches of DNA. One arm contains an autonomous replication sequence (ARS), a centromere (CEN) and a selectable marker (*trp1*). The other arm contains a second selectable marker (*ura3*) for detecting the transfromants.

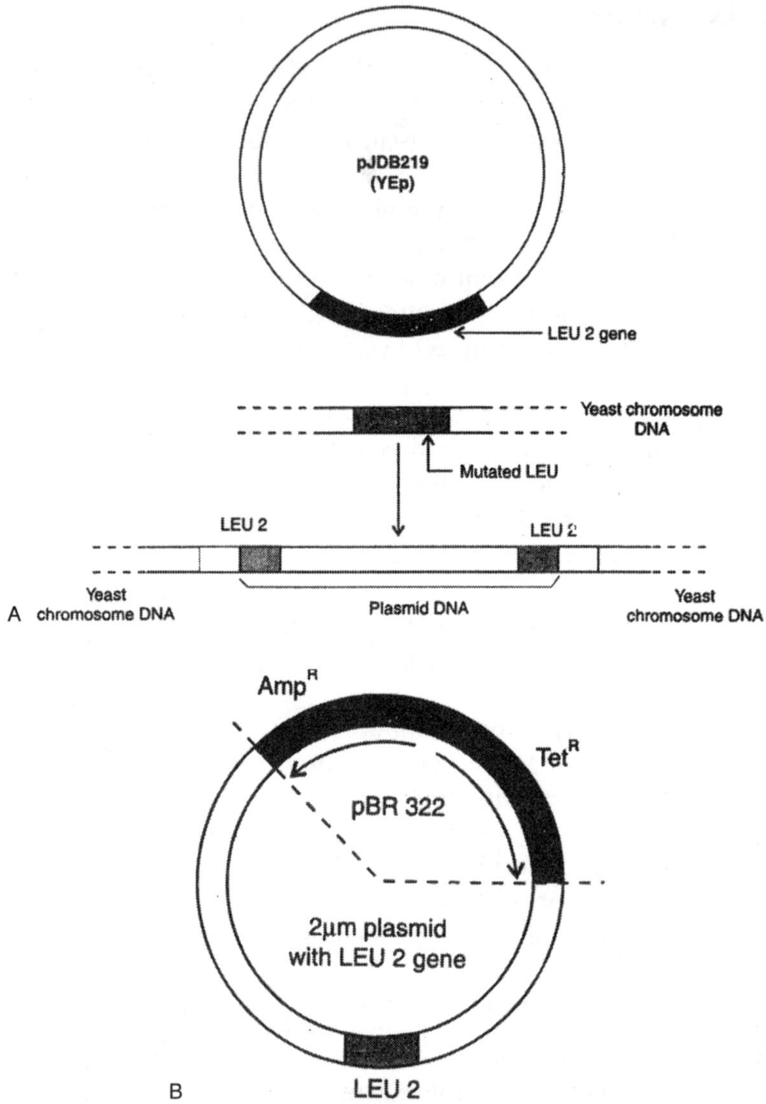

FIGURE 4.11 Structure of Yeast Episomal Plasmid; Homologus Recombination between LEU2 Gene of pJDB 219 and the Mutated LEU2 Gene on the Yeast Chromosome

Bacterial Artificial Chromosomes (BAC): These have been developed to hold much larger pieces of DNA than a plasmid. BAC vectors were originally created from part of an unusual plasmid present in some bacteria called the F' plasmid. The F' plasmid allows bacteria to pass its genome to another bacterium when it is under a lot of stress. F' had been studied extensively and it was found that it could hold up to a million base pairs of DNA from another bacteria.

In addition to this, the F' has origins of replication and bacteria have a way to control how F' is copied. In 1992, Hiroaki Shizuya was the first person to construct a BAC vectors. He took the parts of F' that were important, and designed it into a vector. Figure 4.12 shows the fragments up to 300 kb.

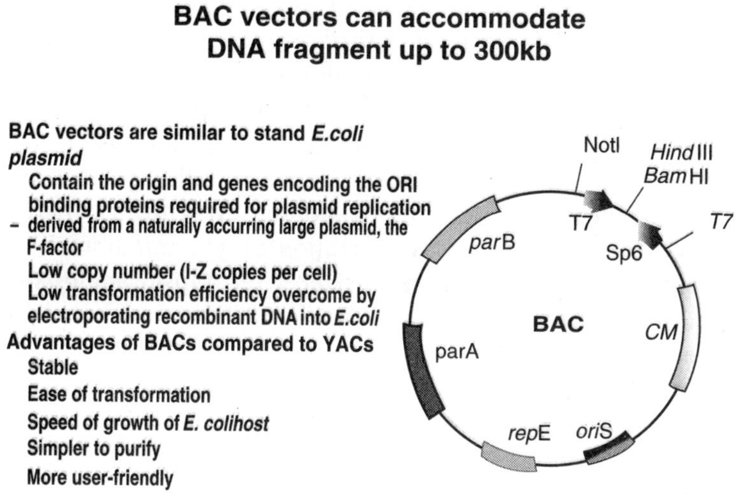

FIGURE 4.12 DNA Fragments up to 300kb

BAC vectors are able to hold up to 350 kb of DNA and have all the tools that a vector needs to work properly, like replication origins, antibiotic resistance genes, and convenient places where clone DNA can insert itself. With these vectors it is possible to study larger genes, several genes at once, or entire viral genomes.

Gene Expression Analysis

The delivery of a gene into a host is only one part of the process of genetic engineering. The most crucial aspect of the entire process is the synthesis of a functional protein by the DNA that is inserted into the host. Toward the end, a number of techniques and processes are employed, under the ambit of gene expression analysis. The main aim of this analysis is to first track the inserted DNA into the host, check for the presence of a complete transcript (mRNA), and finally to determine the presence of a functional protein.

Cloning a Gene

A lot of media attention has been generated over the past decade vis-à-vis the cloning technique. While it has raised a number of ethical and social issues, the technique has also emerged as a powerful tool for the treatment of various ailments using stem cell therapies.

Transformation or the ability to introduce foreign DNA into a host is central to the recombinant technology. Transformation is the process in which the exogenous DNA is taken up by the host cell and is used in the process of cloning. The process of cloning as outlined below consists of various steps. First, the DNA of interest is spliced using the appropriate vector and integrated into a plasmid. The recombinant DNA is then transformed in a host (usually bacterial) and the clones are generated. Although transformation is usually efficient enough for routine cloning experiments, there are some specialized procedures, such as generation of genomic libraries for which higher efficiency is desired. In such instances, the transformation procedure is circumvented by packaging the recombinant DNA into virus particles as described earlier in the chapter. Figure 4.13 shows the schematic representation of cloning a gene.

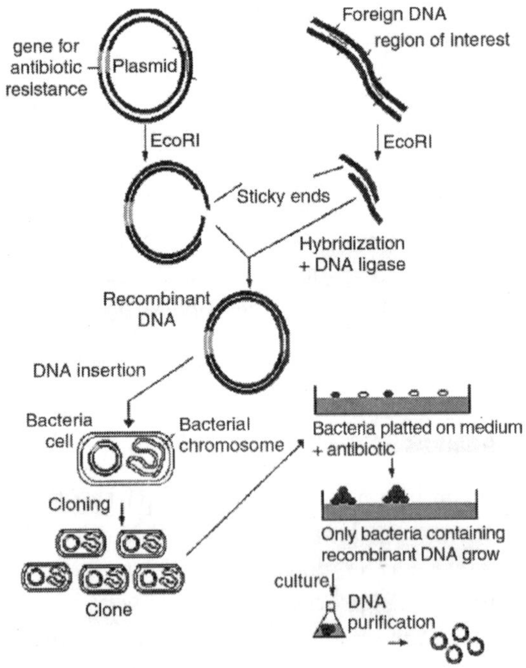

FIGURE 4.13 Cloning a Gene

Nucleic Acid Probes and Hybridization

The delivery and acceptance of foreign DNA by the host is of utmost importance, and thus various probes are used to track it down. The nucleic acid probes are based on the principle of complementation, i.e., by raising specific probes against the DNA that is inserted into the host one can determine if the transfer is successful.

Nucleic acid probes are designed to identify and detect specific target DNA molecules. The soluble probe binds and hybridizes to the target DNA that is immobilized on a nylon or nitrocellulose membrane. Hybridization is used for a variety of biotechnological applications including detection of cloned DNA and diagnosis of genetic diseases. Nucleic acid hybridization technique is based on the ability of single-stranded probe nucleic acid (DNA or RNA) to anneal to complementary single-stranded target sequences (DNA or RNA) within a range of non-complementary nucleic acid residues. The different methods are:

- Southern Blotting: To detect DNA sequences
- Northern Blotting: To detect RNA sequences
- Western blotting: To detect the protein of interest

Figure 4.14 shows southern blotting.

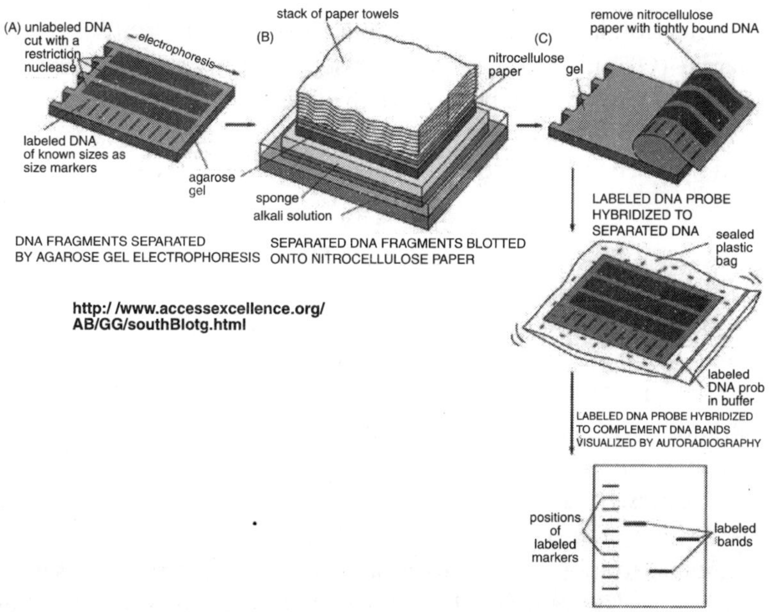

FIGURE 4.14 Southern Blotting

This has emerged as a very powerful technique for the analysis and characterization of proteins. As will be described later on the progression of DNA. Protein is the crux of gene manipulation, since failure of the DNA to synthesize functional protein can nullify the entire process. In this method, the proteins are first blotted onto synthetic membranes. These are then probed using immunological detection (by using protein-specific antibodies) methods.

The techniques mentioned earlier are used in the gene expression analysis. However, certain advanced techniques such as the PCR are in vogue, which will be discussed in the subsequent sections.

Polymerase Chain Reaction (PCR): DNA Copiers

Gene expression, or the ability of the gene to successfully convey the hereditary information, is of utmost importance in all gene manipulation studies. Blotting techniques such as the ones mentioned earlier are not only time and labor intensive but are also limited in their scale of operation. The biggest breakthrough in gene expression studies came in the form of polymerase chain reaction. This is a technique that allows the selective amplification of any fragment of DNA, provided the sequences flanking the same are known. The technique was invented by Kary Mullis, who shared the Nobel Prize for Chemistry in 1993.

The PCR technique relies on the sequence of base pairs along the length of the two strands that make the complete DNA molecule. In the DNA, there are four deoxynucleotides derived from the four bases—Adenosine (A), Thymidine (T), Guanine (G) and Cytidine (C) (discussed in detail later). The two strands of the DNA are held together by means of hydrogen bonds between these bases.

PCR involves three main steps namely, denaturation, annealing and extension by an enzyme called DNA polymerase. The three steps are detailed as follows:

1. Denaturation: The double stranded DNA is heated to 95–98°C and it separates into two complementary single strands.
2. Annealing: At 60°C, the synthetic oligonucleotide primers (chemically synthesized short-chain nucleotides), short sequences of nucleotides around 15–20 base-pairs long, are added and bound to the single strands in places where the strand's DNA complements their own.
3. Extension: At 37–40°C, the primers are extended by the DNA polymerase in the presence of all the four dNTPs (dATP, dGTP, dCTP and dTTP) resulting in the synthesis of new DNA strands complementary to the template strands. The completion of the three steps comprises a cycle and the real power of the technique is that, with 30–35 cycles it leads to massive amplification of DNA, which can be used for various purposes.

Figure 4.15 shows the outline of the PCR protocol.

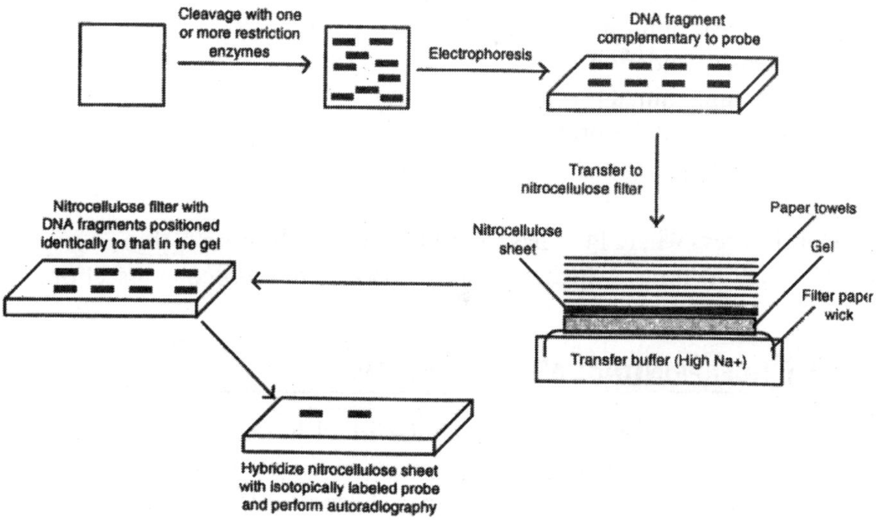

FIGURE 4.15 Use of PCR to Amplify DNA Molecules In Vitro

Figure 4.16 shows the temperature gradients in PCR cycle.

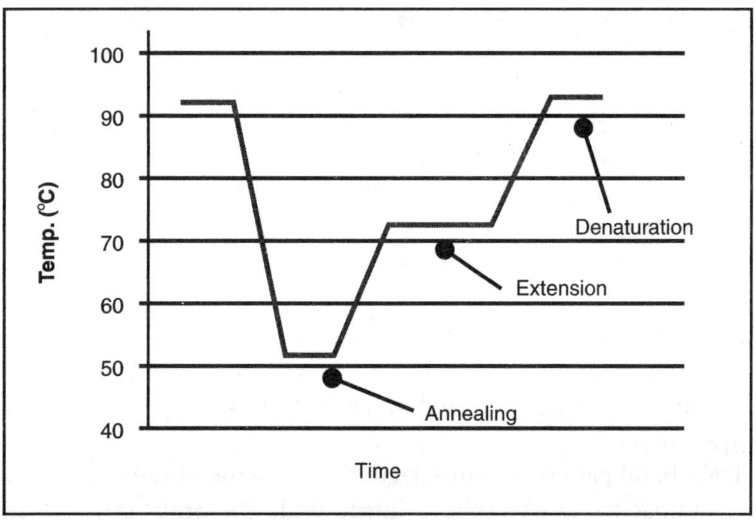

FIGURE 4.16 Temperature Gradients in the PCR Cycle

PCR has many applications in science. Some of them are:

- Molecular biology/genetic engineering
- Infectious and parasitic diseases
- Diagnosis of human genetic diseases
- Forensic validation
- Plant and animal breeding
- Environmental monitoring
- Archaeology

One of the areas where PCR has found massive application is that of genetic (DNA fingerprinting), which will be discussed in the subsequent section.

DNA Fingerprinting: Molecular IDs

The basic composition of DNA is similar in all human beings. The point of difference, if any, lies in the arrangement of the base pairs that make up the DNA. With the increase in the number of genes and the complexity of the organism, the sequence of arrangement of these base pairs also varies. It is this variation in sequence arrangement of the base pairs that form the crux of this technique.

DNA is made up of coding sequences called exons and the non-coding sequences known as introns. The introns are highly repetitive in nature and are dispersed throughout the organism. Currently, there are no functional attributes for introns, making them seemingly useless. However, it has been found that they contain repeated sequences of base pairs. These sequences are called Variable Number Tandem Repeats (VNTRs).

VNTRs differ among the human beings and are the targets for the technique. Determination of a specific VNTR in an individual is performed by a Southern Blot, which is probed, through a hybridization reaction, with a radioactive probe of the VNTR. The pattern, which results from this process, is called a DNA fingerprint.

The process of DNA fingerprinting is described:

1. DNA sample is extracted from blood or cell sample.
2. DNA is cut into fragments using a restriction enzyme.
3. The fragments are separated into bands by electrophoresis through an agarose gel.
4. DNA band pattern is transferred to a nylon membrane.
5. A radioactive DNA probe is introduced. The probe binds to specific DNA sequences on the nylon membrane.

6. The excess probe material is washed away leaving behind the unique DNA band pattern.
7. The radioactive DNA pattern is transferred to an x-ray film using direct exposure. When developed, the resultant visible pattern is the DNA fingerprint.

Applications of DNA Fingerprinting

There are many applications for this technique, some of them being:

- Paternity and maternity determination: Since VNTRs are inherited, their patterns can be used to establish paternity and maternity. The specificity of the patterns is so accurate that the parent's VNTR pattern can be reconstructed using the children's VNTR patterns and vice-versa. This has been used in various paternity dispute cases and conviction in rape cases resulting in unwanted pregnancy.
- Criminal identification and forensics: Comparison of VNTR patterns from DNA isolated from blood, hair, skin cells, or other genetic evidence left at the scene of a crime, with the DNA of a criminal suspect is used determine guilt or innocence. In addition to this, the identity of the patient can be determined on the basis of the VNR profile.

Shortcomings of DNA Fingerprinting

Like most techniques, DNA fingerprinting comes with its share of fallacies. The most important consideration in this technique is that the VNTR profile, though unique, is not entirely deterministic. The technique merely offers a probability of the person's identity based on the VNTR profile. In addition, the VNTR profiles of individuals can also match to a great extent since the probabilities of that pattern occurring in more than a single individual is equally distributed, though at a low significance value.

Technically, the application suffers from the serious consideration of the DNA samples that are used. It is observed that at times the amount of DNA available is so low that it has to be amplified using techniques like PCR. During this step DNA contaminated with other DNA samples can complicate the entire VNTR pattern. Hence the purity and quantity of the samples are very crucial for the success of this procedure.

Gene Engineering Product Optimization

Many commercially important biotechnological processes have been developed using microorganisms found in the environment. These processes are distinct from natural processes and have to be engineered to ensure product optimization.

Subtilisin

One good example of enzyme optimization is that of subtilisin. This is used in 95% of washing detergent formulations and allows protein stains to be removed more effectively at lower temperatures than are usually needed for laundry processing. The ideal requirements for enzyme activities are a temperature of around 70°C and a pH range of around 8–11. Based on an extensive knowledge of its catalytic activity and three-dimensional structure, subtilisin was engineered by site directed mutagenesis to produce variants with improved characteristics. In this method, the specific replacements of nucleotides in a sequence of DNA are performed to modify the activity of genes or their products (subtilisin, in this case).

DNA Shuffling/Sexual PCR

This is a relatively new method used to facilitate the rapid evolution of proteins. The principle involves mixing randomly fragmented DNA encoding closely related genes and then using PCR to reassemble them into full length fragments, with the individual fragments acting as primers. These PCR products are used to generate a library of chimeric genes for the selection of proteins with modified or improved characteristics. This has been used successfully in case of the *E. coli* proteins.

BIOTECHNOLOGY: THE ULTIMATE PANACEA

Biotechnology is an interdisciplinary pursuit today. A logical extension of this is that there has been a quantitative extension in the approaches to problems. Biotechnology has achieved this pursuit to a large extent by marshalling concepts and methodologies from a number of separate disciplines and applying them to specific problems in other areas. This is seen in the number of ventures, both academic and commercial, that have arisen in the last decade. The industries where biotechnology is involved are:

- Bioprocess technology: Historically, this has been the most important area of biotechnology, including brewing, antibiotics, and mammalian cell culture. Extensive development in progress with new products consisting of novel polysaccharides, medically important drugs, solvents, and protein fortified foods. The area has also spawned an industry revolving around the production and design of fermenters for the bioproducts.
- Enzyme technology: The technology has been used in the catalysis of extremely specific chemical reactions; immobilization of enzymes; to create specific molecular converters (bioreactors).

- Waste technology: This area of technology has a long historical significance. However, it is only recently that this has been combined with processes involving conservation and recycling of resources, foods and fertilizers and biological fuels.
- Environmental technology: This area has received major attention in the last decade. Various issues like pollution control, removal of toxic wastes, recovery of metals from mining wastes and oil spills are being tackled using biotechnology.
- Renewable resources technology: With the natural resources dwindling at a rapid pace, a need has been felt for alternative fuel resources. Biotechnology has offered an alternative by providing fuel sources like lignocellulose, methanol and hydrogen that are both economical and renewable.
- Plant and animal agriculture: The world population has been increasing exponentially, and there is a crisis in terms of resource availability. Resources such as land and food are getting scarce thus; necessitating the application of advanced techniques that are capable of high productivity using low resources. Biotechnology has offered a viable solution by the introduction of genetically engineered crops and transgenic animals. Ethical issues notwithstanding, the future of animal and plant agriculture dwells heavily on the technology.
- Healthcare: Biotechnology and genetic engineering have resulted in the research and development of various novel drugs and molecules that are capable of curing hitherto fatal diseases. Technologies involving gene therapy are already offering patients a new hope of survival.

Table 4.3 showcases the different fields of technology, where biotechnology is used extensively.

TABLE 4.3 Fields Using Biotechnology

Field of Technology	Explanation
Therapeutics	Pharmaceutical products for the cure or control of human diseases including various antibiotics, vaccines and gene therapy.
Diagnostics	Clinical testing and diagnosis, food, environment and agriculture.
Forestry/Horticulture	Novel crops or animal varieties, pesticides.
Food	Wide range of food products, fertilizers, beverages and ingredients.

(Continued)

TABLE 4.3 *(Continued)*

Field of Technology	Explanation
Environment	Waste treatment, bioremediation and energy production.
Chemical intermediates	Reagents including enzymes, DNA/RNA, specialty chemicals.
Equipment	Hardware, bioreactors, software and consumables supporting biotechnology.

It is hardly surprising that with such a diverse set of applications biotechnology has emerged as the technology of the century. In such a short span of time it has shown an amazing level of promise and given solutions that have made a difference. Whether it acts as a panacea for mankind is yet to be seen.

TERMS TO KNOW

Adenoviruses	a class of viruses amongst which are the viruses responsible for the common cold.
Attenuated vectors	Microorganisms or pieces of DNA such as plasmids which have certain infective components removed by recombinant DNA technology to destroy their virulence, thereby making them safer to use.
Bacteriophage	Any viruses that infect bacteria, also known as *phage.*
Expression	(of a gene) The central dogma of molecular biology says that genes (DNA) are transcribed into a message (RNA) which is translated into a polypeptide or protein, which in turn gives structural and functional features to a cell.
Gel electrophoresis	Analytical tool to separate substances, eg DNA fragments, from each other by having them move through a gel driven by an electric field.
Gene amplification	The process whereby genes or a sequence of DNA in the genome is greatly increased in number of copies.
Gene cloning	The technique of making many copies of a gene.
Gene expression	the production by a cell of the protein or polypeptide for which the specific gene codes.
Genetic fingerprinting	a technique which enables genetic relationships between close relatives, or the identity of individuals to be established-usually beyond reasonable doubt.
Plasmid	A piece of parasitic genetic material found in a cell that can propagate itself using the cell's energetic resources
Provirus	A virus that has inserted its genome or a complementaly copy of its genome into the host cell genome.
Reverse Transcription	The reverse of transcription-making a copy of complementaly DNA (eDNA) from an RNA sequence-catalyzed by the enzyme, reverse transcriptase.

Transgenic	Describes an organism which has had DNA (genes) foreign to that organism put in it by genetic engineering **(Transgenesis)** in the laboratory. Related words: transgene, transgenesis.
M13 vector	a filamentous, male-specific coliphage (*E. coli* infecting) composed of a single-stranded circular DNA genome and a protein coat.
Ti Plasmids	The virulent strains of bacteria *Agrobacterium tumefaciens,* known to cause crown gall disease in plants, contain large plasmids called the Ti-plasmids.

5 Bioinformatics

- INTRODUCTION
- INFORMATION NETWORKING
- PROTEIN AND GENOME RESOURCES
- SEQUENCE ANALYSIS
- SEQUENCE COMPARISON
- SEARCHING SEQUENCE DATABASES
- SECONDARY DATABASE SEARCH

INTRODUCTION

The quest to decipher the origins of life and the factors that determine them has been one that has spanned ages. Two events that have had a profound influence in the area of biology are the unraveling of the DNA helix and the invention of the World Wide Web. These, along with a host of other technologies, have revolutionized the way science is perceived. Armed with the knowledge that the molecular basis of transmission of hereditary information was through DNA, scientists, during the last quarter century, have been trying voraciously to gather information-first from the genes, then entire genomes (the collection of all the genes in an organism). This has resulted in a deluge of information.

As is evident, biology in the 21st century has moved out of the confines of laboratories and has become more information-driven. The information that drives research in biology is in the form of DNA sequences, RNA expression, protein interactions and molecular confirmations. These form the basis in formulating hypothesis that helps scientists to design large-scale laboratories and experiments. In turn, the lab work ends with the accumulation of further data that must be sifted. It is in this context that bioinformatics has assumed a role of paramount importance.

Bioinformatics represents a new, fast emerging area of science that uses computational approaches to answer biological queries. This requires the investigators to make use of large, complex data sets in a raucous fashion to reach logical and valid biological conclusions. In a sense, bioinformatics is a discipline that is situated at the interface between computer science and biology, such as molecular biology and genetics. Initially, the term dealt only with specific tasks and activities related to the storage of biological data in databases. As the field evolved, the term has grown and includes various algorithms (computer software platforms) and allied techniques and technologies. Due to lack of a consensus that stands for bioinformatics, it is difficult to slot it under a particular niche.

However, bioinformatics can be described as the application of information technology techniques, algorithms and tools used in the context of biological problems. The techniques that are currently employed draw their origins in a number of areas, such as computer sciences, statistics and mathematics.

The genesis of genome projects and the accompanying deluge of sequence information have placed sequence analysis in the driver's seat. In the mid-1980s, the United States initiated a number of projects to construct detailed genetic and physical maps of the human genome, to determine its complete nucleotide sequence, and to localize its estimated 100,000 genes. The draft of the same has been released recently marking the end of one of the most intensive and massive efforts in biology. It was the Human Genome project that gave bioinformatics the fillip it needed since work on this scale required the development of new computational methods for analyzing and storing the massive amount of data. The challenges that bioinformatics faces today; are the rationalization of the mass of sequence data and provide incisive tools for analysis of the same.

The fact remains that mere acquisition of sequence data conveys little more about the intricate biology of the systems from which they are derived, than a telephone directory which might reveal very little about the people and firms listed in it. The task that biologists face is to decode this seemingly cryptic language. Bioinformatics provides both the cryptic language and the means to decipher the same. The message that one has to get is that Nature has its own complex rules, which we are poor at comprehending and which cannot be encapsulated within computer programs or algorithms. Bioinformatics cannot "do" biology. At best, it can provide infallible models of biological systems. It certainly does not provide any final answers, what it provides is an aid to narrow down options from a sea of data.

Computer stimulation of experimental biology is another interesting application of bioinformatics, which is called "*in-silico*" testing. This is an area that is expanding fast due to its usefulness in the area of drug testing and clinical trials. A subfield of particular interest is genomics. The field of genomics encompasses investigations into the structure and function of very large number of genes simultaneously. Under the ambit of genomics, there are a number of subsets. Structural genomics deals with the genetic mapping, physical mapping, and sequencing of genes for entire organisms. Comparative genomics deals with the extending information gained form the study of some organisms in relation to other organisms. Functional genomics is concerned with the role that individual genes or subsets of genes play in the development and life of organisms.

INFORMATION NETWORKING

One of the driving forces behind the evolution and success of biotechnology has been the Internet and the World Wide Web. Together along with a global network of biological information and service providers they have revolutionized the area of biotechnology. Computers emerged to become the preferred medium of storage of huge databases and these machines also provided the advantage of an error-free working environment.

European Molecular Biology Network (EMBnet)

The mid-1980s saw an explosion of biological data and databases proliferating the Internet. This, in turn, hiked up the demand for more efficient means of access to more up-to-date data. It was against this background that a network to link all the European Laboratories using biocomputing and bioinformatics in molecular biology research was set up in 1988. The network known as the EMBnet was envisaged as a way of providing information and services via designated nodes operating in their local languages. Currently, EMBnet has 41 nodes spread in over 30 countries and supports thousands of users all over the world. It has easy access to more than 120 prime biological and bibliographic databases. The nodes are divided into the national and specialist nodes. The national nodes are biocomputing centers appointed by the respective governments in various countries. These are concerned predominantly with providing databases and software tools for biological applications. The specialist nodes, on the other hand, are centers that have an extensive knowledge of specific aspects of computational molecular biology. These are typically responsible for, and devote themselves to the production and maintenance of biologically important databases and software.

National Center for Biotechnology Information

The National Center for Biotechnology Information (NCBI) is the leading American information provider. It was established in the 1988 as a division of the National Library of Medicine (NLM), and is located on the campus of the National Institute of Health (NIH) in Bethesda, Maryland. NLM was an obvious choice considering its immense experience in biomedical database maintenance and as part of the NIH, it could establish a research program in computational biology.

The role of NCBI is in developing new information technologies to aid in the understanding of the molecular and genetic processes that underlie health and disease. Its mandate includes creation of an automated system for storing, analyzing biological information and the development of advanced computational methods in information processing. Since 1992, the major task of NCBI has been the maintenance of GenBank, the NIH DNA databank. The data is regularly updated and exchanged with international databases such as EMBL and DNA Data Bank of Japan (DDBJ).

Protein and Genome Resources

The various genome sequencing projects have lead to an exponential increase in the data that is available across the Internet. This data is largely variable in nature, and ranges from protein sequences, to nucleotide sequences, and ESTs—the list is exhaustive. In such a scenario, it is of utmost importance that one knows the difference in the databases and the types of data they provide. It is here that the role of databases in analyzing sequence information and assembling it into central, shareable resources is of paramount importance. Databases are effectively electronic cabinets and a convenient way of storing and retrieving vast amount of data. The databases vary depending on the nature of information stored (Sequences or structures, 2D or 3D structure images) and the manner in which they are stored.

Protein Resources

In context of protein data, one has to encounter two levels of databases, i.e., the primary and secondary databases. These databases are used to address different aspects of sequence analysis because they store different levels of protein sequences information. It is thus essential to determine when and how to use these different resources depending on the application. The invention of protein sequencing methods and the incidental increase in various protein sequencing projects, saw a tremendous increase in the protein sequence data. To cope with this, there were a number of databases that came forward and catalogued the sequences into different classes based on the nature of data. Some of the most important primary databases that arose due to this are:

- PIR
- MIPS
- SWISS-PROT

- TrEMBL
- NRL-3D

Swiss-Prot is a protein knowledgebase established in 1986 and maintained collaboratively, since 1987, by the Department of Medical Biochemistry of the University of Geneva (now, the Swiss Institute of Bioinformatics) and the EMBL Data Library (now the EMBL Outstation - The European Bioinformatics Institute (EBI)). The Swiss-Prot protein knowledgebase consists of sequence entries. The manner of data storage or the flat file is designed to be user-friendly and, at the same time, to provide exhaustive information.

SwissProt

SwissProt is a curated protein sequence database that strives to provide a high level of annotations (such as the description of the function of a protein, its domains structure, post-translational modifications, variants, etc.), a minimal level of redundancy and high level of integration with other databases. Each protein sequence entry is composed of lines. Different types of lines, each with their own format, are used to record the various data that make up the entry. A sample sequence entry is shown below.

```
ID    GRAA_HUMAN    STANDARD;    PRT;    262 AA.
AC    P12544;
DT    01-OCT-1989 (Rel. 12, Created)
DT    01-OCT-1989 (Rel. 12, Last sequence update)
DT    10-MAY-2005 (Rel. 47, Last annotation update)
DE    Granzyme A precursor (EC 3.4.21.78) (Cytotoxic T-lymphocyte proteinase
DE    1) (Hanukkah factor) (H factor) (HF) (Granzyme 1) (CTL tryptase)
DE    (Fragmentin 1).
GN    Name=GZMA; Synonyms=CTLA3, HFSP;
OS    Homo sapiens (Human).
OC    Eukaryota; Metazoa; Chordata; Craniata; Vertebrata; Euteleostomi;
OC    Mammalia; Eutheria; Euarchontoglires; Primates; Catarrhini; Hominidae;
OC    Homo.
OX    NCBI_TaxID=9606;
RN    [1]
RP    NUCLEOTIDE SEQUENCE.
RC    TISSUE=T-cell;
RX    MEDLINE=88125000; PubMed=3257574;
RA    Gershenfeld H.K., Hershberger R.J., Shows T.B., Weissman I.L.;
RT    "Cloning and chromosomal assignment of a human cDNA encoding a T cell-
RT    and natural killer cell-specific trypsin-like serine protease.";
RL    Proc. Natl. Acad. Sci. U.S.A. 85:1184-1188(1988).
RN    [2]
RP    NUCLEOTIDE SEQUENCE [LARGE SCALE MRNA].
RC    TISSUE=Blood;
RX    MEDLINE=22388257; PubMed=12477932; DOI=10.1073/pnas.242603899;
RA    Strausberg R.L., Feingold E.A., Grouse L.H., Derge J.G.,
RA    Klausner R.D., Collins F.S., Wagner L., Shenmen C.M., Schuler G.D.,
RA    Altschul S.F., Zeeberg B., Buetow K.H., Schaefer C.F., Bhat N.K.,
RA    Hopkins R.F., Jordan H., Moore T., Max S.I., Wang J., Hsieh F.,
RA    Diatchenko L., Marusina K., Farmer A.A., Rubin G.M., Hong L.,
RT    "Generation and initial analysis of more than 15,000 full-length human
RT    and mouse cDNA sequences.";
RL    Proc. Natl. Acad. Sci. U.S.A. 99:16899-16903(2002).
```

```
RN   [3]
RP   NUCLEOTIDE SEQUENCE OF 1-23.
RA   Goralski T.J., Krensky A.M.;
RT   "The upstream region of the human granzyme A locus contains both
RT   positive and negative transcriptional regulatory elements.";
RL   Submitted (NOV-1995) to the EMBL/GenBank/DDBJ databases.
RN   [4]
RP   PROTEIN SEQUENCE OF 29-53.
RX   MEDLINE=88330824; PubMed=3047119;
RA   Poe M., Bennett C.D., Biddison W.E., Blake J.T., Norton G.P.,
RA   Rodkey J.A., Sigal N.H., Turner R.V., Wu J.K., Zweerink H.J.;
RT   "Human cytotoxic lymphocyte tryptase. Its purification from granules
RT   and the characterization of inhibitor and substrate specificity.";
RL   J. Biol. Chem. 263:13215-13222(1988).
RN   [5]
RP   PROTEIN SEQUENCE OF 29-40, AND CHARACTERIZATION.
RX   MEDLINE=89009866; PubMed=3262682;
RA   Hameed A., Lowrey D.M., Lichtenheld M., Podack E.R.;
RT   "Characterization of three serine esterases isolated from human IL-2
RT   activated killer cells.";
RL   J. Immunol. 141:3142-3147(1988).
RN   [6]
RP   PROTEIN SEQUENCE OF 29-39, AND CHARACTERIZATION.
RX   MEDLINE=89035468; PubMed=3263427;
RA   Kraehenbuhl O., Rey C., Jenne D.E., Lanzavecchia A., Groscurth P.,
RA   Carrel S., Tschopp J.;
RT   "Characterization of granzymes A and B isolated from granules of
RT   cloned human cytotoxic T lymphocytes.";
RL   J. Immunol. 141:3471-3477(1988).
RN   [7]
RP   3D-STRUCTURE MODELING.
RX   MEDLINE=89184501; PubMed=3237717;
RA   Murphy M.E.P., Moult J., Bleackley R.C., Gershenfeld H.,
RA   Weissman I.L., James M.N.G.;
RT   "Comparative molecular model building of two serine proteinases from
RT   cytotoxic T lymphocytes.";
RL   Proteins 4:190-204(1988).
RN   [8]
RP   X-RAY CRYSTALLOGRAPHY (2.4 ANGSTROMS) IN COMPLEX WITH A TRIPEPTIDE CMK
RP   INHIBITOR.
RX   PubMed=12819769; DOI=10.1038/nsb944;
RA   Bell J.K., Goetz D.H., Mahrus S., Harris J.L., Fletterick R.J.,
RA   Craik C.S.;
RT   "The oligomeric structure of human granzyme A is a determinant of its
RT   extended substrate specificity.";
RL   Nat. Struct. Biol. 10:527-534(2003).
RN   [9]
RP   X-RAY CRYSTALLOGRAPHY (2.5 ANGSTROMS).
RX   PubMed=12819770; DOI=10.1038/nsb945;
RA   Hink-Schauer C., Estebanez-Perpina E., Kurschus F.C., Bode W.,
RA   Jenne D.E.;
RT   "Crystal structure of the apoptosis-inducing human granzyme A dimer.";
RL   Nat. Struct. Biol. 10:535-540(2003).
CC   -!- FUNCTION: This enzyme is necessary for target cell lysis in cell-
CC   mediated immune responses. It cleaves after Lys or Arg. May be
CC   involved in apoptosis.
CC   -!- CATALYTIC ACTIVITY: Hydrolysis of proteins, including fibronectin,
CC   type IV collagen and nucleolin. Preferential cleavage: Arg-|-Xaa,
CC   Lys-|-Xaa >> Phe-|-Xaa in small molecule substrates.
CC   -!- SUBUNIT: Homodimer; disulfide-linked.
CC   -!- INTERACTION:
CC   Self; NbExp=1; IntAct=EBI-519800, EBI-519800;
CC   -!- SUBCELLULAR LOCATION: Secreted; cytoplasmic granules.
CC   -!- SIMILARITY: Belongs to the peptidase S1 family. Granzyme
```

```
CC   subfamily.
CC   -!- SIMILARITY: Contains 1 peptidase S1 domain.
CC   -----------------------------------------------------------------------
CC   This Swiss-Prot entry is copyright. It is produced through a collaboration
CC   between  the Swiss Institute of Bioinformatics  and the  EMBL outstation -
CC   the European Bioinformatics Institute.  There are no  restrictions on  its
CC   use as long as its content is in no way modified and this statement is not
CC   removed.
CC   -----------------------------------------------------------------------
DR   EMBL; M18737; AAA52647.1; -; mRNA.
DR   EMBL; BC015739; AAH15739.1; -; mRNA.
DR   EMBL; U40006; AAD00009.1; -; Genomic_DNA.
DR   PIR; A31372; A31372.
DR   PDB; 1HF1; Model; @=29-262.
DR   PDB; 1OP8; X-ray; A/B/C/D/E/F=29-262.
KW   3D-structure; Apoptosis; Cytolysis; Direct protein sequencing;
KW   Glycoprotein; Hydrolase; Protease; Serine protease; Signal; T-cell;
KW   Zymogen.
FT   SIGNAL        1    26
FT   PROPEP       27    28     Activation peptide.
FT   CHAIN        29   262     Granzyme A.
FT   DOMAIN       29   259     Peptidase S1.
FT   ACT_SITE     69    69     Charge relay system (By similarity).
FT   ACT_SITE    114   114     Charge relay system (By similarity).
FT   ACT_SITE    212   212     Charge relay system (By similarity).
FT   CARBOHYD    170   170     N-linked (GlcNAc...) (Potential).
FT   DISULFID     54    70
FT   DISULFID    148   218
FT   DISULFID    179   197
FT   DISULFID    208   234
SQ   SEQUENCE    262 AA;  28969 MW;  DA87363A0D92BAF4 CRC64;
     MRNSYRFLAS  SLSVVVSLLL  IPEDVCEKII  GGNEVTPHSR  PYMVLLSLDR  KTICAGALIA
     KDWVLTAAHC  NLNKRSQVIL  GAHSITREEP  TKQIMLVKKE  FPYPCYDPAT  REGDLKLLQL
     TEKAKINKYV  TILHLPKKGD  DVKPGTMCQV  AGWGRTHNSA  SWSDTLREVN  ITIIDRKVCN
     DRNHYNFNPV  IGMNMVCAGS  LRGGRDSCNG  DSGSPLLCEG  VFRGVTSFGL  ENKCGDPRGP
     GVYILLSKKH  LNWIIMTIKG  AV
//
```

This is a sample flat file entry. Some material has been excised for convenience.

Some of the lines and their formats have been discussed in Table 5.1:

TABLE 5.1 Description of Swiss-Prot Entries

Line(s)	Description	General Form
ID (IDentification)	Always the first line of an entry	ID ENTRY_NAME DATA_CLASS; MOLECULE_TYPE; SEQUENCE_LENGTH.
AC (ACcession number)	Lists accession number(s) associated with an entry.	AC AC_number_1; [AC_number_2;]... [AC_number_N;]

(Continued)

TABLE 5.1 *(Continued)*

Line(s)	Description	General Form
DT (DaTe)	Shows date of creation and last modification of database entry.	DT DD-MMM-YYYY
DE (DEscription)	Contain general information about sequence stored. This information is generally sufficient to identify the protein.	DE Description
GN (Gene Name)	Indicates name(s) of gene(s) that code for the stored protein sequence. This line contains Gene names, ordered locus names (position on the chromosome), and the ORF name (sequencing names).	GN Name=<name>; Synonyms=<name1> [,<name2>...]; OrderedLocusNames= <name1> [,<name2> ...]; GN ORFNames=<name1> [, <name2>...];
OS (Organism Species)	Specifies organism(s) that was (were) the source of the stored sequence. In the rare case where all the species information will not fit on a single line, more than one OS line is used. The last OS line is terminated by a period.	OS Name and Species of the organism according to established nomenclatures eg: *Oryza sativa*
OG (OrGanelle)	Indicates if the gene coding originates from any of the cell organelles.	OG Chloroplast.
OC (Organism Classification)	Contain taxonomic classification of the source organism.	OC Node [; Node...]. [; Node...].
The OX (Organism taxonomy cross-reference)	Indicates identifier of a specific organism in a taxonomic database.	OX Taxonomy_database_ Qualifier=Taxonomic code[, Taxonomic code...];
Reference (RN, RP, RC, RX, RG, RA, RT, RL)	Contain literature citations. The citations indicate sources from which the data has been abstracted. The reference lines for a given citation occur in a block,	RN indicates reference number RP indicates reference position

(Continued)

TABLE 5.1 *(Continued)*

Line(s)	Description	General Form
	and are always in the order RN, RP, RC, RX, RG, RA, RT and RL. Within each such reference block, the RN line occurs once, the RC, RX and RT lines occur zero or more times, and the RP, RG/RA and RL lines occur one or more times. If several references are given, there will be a reference block for each.	RA indicates reference authors lines RT indicates reference titles RL indicates reference location line
CC	Conveys any useful information. These comments always appear below the last reference line and are grouped together in comment blocks; a block is made up of one or more comment lines. The first line of a block starts with the characters '-!-'.	`CC -!- TOPIC: First line of a comment block;`
DR (Database cross-Reference)	Used as pointers to information related to entries and found in data collections other than Swiss-Prot.	`DR DATABASE_ IDENTIFIER; PRIMARY_ IDENTIFIER; SECONDARY_ IDENTIFIER [; TERTIARY_ IDENTIFIER] [; QUATERNARY_ IDENTIFIER`
KW (KeyWord)	Provides information used to generate indexes of sequence entries based on functional, structural, or other categories. The keywords chosen for each entry serve as a reference for the sequence.	`KW Keyword [; Keyword...].`
FT (Feature Table)	Provides precise but simple means for annotation of the	No specific format and consists of descriptive

(Continued)

TABLE 5.1 *(Continued)*

Line(s)	Description	General Form
	sequence data. The table describes regions or sites of interest in the sequence. The table lists posttranslational modifications, binding sites, enzyme active sites, local secondary structure or other characteristics reported in the cited references. Sequence conflicts between references are also included in the feature table.	details about the various biological features associated with the sequence.
SQ (SeQuence header)	Marks the beginning of the sequence data and gives a quick summary of its content.	SQ SEQUENCE XXXX AA; XXXXX MW; XXXXXXXXXXXXXXXX CRC64;

The structure of the SWISS-PROT entries coupled with its quality of annotations makes it the most widely used protein sequence database in the world.

TrEMBL

The TrEMBL (Translated EMBL) database is a repository for protein data. TrEMBL contains the translations of all coding sequences (CDS) present in the EMBL Nucleotide Sequence Database, which are not yet integrated into SWISS-PROT. TrEMBL is split into two main sections:

- **SP-TrEMBL** (SwissProt TrEMBL) contains the entries, which should eventually be incorporated into SwissProt and can be considered as a preliminary section of SwissProt as all SP-TrEMBL entries have been assigned SwissProt accession numbers.
- **REM-TrEMBL** (REMaining TrEMBL) contains the entries that we do not want to include in SwissProt. REM-TrEMBL entries have no accession numbers.

Nucleotide Resources

GenBank, the database from the National Center for Biotechnology Information (NCBI), incorporates sequences from various sources that comprise individual author submissions and public databases. The GenBank is divided into smaller discrete divisions as shown in Table 5.2:

TABLE 5.2: GenBank subdivisions

Division	Sequence Subset
PRI	Primates
ROD	Rodents
MAM	Mammals
VRT	Vertebrates
INV	Invertebrates
PLN	Plant, Algal, Fungal
BCT	Bacterial
RNA	Structural RNA
VRL	Viral
PHG	Bacteriophage
SYN	Synthetic
UNA	Unannotated
EST	Expressed Sequence Tags
PAT	Patent
STS	Sequence Tag Sites
GSS	Genome Survey Sequence
HTG	High Through-put Sequence

This superficial separation is useful since it facilitates fast, specific searches by restricting searches to particular databases. Information can be received from GenBank by using the Entrez integrated retrieval system, which combines data from the principal DNA and protein sequence databases. The GenBank release includes various sections as described below:

```
LOCUS       HUMCYCLOX               3387 bp    mRNA    linear   PRI 31-DEC-1994
DEFINITION  Homo sapiens cyclooxygenase-2 (Cox-2) mRNA, complete cds.
ACCESSION   M90100
VERSION     M90100.1  GI:181253
KEYWORDS    cyclooxygenase-2; prostaglandin synthase.
SOURCE      Homo sapiens (human)
  ORGANISM  Homo sapiens
            Eukaryota; Metazoa; Chordata; Craniata; Vertebrata; Euteleostomi;
            Mammalia; Eutheria; Euarchontoglires; Primates; Catarrhini;
            Hominidae; Homo.
REFERENCE   1  (bases 1 to 3387)
  AUTHORS   Hla,T. and Neilson,K.
```

```
  TITLE     Human cyclooxygenase-2 cDNA
  JOURNAL   Proc. Natl. Acad. Sci. U.S.A. 89 (16), 7384-7388 (1992)
   PUBMED   1380156
COMMENT     Original source text: Homo sapiens umbilical vein cDNA to mRNA.
FEATURES             Location/Qualifiers
     source          1..3387
                     /organism="Homo sapiens"
                     /mol_type="mRNA"
                     /db_xref="taxon:9606"
                     /cell_type="endothelial"
                     /tissue_type="umbilical vein"
     gene            1..3387
                     /gene="Cox-2"
     5'UTR           1..97
                     /gene="Cox-2"
     CDS             98..1912
                     /gene="Cox-2"
                     /EC_number="1.14.99.1"
                     /codon_start=1
                     /product="cyclooxygenase-2"
                     /protein_id="AAA58433.1"
                     /db_xref="GI:181254"
                     /translation="MLARALLLCAVLALSHTANPCCSHPCQNRGVCMSVGFDQYKCDC
                     TRTGFYGENCSTPEFLTRIKLFLKPTPNTVHYILTHFKGFWNVVNNIPFLRNAIMSYV
                     LTSRSHLIDSPPTYNADYGYKSWEAFSNLSYYTRALPPVPDDCPTPLGVKGKKQLPDS
                     NEIVGKLLLRRKFIPDPQGSNMMFAFFAQHFTHQFFKTDHKRGPAFTNGLGHGVDLNH
                     IYGETLARQRKLRLFKDGKMKYQIIDGEMYPPTVKDTQAEMIYPPQVPEHLRFAVGQE
                     VFGLVPGLMMYATIWLREHNRVCDVLKQEHPEWGDEQLFQTSRLILIGETIKIVIEDY
                     VQHLSGYHFKLKFDPELLFNKQFQYQNRIAAEFNTLYHWHPLLPDTFQIHDQKYNYQQ
                     FIYNNSILLEHGITQFVESFTRQIAGRVAGGRNVPPAVQKVSQASIDQSRQMKYQSFN
                     EYRKRFMLKPYESFEELTGEKEMSAELEALYGDIDAVELYPALLVEKPRPDAIFGETM
                     VEVGAPFSLKGLMGNVICSPAYWKPSTFGGEVGFQIINTASIQSLICNNVKGCPFTSF
                     SVPDPELIKTVTINASSSRSGLDDINPTVLLKERSTEL"
     sig_peptide     98..148
                     /gene="Cox-2"
     mat_peptide     149..1909
                     /gene="Cox-2"
                     /product="cyclooxygenase-2"
                     /EC_number="1.14.99.1"
     3'UTR           1913..3387
                     /gene="Cox-2"
     polyA_signal    3369..3374
                     /gene="Cox-2"
ORIGIN
        1 gtccaggaac tcctcagcag cgcctccttc agctccacag ccagacgccc tcagacagca
       61 aagcctaccc ccgcgccgcg ccctgcccgc cgctgcgatg ctcgcccgcg ccctgctgct
      121 gtgcgcggtc ctggcgctca gccatacagc aaatccttgc tgttcccacc catgtcaaaa
      181 ccgaggtgta tgtatgagtg tgggatttga ccagtataag tgcgattgta cccggacagg
      241 attctatgga gaaaactgct caacaccgga atttttgaca agaataaaat tatttctgaa
      301 acccactcca aacacagtgc actacatact tacccacttc aagggatttt ggaacgttgt
      361 gaataacatt cccttccttc gaaatgcaat tatgagttat gtgttgacat ccagatcaca
      421 tttgattgac agtccaccaa cttacaatgc tgactatggc tacaaaagct gggaagcctt
      481 ctctaacctc tcctattata ctagagccct tcctcctgtg cctgatgatt gcccgactcc
      541 cttgggtgtc aaaggtaaaa agcagcttcc tgattcaaat gagattgtgg gaaaattgct
      601 tctaagaaga aagttcatcc ctgatcccca gggctcaaac atgatgtttg cattctttgc
      661 ccagcacttc acgcatcagt tttcaagac agatcataag cgagggccag ctttcaccaa
      721 cgggctgggc catggggtgg acttaaatca tatttacggt gaaactctgg ctagacagcg
      781 taaactgcgc cttttcaagg atggaaaaat gaaatatcag ataattgatg gagagatgta
      841 tcctcccaca gtcaaagata ctcaggcaga gatgatctac cctcctcaag tccctgagca
```

```
 901 tctacggttt gctgtggggc aggaggtctt tggtctggtg cctggtctga tgatgtatgc
 961 cacaatctgg ctgagggaac acaacagagt atgcgatgtg cttaaacagg agcatcctga
1021 atggggtgat gagcagttgt tccagacaag caggctaata ctgataggag agactattaa
1081 gattgtgatt gaagattatg tgcaacactt gagtggctat cacttcaaac tgaaatttga
1141 cccagaacta cttttcaaca aacaattcca gtaccaaaat cgtattgctg ctgaatttaa
1201 caccctctat cactggcatc cccttctgcc tgacaccttt caaattcatg accagaaata
1261 caactatcaa cagtttatct acaacaactc tatattgctg gaacatggaa ttacccagtt
1321 tgttgaatca ttcaccaggc aaattgctgg cagggttgct ggtggtagga atgttccacc
1381 cgcagtacag aaagtatcac aggcttccat tgaccagagc aggcagatga aataccagtc
1441 ttttaatgag taccgcaaac gctttatgct gaagccctat gaatcatttg aagaacttac
1501 aggagaaaag gaaatgtctg cagagttgga agcactctat ggtgacatcg atgctgtgga
1561 gctgtatcct gcccttctgg tagaaaagcc tcggccagat gccatctttg gtgaaaccat
1621 ggtagaagtt ggagcaccat tctccttgaa aggacttatg ggtaatgtta tatgttctcc
1681 tgcctactgg aagccaagca cttttggtgg agaagtgggt tttcaaatca tcaacactgc
1741 ctcaattcag tctctcatct gcaataacgt gaagggctgt ccctttactt cattcagtgt
1801 tccagatcca gagctcatta aaacagtcac catcaatgca agttcttccc gctccggact
1861 agatgatatc aatcccacag tactactaaa agaacgttcg actgaactgt agaagtctaa
1921 tgatcatatt tatttattta tatgaaccat gtctattaat ttaattattt aataatattt
1981 atattaaact ccttatgtta cttaacatct tctgtaacag aagtcagtac tcctgttgcg
2041 gagaaaggag tcatacttgt gaagactttt atgtcactac tctaaagatt ttgctgttgc
2101 tgttaagttt ggaaaacagt ttttattctg ttttataaac cagagagaaa tgagttttga
2161 cgtcttttta cttgaatttc aacttatatt ataaggacga aagtaaagat gtttgaatac
2221 ttaaacacta tcacaagatg ccaaaatgct gaaagttttt acactgtcga tgtttccaat
2281 gcatcttcca tgatgcatta gaagtaacta atgtttgaaa ttttaaagta cttttgggta
2341 tttttctgtc atcaaacaaa acaggtatca gtgcattatt aaatgaatat ttaaattaga
2401 cattaccagt aatttcatgt ctacttttta aaatcagcaa tgaaacaata atttgaaatt
2461 tctaaattca tagggtagaa tcacctgtaa aagcttgttt gatttcttaa agttattaaa
2521 cttgtacata taccaaaaag aagctgtctt ggatttaaat ctgtaaaatc agatgaaatt
2581 ttactacaat tgcttgttaa aatattttat aagtgatgtt cctttttcac caagagtata
2641 aacctttta gtgtgactgt taaaacttcc ttttaaatca aaatgccaaa tttattaagg
2701 tggtggagcc actgcagtgt tatctcaaaa taagaatatc ctgttgagat attccagaat
2761 ctgtttatat ggctggtaac atgtaaaaac cccataaccc cgccaaaagg ggtcctaccc
2821 ttgaacataa agcaataacc aaaggagaaa agcccaaatt attggttcca aatttagggt
2881 ttaaactttt tgaagcaaac tttttttag ccttgtgcac tgcagacctg gtactcagat
2941 tttgctatga ggttaatgaa gtaccaagct gtgcttgaat aacgatatgt tttctcagat
3001 tttctgttgt acagtttaat ttagcagtcc atatcacatt gcaaaagtag caatgacctc
3061 ataaaatacc tcttcaaaat gcttaaattc atttcacaca ttaattttat ctcagtcttg
3121 aagccaattc agtaggtgca ttggaatcaa gcctggctac ctgcatgctg ttccttttct
3181 tttcttcttt tagccatttt gctaagagac acagtcttct caaacacttc gtttctccta
3241 ttttgtttta ctagttttaa gatcagagtt cactttcttt ggactctgcc tatattttct
3301 tacctgaact tttgcaagtt ttcaggtaaa cctcagctca ggactgctat ttagctcctc
3361 ttaagaagat taaaaaaaaa aaaaaag
//
```

Some of the lines have been discussed in Table 5.3:

TABLE 5.3 GenBank entries and descriptions

Line	Explanation
Locus	Introduces a short label for the entry suggesting the function of the sequence and data regarding the number of bases, source of sequence data, section of database (PRI) and the date of submission.
Definition	Contains concise description of the sequence.

(Continued)

TABLE 5.3 *(Continued)*

Line	Explanation
Accession	Gives the accession number, which is a unique constant code, assigned to each entry.
NID	Provides a unique reference of the current version of the sequence information, allowing the sequence to be revised without changing the locus name and accession number.
Keywords	Introduces a list of short phrases assigned by the author that describes the gene products and other relevant information about the entry.
Source	Provides information about the tissue from which the data has been derived, the sub keyword *Organism* illustrates the classification of the source organism.
Reference	Indicates the portion of the sequence data to which the cited literature refers. The rest of the sub keywords, such as Author, Title, and Journal are a format for citation.
Features	Provides a unique structure and describes the properties of the sequence in detail.
Base count	Provides details about the frequency of occurrence of the different base types (Number of times A, T, G, C occur).
Origin	Provides location of the first base of the sequence. The entry is terminated with a // marker.

SEQUENCE ANALYSIS

In most scenarios, DNA sequences are the choice of analysis to decipher complex biological problems. This is in contrast to what is generally believed in biology that the real information lies in knowing what a protein does. However, the mere availability of a DNA sequence might not offer a lot of insights. Furthermore, the most sensitive comparisons between various sequences can be made at the protein level, rather than the nucleotide one. A couple of decades earlier, the primary means of analysis revolved around a low throughput method of chemical protein sequencing. This is still relevant in examining and confirming the sequence of protein molecules from a genetically engineered construct. However, in the current scenario, when time and throughput are of vital importance, this method falls short in meeting the demand. It is in this context that the advent of high-throughput automated fluorescent DNA sequencing technology has led to the rapid accumulation of sequence information. It also provides the basis for abundant computationally derived protein sequence data. Analysis of this type of data helps to undermine a number of aspects of research like the detection of phyllogenetic information, genetic engineering using restriction mapping and determination of gene structure through intron/exon prediction. It is in this context that the analyses of DNA sequence have gained a lot of importance in the current decade.

Genes and DNA Sequences

DNA sequence databases typically contain genomic sequence data, which includes information at the level of the untranslated sequence, introns and exons, mRNA, cDNA and translations. This means that the DNA databases contain a variety of different types of data that have to be analyzed. A cursory look at each one of these is vital to understand sequence analysis.

Untranslated regions occur in both RNA and DNA and are portions of the CDS (sequence that codes for an mRNA) that are not translated into protein. The untranslated region particularly at the 3' end is highly specific both to the gene and the species from which the sequence is retrieved.

Given a DNA sequence and knowing the genetic code, it is possible to translate it into protein by looking up successive codons, in a genetic code table such as Table 5.4:

TABLE 5.4 Genetic Code Table

1st Position (5' end)	**2nd Position**				**3rd Position (3' end)**
	U	**A**	**C**	**G**	
U	Phe	Ser	Tyr	Cys	U
	Phe	Ser	Tyr	Cys	C
	Leu	Ser	Stop	Stop	A
	Leu	Ser	Stop	Trp	G
C	Leu	Pro	His	Arg	U
	Leu	Pro	His	Arg	C
	Leu	Pro	Gln	Arg	A
	Leu	Pro	Gln	Arg	G
A	Ile	Thr	Asn	Ser	U
	Ile	Thr	Asn	Ser	C
	Ile	Thr	Lys	Arg	A
	Met	Thr	Lys	Arg	G
G	Val	Ala	Asp	Gly	U
	Val	Ala	Asp	G;y	C
	Val	Ala	Glu	Gly	A
	Val	Ala	Glu	Gly	G

Standard Genetic Code

Synthesis of a peptide always starts from methionine (Met), coded by AUG. The stop codon (UAA, UAG or UGA) signals the end of a peptide. This table applies to mRNA sequences. For DNA, U should be replaced by T. In a DNA molecule, the sequence from an initiating codon (ATG) to a stop codon (TAA, TAG or TGA) is called an Open Reading Frame (ORF), which is likely (but not always) to encode a protein or polypeptide. This is known as the conceptual translation. It is not known whether the first base marks the start of the CDS. For this reason, it is important to carry out a six-frame translation. These are the three forward frames, which are achieved by beginning to translate at the first, second and third bases respectively. The three reverse frames are determined by reversing the DNA sequence and again beginning on the first, second and third bases. Thus, for any sequence of DNA, the result of a six-frame translation is six potential protein sequences. The genetic code is as shown in Table 5.5, but is arranged according to the physical properties of amino acids.

TABLE 5.5 Genetic Codes

Lys	AAA	**Asn**	AAC	**Gly**	GGA	**Met**	AUG
	AAG		AAU		GGC		
					GGG		
Arg	AGA	**Gln**	CAA		GGU	**Pro**	CCA
	AGG		CAG				CCC
	CGA			**Ala**	GCA		CCG
	CGC				GCC		CCU
	CGG				GCG		
	CGU	**Ser**	UCA		GCU		
			UCC			**Tyr**	UAC
His	CAC		UCG	**Val**	GUA		UAU
	CAU		UCU		GUC		
			AGC		GUG	**Phe**	UUC
			AGU		GUU		UUU
Asp	GAC	**Thr**	ACA	**Leu**	CUA	**Trp**	UGG
	GAU		ACC		CUC		
			ACG		CUG		
Glu	GAA		ACU		CUU	**Cys**	UGC
	GAG				UUA		UGU
					UUG		
				Ile	AUA	**Stop**	UAA
					AUC		UAG
					AUU		UGA

The Open Reading Frame (ORF) is generally the longest frame that is uninterrupted by a stop codon and also the one that is eventually transcribed. There are many approaches to detect the ORFs. However, finding the end of an ORF is easier than finding its beginning. This is because the ORFs overlap with each other. This method is routinely used based on the fact that long ORFs rarely occur by chance.

Another very important tool that is used is the codon preference that varies from protein to protein and species to species. In addition, many species show a general preference for G or C over A or T in the third base position of a codon (wobble). This can be used in order to scan ORFs.

Introns and Exons

The eukaryotic genes are made up of regions that contribute towards the CDS, known as exons and those that do not contribute, known as introns. These have a major impact on the gene products and their lengths because all exons might not be represented in the final transcribed mRNA. This is due to the presence of an mRNA editing mechanism which results in different translated polypeptides, forming splice variants, or alternately, spliced forms of the same protein. This results in varying results for sequence searches across different databases.

Sequence Comparison

One of the routine exercises that are part of sequence analysis is the determination of similarity between two sequences. There is no unique, precise, and universally applicable notion of similarity. Proteins and DNA can be similar with respect to their function, their structure, or their primary sequence of amino or nucleic acids. The general rule is that sequence determines shape, and shape determines function. So, when we study sequence similarity, we eventually hope to discover or validate similarity in shape and function. Similarity has both quantitative and qualitative aspect. A similarity measure gives a quantitative answer, saying that two sequences show a certain degree of similarity. An alignment is a mutual arrangement of two sequences, which is a sort of qualitative answer; it exhibits where the two sequences are similar, and where they differ. An optimal alignment is one that exhibits the most correspondences and the least differences. There are various methods used for comparing sequences, of which the most important ones are covered in the subsequent section.

Segment Methods

Segment methods compare all overlapping segments of a predetermined length (e.g., 10 amino acids) with one another. This is the approach used in dotplots. One sequence is represented on each axis and significant matching regions are distributed along diagonals in the matrix. A dot is placed at each position p(x,y) where, the characters are matched. This is generally useful for quick visual comparisons however it is difficult to compare more than two sequences at a time by this method. Figure 5.1 illustrates the dot plot method.

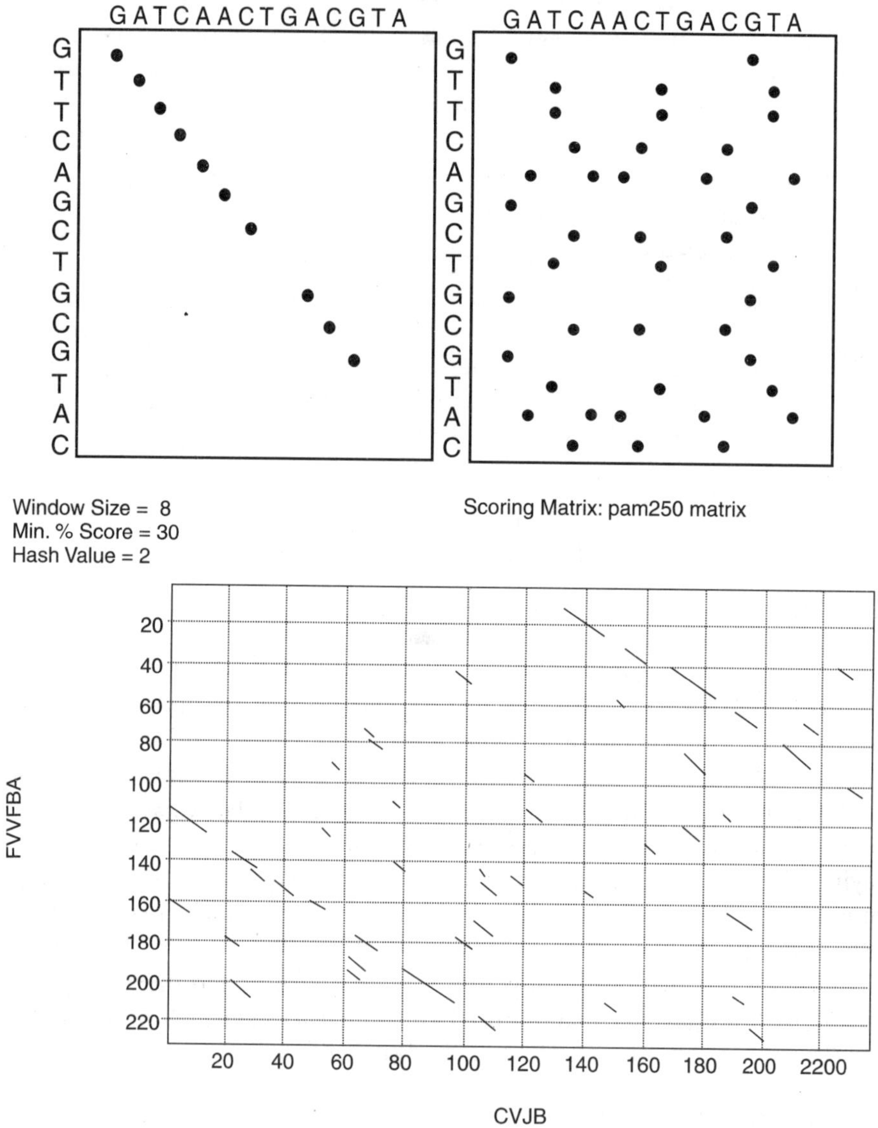

FIGURE 5.1 Dot Plot Method
Source: Simple Dot Plot, Dot Plot with 4 Bases 75% Identity, Sample Dot Plot for Multiple Sequences

Dynamic Programming

Dynamic programming is ideal for both global alignments (GAP is the common version) and best local alignments (BESTFIT). It is a method where one can build a solution to a problem by solving smaller but similar subset problems. Two of the important dynamic programming algorithms are studied here.

Global Alignment

The simultaneous alignment of many nucleotide or amino acid sequences is now an essential tool in molecular biology. Multiple alignments are used:

- To find diagnostic patterns to characterise protein families.
- To detect or demonstrate homology between new sequences and existing families of sequences.
- To help predict the secondary and tertiary structures of new sequences
- To suggest oligonucleotide primers for PCR.
- As an essential prelude to molecular evolutionary analysis.

Conceptually, the method is similar to the dot plot method. In this approach, a maximum match between two sequences is defined as the largest number of amino acids from one protein that can be matched with those of another protein, while allowing for all possible deletions. A penalty is introduced for the same to account for the gap. In this approach, sequences are compared by constructing a 2D matrix.

In 1970, Needleman and Wunsch proposed a maximum match pathway that is obtained computationally by applying an algorithm, where:

- Cells representing identities are scored 1.
- Cells representing mismatches are scored 0.

The 2 D array is then populated with these values. An operation of successive summation of cells is then performed. In the process, each cell in the matrix is examined along with the maximum score along any path leading to the cell. This is then added to its present contents and the summation continues. Once this is completed, the maximum match pathway is constructed. An alignment is then generated by working through the matrix, starting at the highest scoring element at the N-termini, and following the pattern of the high score through the C-termini. This is the reason why the method produces a global alignment as opposed to local alignment technique that is described later. The process of global alignment is depicted in the following figures.

1. Fill the matrix **S** with 1 if match, else 0.

	A	B	C	N	J	R	Q	C	L	C	R	P	M
A	1	0	0	0	0	0	0	0	0	0	0	0	0
J	0	0	0	0	1	0	0	0	0	0	0	0	0
C	0	0	1	0	0	0	0	1	0	1	0	0	0
J	0	0	0	0	1	0	0	0	0	0	0	0	0
N	0	0	0	1	0	0	0	0	0	0	0	0	0
R	0	0	0	0	0	1	0	0	0	0	1	0	0
C	0	0	1	0	0	0	0	1	0	1	0	0	0
K	0	0	0	0	0	0	0	0	0	0	0	0	0
C	0	0	1	0	0	0	0	1	0	1	0	0	0
R	0	0	0	0	0	1	0	0	0	0	1	0	0
B	0	1	0	0	0	0	0	0	0	0	0	0	0
P	0	0	0	0	0	0	0	0	0	0	0	1	0

2. Score on the basis of algorithm.

	A	B	C	N	J	R	Q	C	L	C	R	P	M
A	1	0	0	0	0	0	0	0	0	0	0	0	0
J	0	0	0	0	1	0	0	0	0	0	0	0	0
C	0	0	1	0	0	0	0	1	0	1	0	0	0
J	0	0	0	0	1	0	0	0	0	0	0	0	0
N	0	0	0	1	0	0	0	0	0	0	0	0	0
R	0	0	0	0	0	1	4	3	3	2	2	0	0
C	3	3	4	3	3	3	3	4	3	3	1	0	0
K	3	3	3	3	3	3	3	3	3	2	1	0	0
C	2	2	3	2	2	2	2	3	2	3	1	0	0
R	2	1	1	1	1	2	1	1	1	1	2	0	0
B	1	2	1	1	1	1	1	1	1	1	1	0	0
P	0	0	0	0	0	0	0	0	0	0	0	1	0

3. Trace back from the left-top corner and always select the maximum value from the outmost column and row. Jump to the next maximum in next row or column.

	A	B	C	N	J	R	Q	C	L	C	R	P	M
A	8	7	6	6	5	4	4	3	3	2	1	0	0
J	7	7	6	6	6	4	4	3	3	2	1	0	0
C	6	6	7	6	5	4	4	4	3	3	1	0	0
J	6	6	6	5	6	4	4	3	3	2	1	0	0
N	5	5	5	6	5	4	4	3	3	2	1	0	0
R	4	4	4	4	4	5	4	3	3	2	2	0	0
C	3	3	4	3	3	3	3	4	3	3	1	0	0
K	3	3	3	3	3	3	3	3	3	2	1	0	0
C	2	2	3	2	2	2	2	3	2	3	1	0	0
R	2	1	1	1	1	2	1	1	1	1	2	0	0
B	1	2	1	1	1	1	1	1	1	1	1	0	0
P	0	0	0	0	0	0	0	0	0	0	0	1	0

4. Present alignment according to the scores. This is how it would appear:

```
ABCNJ–RQCLCR–PM
*  *  *  *  *  ** *
AJC-JNR-CKCRBP-
```

or as

```
ABC–NJRQCLCR–PM
*  *  *  *  *  ** *
AJCJN–R–CKCRBP-
```

both with 8 asterisks to denote the 8 matches. Note that in this particular case, gaps are given the same penalty as a mismatch. They simply do not add to the score.

Local Alignment

In 1981, Smith and Waterman described a method commonly called as the Smith-Waterman algorithm for finding regions of similarity in two sequences that are distantly related to each other. A key element of this method is that each cell in the matrix defines the endpoint of a potential alignment, whose similarity is determined by the value stored in the cell. The algorithm begins by filling the edge elements with 0.0 values since they represent the ends of alignments of length zero and, consequently, their similarity score will be zero. This is an important difference in the scoring methodology as compared to the Needleman-Wunsch technique. The next step is to populate the remaining cells in the matrix. Once the matrix is complete the highest score is located and the alignment represented. The process of local alignment is depicted in the following figures:

1. Fill matrix **S** with 0.0 to begin initialization.

	H	E	A	G	A	W	G	H	E	E
P	0.00	0.00	0.00	0.00	0.00	0.00	0.00	0.00	0.00	0.00
A	0.00									
W	0.00									
H	0.00									
E	0.00									
A	0.00									
E	0.00									

2. Continue initializing matrix S by adding Match/Mismatch points. The row (-1st) and column (-1st) are not shown, because they will take no effect on the 0th row and 0th column after applying to the formula.
3. Compute the score matrix and trace back from the highest cell in the matrix by following the arrows till cell with value 0(not included).

	H	E	A	G	A	W	G	H	E	E
P	0.00	0.00	0.00	0.00	0.00	0.00	0.00	0.00	0.00	0.00
A	0.00	0.00	1.00	0.00	1.00	0.00	0.00	0.00	0.00	0.00
W	0.00	0.00	0.00	0.67	0.00	2.00	0.67	0.33	0.00	0.00
H	0.00	0.00	0.00	0.00	0.33	0.67	1.67	1.67	0.33	0.00
E	0.00	1.00	0.00	0.00	0.00	0.33	0.33	1.33	2.67	1.33
A	0.00	0.00	2.00	0.67	1.00	0.00	0.00	0.00	1.33	2.33
E	0.00	1.00	0.67	1.67	0.33	0.67	0.00	0.00	1.00	2.33

4. Represent the alignment.

Table 5.6 depicts the comparison of the global and local alignments:

TABLE 5.6 Comparing global and local alignments

Needleman-Wunsch	Smith-Waterman
Global alignments.	Local alignments.
Residue alignment score should be greater than or equal to 0, or in other words, positive.	Residue alignment score may or may not be negative.
No gap penalty required top work effectively.	Gap penalty required to work effectively.
Score cannot decrease between two cells. It should stay level of a pathway.	Score can increase, decrease or between two cells of a pathway.

SEARCHING SEQUENCE DATABASES

Sequence alignments provide a powerful way to compare novel sequences with previously characterized genes. Both functional and evolutionary information can be inferred from well-designed queries and alignments. Alignment techniques are helpful in studying evolutionary characteristics and have emerged as powerful tools for structure and function predication. There are two methods of searching and aligning sequences.

FastA

This is an algorithm described by Lipman and Pearson in 1985 and is based on the idea of identifying short words, or K-tuples, common to both sequences under comparison. The sizes of the K-tuples used are around 1 or 2 residues in case of protein searches, while larger k-tuples (up to 6 bases) are used in case of DNA searches. FastA uses a heuristic approach to join k-tuples that lie close to each other on the same diagonal. The regions formed in this way contain mismatches lying between matching k-tuples. If a significant number of matches are found, FastA uses a dynamic programming algorithm to compute gapped alignments that incorporate the ungapped regions.

Practically, FastA is a family of programs, allowing queries of DNA versus a protein database, or vice versa. In these variants, there is further distinction that regards the location of gaps. One may assume that gaps occur only in the codon frames corresponding to amino-acid insertion. Alternatively, one can also assume gap location to be arbitrary, accounting for insertion or deletion of nucleotides. The different FastA variants are summarized in Table 5.7:

TABLE 5.7 FastA Variants

FastA Variant	Query Sequence	Database Sequence	Gap Location
FastA	DNA Protein	DNA Protein	Not an issue
FastX	DNA	Protein	Between codons
FastY			Arbitrary
tFastA	Protein	DNA	Not an issue
tFastX	Protein	Translated DNA	Between codons
tFastX			Arbitrary
FastS	Linked Peptides	Protein	Not an issue
FastF	Mixed Peptides	Protein	Not an issue

This is the FastA output format:

```
ggamma.pep
PIR1:HGCZG

P1;HGCZG - hemoglobin gamma-G chain - chimpanzee
N;Alternate names: hemoglobin gamma-1 chain
C;Species: Pan troglodytes (chimpanzee)
```

```
C;Date: 31-May-1996 #sequence_revision 21-Jan-1997 #text_change 14-Nov-1997
C;Accession: I36939; I61853
R;Slightom, J.L.; Chang, L.Y.; Koop, B.F.; Goodman, M. . . .

SCORES   Init1: 971   Initn: 971   Opt: 971   z-score: 1145.0 E(): 6.2e-57
Smith-Waterman score: 971;   100.0% identity in 147 aa overlap

                   10        20        30        40        50        60
ggamma.pep   MGHFTEEDKATITSLWGKVNVEDAGGETLGRLLVVYPWTQRFFDSFGNLSSASAIMGNPK
             ||||||||||||||||||||||||||||||||||||||||||||||||||||||||||||
HGCZG        MGHFTEEDKATITSLWGKVNVEDAGGETLGRLLVVYPWTQRFFDSFGNLSSASAIMGNPK
                   10        20        30        40        50        60
                   70        80        90       100       110       120
ggamma.pep   VKAHGKKVLTSLGDAIKHLDDLKGTFAQLSELHCDKLHVDPENFKLLGNVLVTVLAIHFG
             ||||||||||||||||||||||||||||||||||||||||||||||||||||||||||||
HGCZG        VKAHGKKVLTSLGDAIKHLDDLKGTFAQLSELHCDKLHVDPENFKLLGNVLVTVLAIHFG
                   70        80        90       100       110       120
                  130       140
ggamma.pep   KEFTPEVQASWQKMVTGVASALSSRYH
             |||||||||||||||||||||||||||
HGCZG        KEFTPEVQASWQKMVTGVASALSSRYH
                  130       140
```

Basic Local Alignment Tool (BLAST)

This algorithm was described by Altschul and team, in 1990. BLAST is a set of similarity search programs designed to explore all of the available sequence databases regardless of whether the query is protein or DNA. BLAST programs have been designed for speed, with minimal sacrifice of sensitivity to distant sequence relationships. The scores assigned in a BLAST search have a well-defined statistical interpretation, making real matches easier to distinguish from random background hits. BLAST uses a heuristic algorithm that seeks local as opposed to global alignments and is therefore able to detect relationships among sequences that share only isolated regions of similarity.

Output format for BLAST.

```
Smallest
                                                                      Sum
                                                            High   Probability
Sequences producing High-scoring Segment Pairs:
                                                            Score  P(N)      N
gnl|PID|e288614        (Z83122) R11A5.h [Caenorhabditis ele...  173  1.3e-14   2
gi|1763275             (U73478) acidic nuclear phosphoprote...  134  8.4e-11   1
gi|1769451             (X69465) ryanodine receptor 1 [Sus s...  130  1.2e-09   2
gi|1752736             (D83006) gene required for phosphoyl...  121  8.5e-09   1
sp|P09741|TRT2_RABIT TROPONIN T, CARDIAC MUSCLE ISOFORMS        102  3.8e-08   3
gi|1754694             (U81159) magnesium-dependent calcium...  109  2.1e-06   2
sp|Q08168|HRP_PLABE  58 KD PHOSPHOPROTEIN (HEAT SHOCK-REL...     98  5.1e-06   2
sp|Q10713|MPP1_HUMAN MITOCHONDRIAL PROCESSING PEPTIDASE A...     94  4.2e-05   1
gi|1762998             (U67260) RING-finger protein [Helico...   92  7.4e-05   1
sp|P02547|NFL_PIG    NEUROFILAMENT TRIPLET L PROTEIN (68 ...     93  0.00039   2
gnl|PID|e290386        (Z83731) M04C9.6b [Caenorhabditis el...   64  0.0023    2
gnl|PID|e290387        (Z83731) M04C9.6a [Caenorhabditis el...   64  0.0034    2
sp|P02641|TRT3_RABIT TROPONIN T, FAST SKELETAL MUSCLE ISO...     70  0.0042    3
sp|P50751|TRTC_SHEEP TROPONIN T, CARDIAC MUSCLE ISOFORM (...     69  0.0046    3
```

```
gnl|PID|e283394      (Y08484) swelling dependent chloride...   66  0.0047    2
gi|1763002           (U67262) RING-finger protein [Helico...   78  0.0065    1
sp|P13919|CVCB_PEA   CONVICILIN PRECURSOR                      81  0.0070    2
gi|1763000           (U67261) RING-finger protein [Helico...   77  0.0090    1
gi|1762996           (U67259) RING-finger protein [Helico...   76  0.012     1
gi|1773061           (U82273) FGF homolog [Drosophila mel...   76  0.014     1

WARNING:  Descriptions of 124 database sequences were not reported due to the
          limiting value of parameter V = 20.


>gnl|PID|e288614 (Z83122) R11A5.h [Caenorhabditis elegans]
            Length = 906

 Score = 173 (79.1 bits), Expect = 1.3e-14, Sum P(2) = 1.3e-14
 Identities = 27/73 (36%), Positives = 58/73 (79%)

Query:   131 EGKTGNATDEEEEEEEEEEEGEEEEEEEEEDDDDDDDEDSGAEIQDDDEEGFDDEEEFDDD 190
             E K  ++ +EE++++E+EEE E+EEEE+ED++D+++E++ +E +++D+E  + EEE D++
Sbjct:   576 ELKLDDSDEEEDDDDEDEEEDEDEEEEDEDEEDEEEEENESEEEEEDDEDEESEEESDEE 635

Query:   191 EHDDDDLDNEENE 203
             + ++++ D+ E E
Sbjct:   636 DEEEEEEDDSEPE 648

 Score = 170 (77.8 bits), Expect = 3.3e-14, Sum P(2) = 3.3e-14
 Identities = 28/77 (36%), Positives = 56/77 (72%)

Query:   140 EEEEEEEEEEEGEEEEEEEEEDDDDDDDEDSGAEIQDDDEEGFDDEEEFDDDEHDDDDLDN 199
             +EEE++++E+E E+E+EEEED+D++D+E+   E ++++E+  D+E E + DE D+++ +
Sbjct:   583 DEEEDDDDEDEEEDEDEEEEDEDEEDEEEEENESEEEEEDDEDEESEEESDEEDEEEEEE 642

Query:   200 EENELEELEERVEARKK 216
             +++E EE      + KK
Sbjct:   643 DDSEPEEKPNPKPSTKK 659

 Score = 164 (75.0 bits), Expect = 2.1e-13, Sum P(2) = 2.1e-13
 Identities = 26/63 (41%), Positives = 49/63 (77%)

Query:   139 DEEEEEEEEEEEEGEEEEEEEEEDDDDDDDEDSGAEIQDDDEEGFDDEEEFDDDEHDDDDLD 198
             D+++E+EEE+E+ EEE+E+EED+++++++E    E  D+DEE  ++ +E D++E ++DD +
Sbjct:   587 DDDDEDEEEDEDEEEEDEDEEDEEEEENESEEEEEDDEDEESEEESDEEDEEEEEEDDSE 646

Query:   199 NEE 201
              EE
Sbjct:   647 PEE 649

 Score = 151 (69.1 bits), Expect = 1.2e-11, Sum P(2) = 1.2e-11
 Identities = 24/85 (28%), Positives = 56/85 (65%)

Query:   142 EEEEEEEEEGEEEEEEEEEDDDDDDDEDSGAEIQDDDEEGFDDEEEFDDDEHDDDDLDNEE 201
             ++ +EEE++ +E+EEE+ED++++D+++   E ++++ E  ++++E ++ E + D+ D EE
Sbjct:   580 DDSDEEEDDDDEDEEEDEDEEEEDEDEEDEEEEENESEEEEEDDEDEESEEESDEEDEEE 639

Query:   202 NELEELEERVEARKKTTEKQSAAAL 226
              E ++ E   +   K + K+ A  L
Sbjct:   640 EEEDDSEPEEKPNPKPSTKKQALIL 664

 Score = 149 (68.2 bits), Expect = 2.3e-11, Sum P(2) = 2.3e-11
 Identities = 23/64 (35%), Positives = 48/64 (75%)

Query:   139 DEEEEEEEEEEEEGEEEEEEEEEDDDDDDDEDSGAEIQDDDEEGFDDEEEFDDDEHDDDDLD 198
             D+E+EEE+E+EE E+E+EE+E+++++++ E+   + +D++ E  DEE+ +++E DD + +
Sbjct:   589 DDEDEEEDEDEEEEDEDEEDEEEEENESEEEEEDDEDEESEEESDEEDEEEEEEDDSEPE 648
```

```
Query:   199 NEEN 202
               + N
Sbjct:   649 EKPN 652

 Score = 148 (67.7 bits), Expect = 3.1e-11, Sum P(2) = 3.1e-11
 Identities = 27/70 (38%), Positives = 48/70 (68%)

Query:   126 DLSNVEGKTGNATDEEEEEEEEEEEEGEEEEEEEEEDDDDDDDEDSGAEIQDDDEEGFDDEE 185
             D S+ E    +   +EE+E+EEEE+E EE+EEEEE++ ++++ED   E  +++ +  D+EE
Sbjct:   580 DDSDEEEDDDDEDEEEDEDEEEEDEDEEDEEEEENESEEEEEDDEDEESEEESDEEDEEE 639
```

The original BLAST program suffers from a basic limitation that it can only produce ungapped alignments. It is known that, on many occasions, several ungapped non-overlapping alignments result from a match to a single database sequence. This would not be a true reflection of the sequence similarity between the two sequences. To overcome this a modification, a new algorithm has been recently introduced. The new algorithm seeks only one, rather than all, ungapped alignments that make up a significant match, and hence speeds the initial database search.

SECONDARY DATABASE SEARCH

Primary databases are growing at an alarming rate and the research in this decade is mainly focused on deducing functional features from the sequence data that is available. Sequence search and analysis methodologies are vastly advanced. However, the results derived by these are difficult to interpret and do not provide answers to some of the more sophisticated questions. This is because of the redundancy in primary databases due to irrelevant matches and the presence of repetitive entries. Also, in case of proteins, it was difficult to determine at what level a match was made (single domain, multiple domain or the whole protein). Due to these reasons, the secondary databases gained prominence.

Secondary databases are so called, because they contain the results of analyses of the sequences in the primary sources. The information contained in the secondary databases is largely variable owing to the fact that there are several primary databases, and a variety of ways of analyzing the protein sequences. Although this seems impossible due to the lack of any standards or consensus, Swiss-Prot has emerged as one of the most popular primary resources and many secondary databases now use it.

The common principle behind secondary databases is that the homologous sequences may be gathered in multiple alignments, within which are conserved regions that show little or no variation between constituent sequences. These conserved regions are called motifs and reflect some important biological functions. Secondary databases use such conserved motifs to build characteristic signatures that aid family and functional diagnosis of new sequences. Secondary databases use various approaches to determine these signatures.

Some of the main secondary databases are displayed in Table 5.8:

TABLE 5.8 Main Secondary Database

Secondary Database	Primary Database	Stored Information
Prosite	Swiss-Prot	Regular expressions (Patterns)
Profiles	Swiss-Prot	Weighted matrices (Profiles)
Prints	OWL*	Aligned motifs (fingerprints)
Pfam	Swiss-Prot	Hidden Markov Models
Blocks	Prosite/Prints	Aligned motifs (blocks)
Identify	Blocks/Prints	Fuzzy regular expressions

*Swiss-Prot is OWL's highest priority source

Prosite

Prosite is a database of protein families and domains. It is based on the observation that, while there is a huge number of different proteins, most of them can be grouped, on the basis of similarities in their sequences, into a limited number of families. Prosite currently contains patterns and profiles specific for more than a thousand protein families or domains. Each of these signatures comes with documentation providing background information on the structure and function of these proteins.

```
ID   OPSIN; PATTERN.
AC   PS00238;
DT   APR-1990 (CREATED); DEC-2004 (DATA UPDATE); APR-2005 (INFO UPDATE).
DE   Visual pigments (opsins) retinal binding site.
PA   [LIVMFWAC]-[PSGAC]-x-{G}-x-[SAC]-K-[STALIMR]-[GSACPNV]-[STACP]-x(2)-
PA   [DENF]-[AP]-x(2)-[IY].
NR   /RELEASE=47.0,181571;
NR   /TOTAL=198(197); /POSITIVE=195(194); /UNKNOWN=0(0); /FALSE_POS=3(3);
NR   /FALSE_NEG=1; /PARTIAL=4;
CC   /TAXO-RANGE=??E??; /MAX-REPEAT=2;
CC   /SITE=7,retinal;
CC   /VERSION=1;
DR   Q9W6A9, OP1S1_BRARE, T; Q9W6A8, OP1S2_BRARE, T; Q9H1Y3, OPN3_HUMAN , T;
DR   Q9WUK7, OPN3_MOUSE , T; Q9UHM6, OPN4_HUMAN , T; Q9QXZ9, OPN4_MOUSE , T;
DR   P22269, OPS1_CALVI , T; P06002, OPS1_DROME , T; P28678, OPS1_DROPS , T;
DR   Q94741, OPS1_SCHGR , T; P08099, OPS2_DROME , T; P28679, OPS2_DROPS , T;
DR   Q26495, OPS2_SCHGR , T; P04950, OPS3_DROME , T; P28680, OPS3_DROPS , T;
DR   P08255, OPS4_DROME , T; P29404, OPS4_DROPS , T; P17646, OPS4_DROVI , T;
DR   P91657, OPS5_DROME , T; 001668, OPS6_DROME , T; P51471, OPSB_ANOCA , T;
DR   P90680, OPSB_APIME , T; P51472, OPSB_ASTFA , T; P51490, OPSB_BOVIN , T;
DR   P32310, OPSB_CARAU , T; P28682, OPSB_CHICK , T; 013227, OPSB_CONCO , T;
DR   P35357, OPSB_GECGE , T; P03999, OPSB_HUMAN , T; P51491, OPSB_MOUSE , T;
DR   P87365, OPSB_ORYLA , T; P60573, OPSB_PANPA , T; P60015, OPSB_PANTR , T;
DR   Q63652, OPSB_RAT   , T; 013092, OPSB_SAIBB , T; Q25157, OPSC1_HEMSA, T;
DR   Q25158, OPSC2_HEMSA, T; Q90214, OPSD1_ANGAN, T; 015973, OPSD1_PATYE, T;
```

```
DR   Q90215, OPSD2_ANGAN, T; 015974, OPSD2_PATYE, T; 042294, OPSD_ABYKO , T;
DR   P52202, OPSD_ALLMI , T; Q90245, OPSD_AMBTI , T; P41591, OPSD_ANOCA , T;
 OPSD_CORAU , T;
 OPSG2_ASTFA, T;

DO   PDOC00211;
//
```

The following are the explanations of the different lines used in the sequence provided above.

- **The ID (IDentification) line:** is always the first line of an entry. The first item on the ID line is the entry name. This name is a useful means of identifying an entry. The general form of the ID line is: `ID ENTRY_NAME; ENTRY_TYPE.`
- **The AC (ACcession number) line:** lists the accession number associated with an entry. It is always the second line of an entry. An accession number never changes. Accession numbers allow unambiguous citation of database entries. The format of the AC line is: `AC PSnnnnn;` Where 'PS' stands for Prosite and 'nnnnn' is a five digit number.
- **The DT (DaTe) line:** shows the date of entry or last modification of the entry. It is always the third line of an entry. The format of the DT line is: `DT MMM-YYYY (CREATED); MMM-YYYY (DATA UPDATE); MMM-YYYY (INFO UPDATE).`
- **The DE (DEscription) line:** provides descriptive information about the content of the entry. It is always the fourth line of an entry. The format of the DE line is: `DE  Description.`
- **The PA (PAttern) lines:** contain the definition of a Prosite pattern. For instance, PA [AC]-x-V-x(4)-{ED}. This pattern is translated as: [Ala or Cys]-any-Val-any-any-any-any-{any but Glu or Asp} PA <A-x-[ST](2)-x(0,1)-V. This pattern, which must be in the N-terminal of the sequence ('<'), is translated as: Ala-any-[Ser or Thr]-[Ser or Thr]-(any or none)-Val.
- **The MA (MAtrix) lines:** contain the definition of a PROSITE profile (or matrix) entry.
- **The RU (RUle) lines:** contain the definition of a PROSITE rule entry. The format of the RU line is `RU Rule_Description.`
- **The NR (Numerical Results) lines:** contain information relevant to the results of the scan with a pattern on the complete Swiss-Prot knowledgebase. The format of the NR line is `NR /QUALIFIER=data; /QUALIFIER=data;`
- **The CC (Comments) lines:** contain various types of comments. The format of the CC line is `CC /QUALIFIER5data; /QUALIFIER5 data;`

- **The TAXO-RANGE qualifier**: is used to indicate the taxonomic range of a pattern or matrix. The syntax of that qualifier is `/TAXO-RANGE=ABEPV;` where:
    - 'A' stands for archaea
    - 'B' stands for bacteriophages
    - 'E' stands for eukaryotes
    - 'P' stands for prokaryotes (bacteria)
    - 'V' stands for eukaryotic viruses
- **The/MAX-REPEAT qualifier**: is used to indicate the maximum number of times a given pattern or profile has been found in a single protein sequence. The syntax of that qualifier is `/MAX-REPEAT=nn;`
- **The DR (Database Reference) lines**: are used as pointers to the Swiss-Prot entries that are picked up (or missed) by the pattern being described in the entry. The format of the DR line is `DR AC_NB, ENTRY_NAME, C; AC_NB, ENTRY_NAME, C; AC_NB, ENTRY_NAME, C;`
- **The 3D (3D-structure) line**: is used to list the code(s) of the Protein Data Bank (PDB) entries that contain structural data corresponding the sequence region described in a PROSITE entry. The format of the 3D line is `3D name; [name2;...]`
- **The DO (DOcumentation) line**: contains a pointer to the entry in the PROSITE documentation file that describes the entry. The format of the DO line is `DO PDOCnnnnn;` where 'PDOC' stands for PROSITE DOCumentation and 'nnnnn' is a five-digit number.
- **The // (terminator) line**: contains no data or comments. It designates the end of an entry.

Patterns

Patterns are derived with the construction of a multiple alignment and manual inspection conserved regions. Sequence information within the individual motifs is reduced to single consensus expressions, and the resulting patterns are used to search Swiss-Prot (or other primary databases). The results of this search are further sieved to eliminate incorrect or redundant matches to create the optimal pattern. The strict outcome of this type of pattern searching has severe diagnostic limitations. Due to the high level of stringency that is built in to the algorithm, there are chances that one might end up losing a lot of patterns that might be crucial in resolving expressions of certain proteins. Another important limitation of this approach is that it is based on the philosophy of using single motifs to characterize entire proteins. This requires the advance knowledge of the most conserved regions of the sequence alignments. This approach is extremely unreliable in case of proteins that have a number of conserved regions. Since only a single motif is considered the chances of getting fuzzy results are higher.

Profiles

There is an alternative to the motif-based approach of protein family characterization. This method is based on the principle that variable regions between conserved motifs also contain valuable sequence information. In this case, the complete sequence alignment effectively becomes the discriminator. The discriminator, termed a profile, is weighted to indicate where insertions and deletions (INDELS) are allowed, the type of residues allowed and at what positions do the conserved regions belong to. Profiles are also called weight matrices and they provide a sensitive means of detecting distant sequence relationships where only a few residues are well conserved. This enables to overcome the limitations of regular expressions, which would either miss any true positives or catch too many false positives.

Pfam

A basic variation on the theme of profiles is to encode an alignment in the form of a Hidden Markov Model (HMM). These are probabilistic models consisting of a number of interconnecting states; they are essentially linear chains of match, delete, or insert states, which are used to encode sequence conservation within alignments. Figure 5.2 depicts an HMM.

The model in the figure shown is a linear sequence of nodes, each of which includes match (square), insert (diamond), and delete (circle) states. Each match state has a distribution over the appropriate alphabet indicating

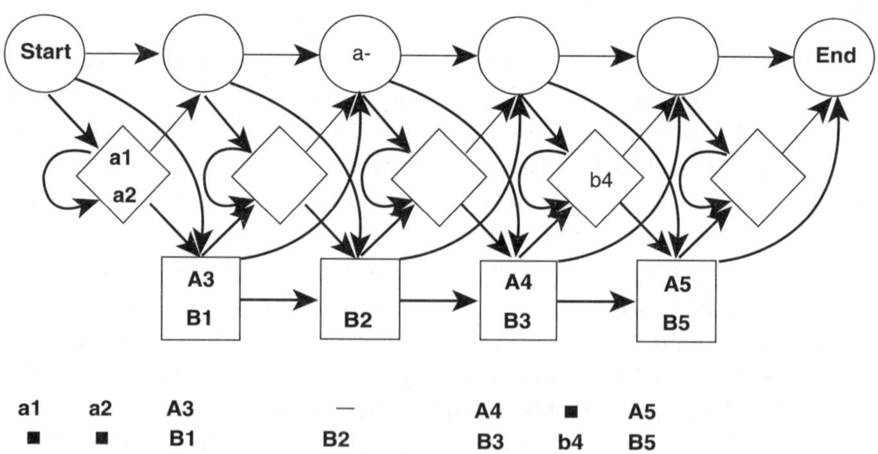

FIGURE 5.2 Hidden Markov Model

which characters are most likely. The chain of match states forms a model of the family, or of columns of a multiple alignment. Some sequences may not have characters in specific positions—delete states enable them to skip through a node without 'using up' any characters. Other sequences may have extra characters, which are modeled with the insert states. As a result, insertions are used when a small number of sequences have positions not found in most other sequences, while delete states are used when a small number of sequences do not have a character in a position found in most other sequences. As with the match states, all transition probabilities (the chance of having a delete or moving to an insertion state) are local, enabling, for example, the system to strongly penalize sequences that delete conserved regions.

A match state is assigned to each conserved column in a sequence alignment; an insert state allows insertions relative to the match states; and delete states allow match positions to be skipped. Thus, building of HMMs requires each position to be categorized under these categories. This is the basis of Pfam databases.

The Pfams are essentially divided into two: one containing the seed and full alignments coupled with minimal annotations and HMMs themselves called PfamA, and the other containing all sequences that are not deposited in Pfam A called Pfam B. The format varies slightly across different databases however they have certain commonalities. The format is as shown:

```
Compulsory fields:
-------------

 AC   Accession number:           Accession number in form PFxxxxx or PBxxxxxx.
 ID   Identification:             One word name for family.
 DE   Definition:                 Short description of family.
 AU   Author:                     Authors of the entry.
 AL   Alignment method of seed:   The method used to align the seed members.
 SE   Source of seed:             The source suggesting the seed members belong to
                                  one family.
 GA   Gathering method:           Search threshold to build the full alignment.
 TC   Trusted Cutoff:             Lowest sequence score and domain score of match
                                  in the full alignment.
 NC   Noise Cutoff:               Highest sequence score and domain score of match
                                  not in full alignment.
 SQ   Sequence:                   Number of sequences in alignment.
 //                               End of alignment.

Optional fields:
-------------

 DC   Database Comment:           Comment about database reference.
 DR   Database Reference:         Reference to external database.
 RC   Reference Comment:          Comment about literature reference.
 RN   Reference Number:           Reference Number.
 RM   Reference Medline:          Eight digit medline UI number.
 RT   Reference Title:            Reference Title.
 RA   Reference Author:           Reference Author
 RL   Reference Location:         Journal location.
```

```
PI   Previous identifier:        Record of all previous ID lines.
KW   Keywords:                   Keywords.
CC   Comment:                    Comments.
```

Blocks

The diagnostic limitations of regular expressions prompted the development of multiple-motif databases, based on the protein families contained in Prosite. In this resource, the motifs, or blocks, are created by automatically detecting the most highly conserved regions of each protein family. This is achieved by a method based on the initial stages of identification of three conserved amino acids. The resulting blocks are then scanned against Swiss-Prot to obtain a measure of a chance match. The blocks are then scored at two levels; the first denotes the level at which 99.5% of the matches are true negatives (which are unrelated), the second is a median value of the true positives (related). The format is as shown:

```
ID   short_identifier; BLOCK
AC   block_number; distance from previous block = (min,max)
DE   description
BL   xxx motif; width=w; seqs=s; 99.5%=n1; strength=n2
sequence_id  (offset) sequence_segment  sequence_weight
 .
 .
 .

//
```

The following are the explanations of the different lines used in the sequence provided above:

- **ID line:** Starts block entry and contains short identifier for the group of sequences from which the block was made. If the block was taken from Prosite, it will be the Prosite group ID. A semi-colon terminates the identifier, and the word "BLOCK" indicates the entry type.
- **AC line:** Contains the block number, a seven-character group number for sequences from which the block was made, followed by a letter (A-Z) indicating the order of the block in the sequences. If
  - The group has only one block, the letter is omitted.
  - The block was made from Prosite group PSnnnnn, the block number is BLnnnnna.
  - The block was converted from Terri Attwood's Prints Database, the block number is PRnnnnna.

`(min,max)` is the minimum and maximum number of amino acids from previous block for sequences in this block. For the first block in the group, the distance is from the beginning of the sequences.

- **DE line:** Contains a description of the group of sequences from which the block was made. If the block was taken from Prosite, it will be a slightly edited version of the Prosite description.
- **BL Line:** Contains information about the block. The components of this line are:

  - *xxx* : amino acids in the spaced triplet found by MOTIF.
  - *w* : width of sequence segments (columns).
  - *s* : number of sequence segments (rows).
  - *n1* : raw calibration score; 99.5th percentile score of true negative sequences. Raw search scores are normalized by dividing by the score of the true negative sequences and multiplying by 1000.
  - *n2* : median normalized score of known true positive sequences as documented in Prosite.

Following the BL line are *s* lines, one for each sequence with a segment in the block. The segments may be clustered with clusters separated by blank lines. Each segment line contains a sequence identifier of 10 or fewer characters, the offset from the beginning of the sequence to the block in parentheses, the sequence segment, and a weight for the segment. The weights are normalized so that the most distant segment has a weight of 100.

- **// line** : Terminates a block entry.

Terms to Know

(Adapted from http://www.library.csi.cuny.edu/~davis/ Bio_326/ bioinfo_glossary.html)

Accession number	Identifier supplied by curators of major biological databases, upon submission of a novel entry that uniquely identifies that sequence (or other) entry.
Annotation	Combination of comments, notations, references, and citations, either in free format or utilizing a controlled vocabulary, together describing all experimental and inferred information about a gene or protein. Annotations can also be applied to the description of other biological systems. Batch, automated annotation of bulk biological sequence is one of the key uses of Bioinformatics tools.
Algorithm	Series of steps defining a procedure or formula for solving a problem that can be coded into a programming language and executed. Bioinformatics algorithms are used to process, store, analyze, visualize and make predictions from biological data.
Consensus sequence	Single sequence delineated from an alignment of multiple constituent sequences that represents a "best fit" for all those sequences. A "voting" or other selection procedure is used to determine which residue (nucleotide or amino acid) is placed at a given position in the event that not all of the constituent sequences have the identical residue at that position.
Data Mining	Ability to query very large databases in order to satisfy a hypothesis ("top-down" data mining); or to interrogate a database in order to generate new hypotheses based on rigorous statistical correlations ("bottom-up" data mining).
Expression (gene or protein)	Measure of the presence, amount, and time-course of one or more gene products in a particular cell or tissue. Expression studies are typically performed at the RNA (mRNA) or protein level in order to determine the number, type, and level of genes that may be up-regulated or down-regulated during a cellular process, in response to an external stimulus, or

	in sickness or disease. Gene chips and proteomics now allow the study of expression profiles of sets of genes or even entire genomes.
Gene Index	Listing of the number, type, label and sequence of all the genes identified within genome of a given organism. Gene indices are usually created by assembling overlapping EST sequences into clusters, and then determining if each cluster corresponds to a unique gene. Methods by which a cluster can be identified as representing a unique gene include identification of long Open Reading Frames (ORFs), comparison to genomic sequence, and detection of SNPs or other features in the cluster that are known to exist in the gene.
Genome	Complete genetic content of an organism.
Genomic DNA (sequence)	DNA sequence obtained from mammalian or other higher-order species, which includes intron and exon sequence (coding sequence), as well as non-coding regulatory sequences, such as promoter, and enhancer sequences.
Homology	(Strict definition) Two or more biological species, systems or molecules that share a common evolutionary ancestor. (General) Two or more gene or protein sequences that share a significant degree of similarity, typically measured by the amount of identity (in the case of DNA), or conservative replacements (in the case of protein), that they register along their lengths. Sequence "homology" searches are typically performed with a query DNA or protein sequence to identify known genes or gene products that share significant similarity and hence might inform on the ancestry, heritage and possible function of the query gene.
Markov chain	Any multivariate probability density whose independence diagram is a chain. The variables are ordered, and each variable "depends" only on its neighbors in the sense of being conditionally independent of the others. Markov chains are an integral component of hidden Markov models.

Motif	Conserved element of a protein sequence alignment that usually correlates with a particular function. Motifs are generated from a local multiple protein sequence alignment corresponding to a region whose function or structure is known. It is sufficient that it is conserved, and is hence likely to be predictive of any subsequent occurrence of such a structural/functional region in any other novel protein sequence.
Pattern	Molecular biological patterns usually occur at the level of the characters making up the gene or protein sequence. A pattern language must be defined in order to apply different criteria to different positions of a sequence. In order to have position-specific comparison done by a computer, a pattern-matching algorithm must allow alternative residues at a given position, repetitions of a residue, exclusion of alternative residues, weighting, and ideally, combinatorial representation.
Protein families	Sets of proteins that share a common evolutionary origin reflected by their relatedness in function, which is usually reflected by similarities in sequence, or in primary, secondary or tertiary structure. Subsets of proteins with related structure and function.
Sequence Tagged Site (STS)	Unique sequence from a known chromosomal location that can be amplified by PCR. STSs act as physical markers for genomic mapping and cloning.
Virtual libraries	Creation and storage of vast collections of molecular structures in an electronic database. These databases may be queried for subsets that exhibit specific physicochemical features, or may be "virtually screened" for their ability to bind a drug target. This process may be performed prior to the synthesis and testing of the molecules themselves.
Coding regions (CDS)	Portion of a genomic sequence bounded by start and stop codons that identifies the sequence of the protein being coded for by a particular gene.

Complexity (of gene sequence)	The term "low complexity sequence" may be thought of as synonymous with regions of locally biased amino acid composition. In these regions, the sequence composition deviates from the random model that underlies the calculation of the statistical significance (P-value) of an alignment. Such alignments among low complexity sequences are statistically but not biologically significant, i.e., one cannot infer homology (common ancestry) or functional similarity.
Data Processing	Data processing is defined as the systematic performance of operations upon data such as handling, merging, sorting, and computing. The semantic content of the original data should not be changed, but the semantic content of the processed data may be changed.
Gaps (affine gaps)	Gap is defined as any maximal, consecutive run of spaces in a single string of a given alignment. Gaps help create alignments that better conform to underlying biological models and more closely fit patterns that one expects to find in meaningful alignment. The idea is to take in account the number of continuous gaps and not only the number of spaces when calculating an alignment. Affine gaps contain a component for gap insertion and a component for gap extension, where the extension penalty is usually much lower than the insertion penalty. This mimics biological reality as multiple gaps would imply multiple mutations, but a single mutation can lead to a long gap quite easily.
Junk DNA	Term used to describe the excess DNA that is present in the genome beyond that required to encode proteins. A misleading term since these regions are likely to be involved in gene regulation, and other as yet unidentified functions.
Modeling	In bioinformatics, modeling usually refers to molecular modeling, a process whereby the three-dimensional architecture of biological molecules is interpreted (or predicted), visually represented, and manipulated in order to

	determine their molecular properties. (General) A series of mathematical equations or procedures, which simulate a real-life process, given a set of assumptions, boundary parameters, and initial conditions.
Selectivity	Selectivity of bioinformatics similarity search algorithms is defined as the significance threshold for reporting database sequence matches. As an example, for BLAST searches, the parameter E is interpreted as the upper bound on the expected frequency of chance occurrence of a match within the context of the entire database search. E may be thought of as the number of matches one expects to observe by chance alone during the database search.
Query (sequence)	DNA, RNA of protein sequence used to search a sequence database in order to identify close or remote family members (homologs) of known function, or sequences with similar active sites or regions (analogs), from whom the function of the query may be deduced.
Domain (protein)	Region of special biological interest within a single protein sequence. However, a domain may also be defined as a region within the three-dimensional structure of a protein that may encompass regions of several distinct protein sequences that accomplishes a specific function. A domain class is a group of domains that share a common set of well-defined properties or characteristics.
Gene families	Subsets of genes containing homologous sequences that usually correlate with a common function.

6 Animal Biotechnology

- INTRODUCTION
- MAMMALIAN CELL LINES AND CHARACTERISTICS
- MICROBIAL V/S ANIMAL CELL CULTURE
- TECHNIQUES OF MAMMALIAN CELL CULTURE
- STEM CELL RESEARCH

INTRODUCTION

The basic principles of animal cell culture date back to the beginning of the present millennium. In 1903, it was first shown that cells could survive and die in vitro. However, it is Ross Harrison who has the distinction of being the pioneer of animal cell and tissue culture. In 1907, he demonstrated this, using frogs. In 1912, Alexis Carell further demonstrated the use of embryo extract in culturing animal cells. Fifty years later, the techniques for animal cell culture were refined from using animal tissue explants to dispersed cell-specific cultures. The technique got a fillip in the 1940s with the discovery of antibiotics that ensured a bacteria-free environment for the cultures. The ultimate boost was in the 1950s when the technique was used for the production of Polio vaccine. Animal tissue culture is basically a technique in which the animal tissue/cells are taken out of the animal's body, grown and maintained in a suitable medium, under controlled conditions. Like most of the techniques, animal cell culture has also had its share of supporters and naysayers. Some of the major advantages of animal cell culture are:

- The nature and behaviour of cells in controlled conditions can be studied and interpreted to enable further research.
- The cell types are homogenous and can be maintained as pure lines for research.
- It is more economical and convenient since the need for costly animal houses and feeds are eliminated.
- It offers an alternative to animal testing by replicating the similar cell constitution and conditions in vitro. This is significant in the context of legal, moral and ethical considerations.

Figure 6.1 shows the application of tissue culture.

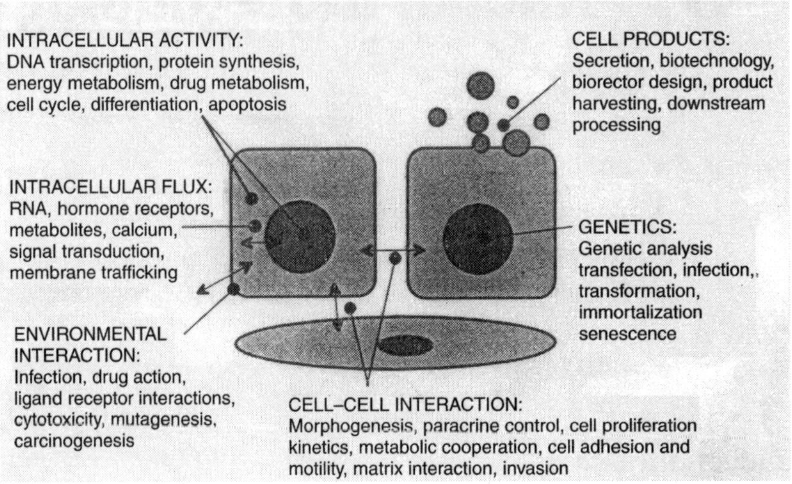

FIGURE 6.1 Tissue Culture Application

The above benefits clearly outweigh any disadvantages that these techniques might pose. The most notable disadvantage is the high initial startup costs in terms of advanced laboratories and resources. Another area of concern is the fact that the technique is time-consuming and dependent on highly skilled labour. However, the fact remains that in the present era this has become an indispensable element in the field of medical and agricultural sciences.

Basic Set-up

The most vital part of any tissue culture technique is the basic infrastructure that contains a host of items that determine the success of the entire technique. The major requirements are:

- Air conditioning, to maintain optimum temperatures at all times.
- "Hot-room" with temperature recorder for incubation of cell lines.
- Microscope room that contains a range of microscopes, from simple microscopes to advanced phase-contrast and fluorescent microscopes.
- Dark room, in order to study the effect of light on various parameters of cell growth.
- Service room, for maintaining and storing stocks.
- Sterilization room, to disinfect all glassware and media components that will be used for the technique.
- Animal houses to maintain live animals for procuring tissues and cells from different organs.

Substrates for Cell Cultures

The type of substrate that has to be used depends on the nature of cells that have to be cultured. These are also called anchorage dependent cells. A large number of substrates are used owing to the nature of different cell types. Some of the substrates are:

- Plastic: Disposable plastic substrates made of polystyrene are commonly used in the laboratories. They are treated by various methods that include gamma irradiation and electric arc supply to develop charges on the surface. This is crucial to anchor cells. Many other materials, such as Teflon, polyvinyl chloride (PVC) and polycarbonate are used to manufacture substrates.
- Glass: This is an important and indispensable substrate in laboratories. Most of the equipments, right from test tubes, slide, coverslips, pipettes, flasks, rods, bottles, Petri plates, to storage bottles are made of glass. It can be sterilized by autoclaving (moist heat), dry heat (oven) and chemical radiations.
- Palladium: This element is used, though not very routinely, as a substrate. It is deposited on agarose and used for the culture of fibroblasts and glial cells.

Treatment of substrates is crucial to experiments. They are first sterilized using earlier mentioned techniques. These are then coated with a layer of chemical/natural substrate in order to ensure cell attachment and proliferation. It is also treated with a monolayer of special type of cells called the feeder layer. This is most often made up of glial cells, fetal cells and other cells.

Culture Media

The principle of animal cell culture lies in maintaining and proliferating animal cells in vitro. This is made possible by simulating all conditions that the cell needs to survive; from the temperature range to factors required for growth and division. The culture media provides these to the cell in the form of stable pH, osmotic pressure and chemical constituents. These are divided into two main categories: natural and synthetic media.

Natural Media

These are natural sources of nutrients and components that the cell would require to grow and proliferate. In spite of the advancement with synthetic media, natural sources are still considered as the first choice due to their compatibility with cells. The most commonly used natural media sources are:

- **Blood plasma:** is a basic component of blood. It is the liquid in which blood cells are suspended. Blood plasma contains proteins, nutrients, metabolic end products, hormones and inorganic electrolytes. These components act as a source of nutrition for the cells and help them to sustain growth and proliferation over a period of time. Unlike the synthetic media, this does not need optimization since all the components are in a predetermined natural ratio. Plasma is derived from whole blood. To prevent clotting, an anticoagulant, such as citrate or heparin, is added to the blood specimen immediately after it is obtained. Plasma is then obtained by a process called centrifugation. It can be stored at a temperature below −80°C for long term use. There are various advantages of this media. Blood plasma:

- Provides a medium for growth and conditioning of freshly transferred cells on the substrate.
- Helps cells in the process of acclimatization by providing it a natural environment.
- Provides the cell protection from excessive damage during subsequent sub cultures.

- **Blood serum:** is the same as blood plasma except that factors that assist in clotting of blood are removed from it. In other words, serum is blood plasma without fibrinogen. It was assumed, for a number of years, that the whole serum is toxic. However, in 1928, this claim was refuted by de Ligneris and Parker when they successfully cultured animal cells in diluted serum. The use of serum, however, is not common owing to a number of disadvantages, such as:
  - Unlike plasma, serum cannot be stored for a long time period. It deteriorates within a year and, further, the quality varies from batch to batch.
  - Serum increases the coat of the medium manifold.
  - Serum is known to stimulate undesirable growth and inhibit the same, in some cases.

Table 6.1 shows the constituents of serum.

TABLE 6.1 Constituents of Serum

Constituent	Range of Concentration
Proteins and Polypeptides	40–80 mg/ml
Albumin	20–50 mg/ml
Fetuin	10–20 mg/ml
Fibronectin	1–10 μg/ml
Globulins	1–15 mg/ml
	(Continued)

TABLE 6.1 *(Continued)*

Constituent	Range of Concentration
Protease inhibitors: α_1-antitrypsin, α_2-macroglobulin	0,5–2.5 mg/ml
Transferrin	2–4 mg/ml
Growth factors:	
EGF, PDGF, IGF-1 and 2, FGF, IL-1, IL-6	1–100 ng/ml
Amino acids	0.01–1.0 μM
Lipids	2–10 mg/ml
Cholesterol	10 μM
Fatty acids	0.1–1.0 μM
Linoleic acid	0.01–0.1 μM
Phospholipids	0.7–3.0 mg/ml
Carbohydrates	1.0–2.0 mg/ml
Glucose	0.6–1.2 mg/ml
Hexosamine	6–1.2 mg/ml
Lactic acid	0.5–2.0 mg/ml
Pyruvic acid	2–10 μg/ml
Polyamines:	
Putrescine, spermidine	0.1–1.0 μM
Urea	170–300 μg/ml
Inorganics	0.14–0.16 M
Calcium	4–7 mM
Chlorides	100 μM
Iron	10–50 μM
Potassium	5–15 mM
Phosphate	2–5 mM
Selenium	0–01μM
Sodium	135–155 mM
Zinc	0.1–1.0 μM
Hormones	0.1–200 nM
Hydrocortisone	10–200 nM

(Continued)

TABLE 6.1 *(Continued)*

Constituent	Range of Concentration
Insulin	1–100 ng/ml
Triiodothyronine	20 nM
Thyroxine	100 nM
Vitamins	10 ng–10 μg/ml
Vitamin A	10–100 ng/ml
Folate	5–20 ng/ml

- **Tissue extracts:** are another natural medium for animal cell culture. In 1912, Carrel demonstarted that embryo tissue extract was capable of promoting and sustaining animal cell growth and proliferation. One of the questions that has confounded scientists for long is the exact component of the tissue that promotes growth. In the recent past, there have been growing evidences that activity was confined largely in the protein portion containing nucleoproteins and glycoproteins. Growth promotion was attributed to the nucleoproteins of the RNA type. Tissue extracts are made from various sources, with the chick embryo being a routine source. The extract can be stored indefinitely, after it is retrieved.

Synthetic Media

These are prepared artificially by the addition of several nutrients (organic and inorganic), vitamins, salts, serum proteins and carbohydrates. Different types of media are prepared based on the types of cells to be cultured. Synthetic media are of two types:

- Serum-containing media
- Serum-free media

As the name suggests, in case of the serum-free media, there is no serum. This is prepared in applications that require selective media. It is also used to counter disadvantages of serum-containing media, which are enumerated above. Some examples of serum-free media are displayed in Table 6.2:

TABLE 6.2 Serum-free Media

Component	Mol. wt.	MCDB 110 Human Lung Fibroblasts [Bettger et al., 1981]	MCDB 131 Human Vascular Endothelium [Knedler and Ham, 1987; Gupta et al., 1997]	MCDB 170 Mammary Epithelium (Hammond et al., 1984]	MCDB 202 Chick Embryo Fibroblasts [McKeehan and Ham, 1976b]	MCDB 302 CHO cells [Hamilton and Ham, 1977]	MCDB 402 3T3 Cells [Shipley and Ham, 1983]	WAJC 404 Prostatic Epithelium [McKeehan et al., 1984; Chaproniere and McKeehan, 1986]	MCDB 153 Keratinocytes [Peehl and Ham, 1986]	Iscove's Lymphoid Cells [Iscove and Melchers, 1978]	LHC-9 Bronchial Epithelium [Lechner and LaVeck, 1985]
Amino acids											
L-alanine	89	1.0E-04	3.0E-05	1.0E-04	1.0E-04		1.0E-04	1.0E-04	2.8E-04	1.0E-04	
L-arginine	211	1.0E-03	3.0E-04	3.0E-04	3.0E-04	1.0E-03	3.0E-04	1.0E-03	1.0E-03	4.0E-04	2.0E-03
L-asparagine	132	1.0E-04	1.0E-04	1.0E-03	1.0E-03	1.1E-04	1.0E-04	1.0E-04	1.0E-04	1.9E-04	1.0E-04
L-aspartic acid	133	1.0E-04	1.0E-04	1.0E-04	1.0E-04	1.0E-04	1.0E-05	3.0E-05	3.0E-05	2.3E-04	3.0E-05
L-cysteine	176	5.0E-05	2.0E-04	7.0E-05	2.0E-04	1.0E-04			2.4E-04		2.4E-04
L-cystine	240			2.0E-04	2.0E-04		4.0E-04	2.4E-04		2.9E-04	
L-glutamic acid	147	1.0E-04	3.0E-05	1.0E-04	1.0E-04	1.0E-04	1.0E-05	1.0E-04	1.0E-04	5.1E-04	1.0E-04
L-glutamine	146	2.5E-03	1.0E-02	2.0E-03	1.0E-03	3.0E-03	5.0E-03	6.0E-03	6.0E-03	4.0E-03	6.0E-03
Glycine	75	3.0E-04	3.0E-05	1.0E-04	1.0E-04	1.0E-04	1.0E-04	1.0E-04	1.0E-04	4.0E-04	1.0E-04
L-histidine	210	1.0E-04	2.0E-04	1.0E-04	1.0E-04	1.0E-04	2.0E-03	8.0E-05	8.0E-05	2.0E-04	1.6E-04
L-isoleucine	131	3.0E-05	5.0E-04	1.0E-04	1.0E-04	3.0E-05	1.0E-03	1.5E-05	1.5E-05	8.0E-04	3.0E-05
L-leucine	131	1.0E-04	1.0E-03	3.0E-04	3.0E-04	1.0E-04	2.0E-03	5.0E-04	5.0E-04	8.0E-04	1.0E-03
L-lysine HCl	183	2.0E-04	1.0E-03	2.0E-04	2.0E-04	2.0E-04	8.0E-04	1.0E-04	1.0E-04	8.0E-04	2.0E-04
L-methionine	149	3.0E-05	1.0E-04	3.0E-05	3.0E-05	3.0E-05	2.0E-04	3.0E-05	3.0E-05	2.0E-04	6.0E-05
L-phenylalanine	165	3.0E-05	2.0E-04	3.0E-05	3.0E-05	3.0E-05	3.0E-04	3.0E-05	3.0E-05	4.0E-04	6.0E-05

(Continued)

TABLE 6.2 *(Continued)*

Component	Mol. wt.	MCDB 110 Human Lung Fibroblasts [Bettger et al., 1981]	MCDB 131 Human Vascular Endothelium [Knedler and Ham, 1987; Gupta et al., 1997]	MCDB 170 Mammary Epithelium (Hammond et al., 1984]	MCDB 202 Chick Embryo Fibroblasts [McKeehan and Ham, 1976b]	MCDB 302 CHO cells [Hamilton and Ham, 1977]	MCDB 402 3T3 Cells [Shipley and Ham, 1983]	WAJC 404 Prostatic Epithelium [McKeehan et al., 1984; Chaproniere and McKeehan, 1986]	MCDB 153 Keratinocytes [Peehl and Ham, 1986]	Iscove's Lymphoid Cells [Iscove and Melchers, 1978]	LHC-9 Bronchial Epithelium [Lechner and LaVeck, 1985]
L-proline	115	3.0E-04	1.0E-04	5.0E-05	5.0E-05	3.0E-04		3.0E-04	3.0E-04	3.5E-04	3.0E-04
L-serine	105	1.0E-04	3.0E-04	3.0E-04	3.0E-04	1.0E-04	1.0E-04	6.0E-04	6.0E-04	4.0E-04	1.2E-03
L-threonine	119	1.0E-04	1.0E-04	3.0E-04	3.0E-04	1.0E-04	5.0E-04	1.0E-04	1.0E-04	8.0E-04	2.0E-04
L-tryptophan	204	1.0E-05	2.0E-05	3.0E-05	3.0E-05	1.0E-05	1.0E-05	1.5E-05	1.5E-05	7.8E-05	3.0E-05
L-tyrosine	181	3.5E-05	1.0E-04	5.0E-05	5.0E-05	4.4E-05	2.0E-04	1.5E-05	1.5E-05	4.6E-04	3.0E-05
L-valine	117	1.0E-04	1.0E-03	3.0E-04	3.0E-04	1.0E-04	2.0E-03	3.0E-04	1.0E-04	8.0E-04	6.0E-04
Vitamins											
Biotin	244	3.0E-08	3.0E-08	3.0E-08	3.0E-08	3.0E-08	3.0E-08	6.0E-08	6.0E-08	5.3E-08	6.0E-08
Choline chloride	140	1.0E-04		1.0E-04	1.0E-04	1.0E-04	1.0E-04	1.0E-04	1.0E-04	2.0E-05	2.0E-04
Folic acid	441		1.0E-04			3.0E-06		1.8E-06	1.8E-06	9.1E-06	1.8E-06
Folinic acid	512	1.0E-09	1.0E-06	1.0E-08	1.0E-08		1.0E-06				
myo-Inositol	180	1.0E-04	4.0E-05	1.0E-04	1.0E-04	1.0E-04	4.0E-05	1.0E-04	1.0E-04	4.0E-05	1.0E-04
Nicotinamide	122	5.0E-05	5.0E-05	5.0E-05	5.0E-05	3.0E-07	5.0E-05	3.0E-07	3.0E-07	3.3E-05	3.0E-07
Pantothenate	238	1.0E-06	5.0E-05	1.0E-06	1.0E-06	1.0E-06	5.0E-05	1.0E-06	1.0E-06	1.7E-05	1.0E-06
Pyridoxal HCl	204									2.0E-05	
Pyridoxine HCl	206	3.0E-07	1.0E-05	3.0E-07	3.0E-07	3.0E-07	1.0E-04	3.0E-07	3.0E-07		3.0E-07
Riboflavin	376	3.0E-07	1.0E-08	3.0E-07	3.0E-07	1.0E-07	1.0E-06	1.0E-07	1.0E-07	1.1E-06	1.0E-07
Thiamin HCl	337	1.0E-06	1.0E-05	1.0E-06	1.0E-06	1.0E-06	1.0E-04	1.0E-06	1.0E-06	1.2E-05	1.0E-06

α-Tocopherol	430	1.4E-07									
Retinoic acid	300										3.3E-07
Retinol acetate	329	4.2E-07									
Vitamin B_{12}	1355	1.0E-07	1.0E-08	1.0E-07	1.0E-07	1.0E-07	1.0E-08	3.0E-07	3.0E-07	9.6E-09	3.0E-07
Antioxidunts											
Dithiothreitol	154	6.5E-06									
Glutathione	307	6.5E-07									
Inorganic salts											
$CaCl_2$	147	1.0E-03	1.6E-03	2.0E-03	2.0E-03	6.0E-04	1.6E-03	1.3E-04	3.0E-05	1.5E-03	1.1E-04
KCl	75	5.0E-03	4.0E-03	3.0E-03	3.0E-03	3.0E-03	4.0B-03	1.5E-03	1.5E-03	4.4E-03	1.5E-O3
KNO_3	160					1.6E-08				7.5E-07	
$MgCl_2$	203					6.0E-04		6.0E-04	6.0E-04		2.2E-02
$MgSO_4$	247	1.0E-03	1.0E-02	1.5E-03	1.5E-03	6.1E-10	8.0E-04			8.1E-04	
NaCl	58	1.1 E 01	1.1E-01	1.2E-01	1.2E-01	1.3E-01	1.2E-01	1.2E-01	1.2E-01	7.7E-02	1.0E-01
$NaHCO_3$	84		1.4E-02			1.4E-02	1.4E-02	1.4E-02	1.4E-02	3.6E-02	1.2E-02
Na_2HPO_4	120	3.0E-03	5.9E-04	5.0E-04	5.0E-04	1.2E-03	5.0E-04	2.0E-03	2.0E-03	1.0E-03	2.0E-03
Trace elements											
$CuSO_4 \cdot 5H_2O$	160	1.0E-09	7.5E-09	1.0E-09	1.0E-09		5.0E-09	1.0E-09	1.1E-08		1.0E-08
FeSO4	278	5.0E-06	1.0E-06	5.0E-06	5.0E-06	3.0E-06	1.0E-06	5.0E-06	5.0E-06		5.4E-04
$MnSO_4 \cdot H_2O$	169	1.0E-09	1.2E-09	5.0E-10	5.0E-10		1.0E-09	1.0E-09	1.0E-09		1.0E-09
$(NH_4)_6Mo_7O_{24}$	1236	1.0E-09	3.0E-09	1.0E-09	1.0E-09	1.0E-08	3.0E-09	1.0E-09	1.0E-09		1.0E-09
$NiCl_2$	238	5.0E-10	4.2E-10	5.0E-12	5.0#-12		3.0E-10	5.0E-10	5.0E-10		5.0E-10
H_2SeO_3	129	3.0E-08	2-3E-05	3.0E-08	3.0E-08	1.3E-08	1.0E-08	3.0E-08	3.0E-08	1.0E-07	3.0E-08
Na_2SiO_3	122	5.0E-07	5.1E-09	5.0E-07	5.0E-07		1.0E-05	5.0E-07	5.0E-07		5.0E-07

(Continued)

TABLE 6.2 *(Continued)*

Component	Mol. wt.	MCDB 110 Human Lung Fibroblasts [Bettger et al., 1981]	MCDB 131 Human Vascular Endothelium [Knedler and Ham, 1987; Gupta et al., 1997]	MCDB 170 Mammary Epithelium (Hammond et al., 1984]	MCDB 202 Chick Embryo Fibroblasts [McKeehan and Ham, 1976b]	MCDB 302 CHO cells [Hamilton and Ham, 1977]	MCDB 402 3T3 Cells [Shipley and Ham, 1983]	WAJC 404 Prostatic Epithelium [McKeehan et al., 1984; Chaproniere and McKeehan, 1986]	MCDB 153 Keratinocytes [Peehl and Ham, 1986]	Iscove's Lymphoid Cells [Iscove and Melchers, 1978]	LHC-9 Bronchial Epithelium [Lechner and LaVeck, 1985]
$SnCl_2$	190	5.0E-10	1.0E-09	5.0E-12	5.0E-12			5.0E-10	5.0E-10		5.0E-10
NH_4VO_3	117	5.0E-09		5.0E-09	5.0E-09	1.0E-08	5.0E-09	5.0E-09	5.0E-09		5.0E-09
$2nSO_4 \cdot 7H_2O$	288	5.0E-07		1.0E-07	1.0E-07	3.0E-06	1.0E-06	5.0E-07	5.0E-07		4.8E-07
Lipids and precursors											
Cholesterol	387	7.6E-06									
Ethanolamine	61			1.0E-04					1.0E-04		
Linoleic acid	280			2.0E-07	2.0E-07	3.0E-07	3.0E-07		1.0E-06		
Lipoic acid	206	1.0E-08	1.0E-08	1.0E-08	1.0E-08	9.8E-07	1.0E-08	1.0E-06			1.0E-06
Phosphoethanolamine	141			1.0E-04							5.0E-07
Soya lecithin μg/ml		6									
Soybean lipid μg/ml										50	
Sphingomyelin μg/ml		1									
Hormones and growth factors											
EOF (ng/ml)		30		10				25	25		5
Epinephrine	183										2.7E-06
Hydrocortisone[1]	362	5.0E-07		1.4E-07				5.0E-07	1.4E-07		2.0E-07
Insulin (μg/ml)		1		5				10	5		5
Prolactin μg/ml		1		5							

PGE_1	355	2.5E-08							2.5E-08		
Triiodothyronine	673										1.0E-08
Nucleosides, etc.											
Adenine SO_4	184	1.0E-05	1.0E-06	1.0E-06	1.0E-06		1.0E-06	1.8E-04	1.8E-04		1.8E-04
Hypoxanthine	136					3.0E-05					
Thymidine	242	3.0E-07	1.0E-07	3.0E-07	3.0E-07		1.0E-06	3.0E-06	3.0E-06		3.0E-06
Energy metabolism											
D-glucose	180	4.0E-03	5.6E-03	8.0E-03	8.0E-03	1.0E-02	5.5E-03	6.0E-03	6.0E-03	2.5E-02	6.0E-03
Phosphoenolpyruvate	190	1.0E-05									
Sodium acetate·$3H_2O$	136						1.0E	3.7E-03	3.7E-03		3.7E-03
Sodium pyruvate	110	1.0E-03	1.0E-03	1.0E-03	5.0E-04	1.1E + 02		5.0E-04	5.0E-04	1.0E-03	5.0E-04
Other components											
Cholera toxin	~90 KDa							2.0E-10			
HEPES, Na salt	260	3.0E-02		3.0E-02	3.0E-02			2.8E-02	2.8E-02	2.5E-02	2.3E-02
Phenol red	376	3.3E-06	3.3E-05	3.3E-06	3.3E-06	3.3E-06	3.3E-05	3.3E-05	3.3E-05	4.0E-05	3.3E-06
Putrescine 2HCl	161	1.0E-09	1.2E-09	1.0E-09	1.0E-09	1.0E-06	1.0E-09	1.0E-06	1.0E-06		1.0E-06
Protein supplements											
BPE $\mu gP/ml^2$		70						25			35
BSA, mg/ml										0.5-10	
Dialyzed FBS, $\mu gP/ml$									1		
Transferrin, Fe^{3+}				5						30–300	10
saturated ($\mu g/ml$)											
Gas phase											
CO_2	44	2%	5%	2%	2%	5%	5%	5%	5%	10%	5%

Regardless of which media is used, certain basic components are mandatory for the survival of the cell. These are:

- **Inorganic salts:** perform several functions. They retain the osmotic balance of cells and help to regulate membrane potential. This is done by the provision of sodium, potassium and calcium ions that are required in the cell matrix for cell attachment and as enzyme cofactors. Table 6.3 shows a list of balanced salt solutions.
- **Carbohydrates:** Is the main source of energy, and are generally present in the form of sugars. The major sugars used are, glucose and galactose, maltose and fructose. The concentration of sugar varies from basal media containing 1g/l to 4.5g/l in some complex media. The media containing higher concentration of sugars are able to support the growth of a wider range of cell types.
- **Amino acids:** are the building blocks that eventually make proteins.
- **Vitamins:** are necessary for cell growth and proliferation. While the presence of B12 is essential for some lines, many media are enriched with vitamins to make them more suitable for a wider range of cell lines. Serum is an important source of vitamins in cell culture. Some media also have increased levels of vitamins A and E. The commonly used vitamins in media include riboflavin, thiamine and biotin.
- **Fatty acids and lipids:** are important components of serum that are needed even in serum-free media. Cholesterol and steroids are some of the fatty acids and lipids that are essential for specialized cells.
- **Proteins and peptides:** are particularly important in serum-free media. The most common proteins and peptides include albumin, transferrin, fibronectin and fetuin. These molecules are used to replace those that are normally present in the media, through the addition of serum to the medium.
- **Serum:** is a complex mix of albumins, growth factors and growth inhibitors and is probably one of the most important components of cell culture medium. The most commonly used serum is fetal bovine serum. Other types of serum include newborn calf serum and horse serum. The quality, type and concentration of serum affect the growth of cells. For this reason, it is important to screen batches of serum for their ability to support growth of cells. Also, there are other tests that are used to aid the selection of a batch of serum including cloning efficiency, plating efficiency and the preservation of cell characteristics. Serum is also able to increase the buffering capacity of cultures that can be important for slow-growing cells or where the seeding density is low (e.g., cell cloning experiments). It also helps to protect against mechanical damage which may occur in stirred cultures or whilst using a cell

TABLE 6.3 Balanced Salt Solutions

Component		Earle's BSS		Dulbecco's PBS A		Dulbecco's PBS B		Hanks's BSS		Spinner salts (as in S-MEM)	
Inorganic salts	**M.W.**	**g/l**	**mM**	**g/l**	**mM**	**g/l**	**mM**	**g/l**	**mM**	**g/l**	**mM**
$CaCl_2$ (anhydrous)	111	0.02	0.18			0.2	1.80	0.14	1.3		
KCl	74.55	0.4	5.37	0.2	2.7	0.2	2.68	0.4	5.4	0.40	5. 37
KH_2PO_4	136.1			0.2	1.5	0.2	1.47	0.06	0.4		
$MgCl_2 \cdot 6H_2O$	203.3						0.00	0.1	0.5		
$MgSO_4 \cdot 7H_2O$	246.5	0.2	0.81			0.98	3.98	0.1	0.4	0.20	0.81
NaCl	58.44	6.68	114.31	8	136.9	8	136.89	8	136.9	6.80	116.36
$NaHCO_3$	84.01	. 2.2	26.19				0.00	0.35	4.2	2.20	26.19
$Na_2HPO_4 \cdot 7H_2O$	268.1			2.2	8.1	2.16	8.06	0.09	0.3		
$NaH_2PO_4 \cdot H_2O$	138	0.14	1.01							1.40	10.14
Total salt			147.86		149.1		154.88		149.4		158.87
Other components											
D-glucose	180.2	1	5.55					1	5.5	1.00	5.55
Phenol red	354.4	0.01	0.03					0.01	0.0	0.01	0.03
Gas phase		5% CO_2				Air		Air		5% CO_2	
Buffer											
HEPES, Na salt	260.3	13.02	50.00	5.21	20.0	5.21	20	2.08	8.0	13.02	50.00

scraper. A further advantage of serum is the wide range of cell types with which it can be used despite the varying requirements of different cultures in terms of growth factors. In addition, serum is able to bind and neutralize toxins.

TABLE 6.4 Frequently Used Media for Animal Tissue Culture

Component	MEM	DMEM	F12	DMEM/ F12	αMEM
Amino acids					
L-alanine			1.0E-04	5.0E-05	2.8E-04
L-arginine	6.0E-04	4.0E-04	1.0E-03	7.0E-04	6.0E-04
L-asparagine			1.0E-04	5.0E-05	3.3E-04
L-aspartic acid			1.0E-04	5.0E-05	2.3E-04
L-cysteine			2.0E-04	1.0E-04	5.7E-04
L-cystine	1.0E-04	2.0E-04		1.0E-04	1.0E-04
L-glutamic acid			1.0E-04	5.0E-05	5.1E-04
L-glutamine	2.0E-03	4.0E-03	1.0E-03	2.5E-03	2.0E-03
Glycine		4.0E-04	1.0E-04	2.5E-04	6.7E-04
L-histidine	2.0E-04	2.0E-04	1.0E-04	1.5E-04	2.0E-04
L-hydroxy-proline					
L-isoleucine	4.0E-04	8.0E-04	3.0E-05	4.2E-04	4.0E-04
L-leucine	4.0E-04	8.0E-04	1.0E-04	4.5E-04	4.0E-04
L-lysine HCl	4.0E-04	8.0E-04	2.0E-04	5.0E-04	4.0E-04
L-methionine	1.0E-04	2.0E-04	3.0E-05	1.2E-04	1.0E-04
L-phenylalanine	2.0E-04	4.0E-04	3.0E-05	2.2E-04	1.9E-04
L-proline			3.0E-04	1.5E-04	3.5E-04
L-serine		4.0E-04	1.0E-04	2.5E-04	2.4E-04
L-threonine	4.0E-04	8.0E-04	1.0E-04	4.5E-04	4.0E-04
L-trypcophan	4.9E-05	7.8E-05	1.0E-05	4.4E-05	4.9E-05
L-tyrosine	2.0E-04	4.0E-04	3.0E-05	2.1E-04	2.3E-04
L-valine	4.0E-04	8.0E-04	1.0E-04	4.5E-04	3.9E-04

- **Trace elements**: such as zinc, copper, selenium and tricarboxylic acid intermediates are important for cell culture. Selenium is a detoxifier and helps remove oxygen-free radicals. Table 6.4 lists the frequently used media for Animal tissue culture.

CMRL 1066	RPMI 1640	M199	L15	McCoy's 5A	Fischer	MB 752/1
2.8E-04		2.8E-04	2.5E-03	1.5E-04		
3.3E-04	1.1E-03	3.3E-03	2.9E-03	2.0E-04	7.1E-05	3.6E-04
3.8E-04	1.7E-03	3.0E-04	7.6E-05			
2.3E-04	1.5E-04	2.3E-04		1.5E-04		4.5E-04
1.5E-03		5.6E-07	9.9E-04	2.0E-04		5.0E-04
8.3E-05	2.1E-04	9.9E-05			9.9E-05	6.3E-05
5.1E-04	1.4E-04	4.5E-04		1.5E-04		1.0E-03
6.8E-04	2.1E-03	6.8E-04	2.1E-03	1.5E-03	1.4E-03	2.4E-03
6.7E-04	1.3E-04	6.7E-04	2.7E-03	1.0E-04		6.7E-04
9.5E-05	9.7E-05	1.0E-04	1.6E-03	1.0E-04	3.9E-04	8.3E-04
7.6E-05	1.5E-04	7.6E-05		1.5E-04		
1.5E-04	3.8E-04	1.5E-04	9.5E-04	3.0E-04	5.7E-04	1.9E-04
4.6E-04	3.8E-04	4.6E-04	9.5E-04	3.0E-04	2.3E-04	3.8E-04
3.8E-04	2.2E-04	3.8E-04	5.1E-04	2.0E-04	2.7E-04	1.3E-03
1.0E-04	1.0E-04	1.0E-04	5.0E-04	1.0E-04	6.7E-04	3.4E-04
1.5E-04	9.1E-05	1.5E-04	7.6E-04	1.0E-04	4.1E-04	3.0E-04
3.5E-04	1.7E-04	3.5E-04		1.5E-04		4.3E-04
2.4E-04	2.9E-04	2.4E-04	1.9E-03	2.5E-04	1.4E-04	
2.5E-04	1.7E-04	2.5E-04	2.5E-03	1.5E-04	3.4E-04	6.3E-04
4.9E-05	2.5E-05	4.9E-05	9.8E-05	1.5E-05	4.9E-05	2.0E-04
2.2E-04	1.1E-04	2.2E-04	1.7E-03	1.2E-04	3.3E-04	2.2E-04
2.1E-04	1.7E-04	2.1E-04	8.5E-04	1.5E-04	6.0E-04	5.6E-04

(Continued)

TABLE 6.4 *(Continued)*

Component	MEM	DMEM	F12	DMEM/ F12	αMEM
Vitamins					
p-Aminobenzoic acid					
L-Ascorbic acid					2.5E-04
Biotin			3.0E-08	1.5E-08	4.1E-07
Calciferol					
Choline chloride	7.1E-06	2.9E-05	1.0E-04	6.4E-05	7.1E-06
Folic acid	2.3E-06	9.1E-06	2.9E-06	6.0E-06	2.3E-06
myo-inositol	1.1E-05	4.0E-05	1.0E-04	7.0E-05	1.1E-05
Menadione					
Nicotinamide	8.2E-06	3.3E-05	3.3E-07	1.7E-05	8.2E-06
Nicotinic acid					
D-Ca pantothenate	4.2E-06	1.7E-05	2.0E-06	9.4E-06	4.2E-06
Pyridoxal HCl	4.9E-06	2.0E-05	3.0E-07	1.0E-05	4.9E-06
Pyridoxine HCl			3.0E-07	1.5E-07	
Riboflavin	2.7E-07	1.1E-06	1.0E-07	5.8F-07	2.7F-07
Thiamin mono PO_4					
α-tocopherol					
Retinol acetate					
Vitamin B_{12}			1.0E-06	5.0E-07	1.0E-06
Antioxidants					
Glutathione					
Inorganic salts					
$CaCl_2$	1.8E-03	1.8E-03	3.0E-04	1.1E-03	1.8E-03
KCl	5.3E-03	5.3E-03	3.0E-03	4.2E-03	5.3E-04
KH_2PC_4					
$MgCl_2$					1.2E-01
$MgSO_4$	8.1E-04	8.1E-04		4.0E-04	8.1E-04
NaCl	1.2E-01	1.1E-01	1.3E-01	1.2E-01	
$NaHCO_3$	2.6E-02	4.4E-02	1.4E-02	2.9E-02	2.6E-02
NaH_2PO_4	1.0E-03	9.1E-04		4.5E-04	
Na_2HPO_4			1.0E-03	5.0E-04	

CMRL 1066	RPMI 1640	M199	L15	McCoy's 5A	Fischer	MB 752/1
3.6E-07	7.3E-06	3.6E-07		7.3E-06		
2.8E-04		2.8E-07		3.2E-06		9.9E-05
4.1E-08	8.2E-07	4.1E-08		8.2E-07	4.1E-08	8.2E-08
		2.5E-07				
3.6E-06	2.1E-05	3.6E-06	7.1E-06	3.6E-05	1.1E-05	1.8E-03
2.3E-08	2.3E-06	2.3E-08	2.3E-06	2.3E-05	2.3E-05	9.1E-07
2.8E-07	1.9E-04	2.8E-07	1.1E-05	2.0E-04	8.3E-06	5.6E-06
		6.9E-08				
2.0E-07	8.2E-06	2.0E-07	8-2E-06	4.1E-06	4.1E-06	8.2E-06
		2.0E-07		4.1E-06		
4.2E-08	1.1E-06	4.2E-08	4.2E-06	8.4E-07	2.1E-06	4.2E-06
1.2E-07		1.2E-07		2.5E-06	2.5E-06	
1.2E-07	4.9E-06	1.2E-07		2.4E-06		4.9E-06
2.7E-08	5.3E-07	2.7E-08	1.9E-07	5.3E-07	1.3E-06	2.7E-06
				4.8E-06		
		2.3E-08				
		3.5E-07				
	3.7E-09					1.5E-07
3.0E-05	3.0E-06	1.5E-07		1.5E-06		4.5E-05
1.8E-03		1.3E-03	1.3E-03	9.0E-04	6.2E-04	8.2E-04
5.3E-03	5.3E-03	5.3E-03	5.3E-03	5.3E-03	5.3E-03	2.0E-03
		4.4E-04	4.4E-04			5.9E-04
						1.2E-03
8.1E-04	4.0E-04	8.1E-04	1.6E-03	8.1E-04	4.9E-04	8.1E-04
1.2E-01	1.0E-01	1.4E-01	1.4E-01	1.1E-01	1.4E-01	1.0E-01
2.6E-02	2.6E-02	4.2E-03		2.6E-02	1.3E-02	2.7E-02
1.0E-03				4.2E-03	5.7E-04	
	5.6E-03	4.0E-04	1.6E-03		5.0E-04	2.1E-03

(Continued)

TABLE 6.4 *(Continued)*

Component	MEM	DMEM	F12	DMEM/ F12	αMEM
Trace elements					
$CuSO_4 \cdot 5H_2O$			1.6E-08	7.8E-09	
$Fe(NO_3)_3 \cdot 9H_2O$		2.5E-07		1.2E-07	
$FeSO_4 \cdot 7H_2O$			3.0E-06	1.5E-06	
$ZnSO_4 \cdot 7H_2O$			3.0E-06	1.5E-06	
Bases, nucleosides, etc.					
Adenine SO_4					
Adenosine					3.7E-05
AMP					
ATP					
Cytidine					4.1E-05
Deoxyadenosine					4.0E-05
Deoxycytidine					4.2E-05
Deoxyguanosine					3.7E-05
2-Deoxyribose					
DPN					
FAD					
Glucuronate, Na					
Guanine					
Guanosine					3.5E-05
Hypoxanthine			3.0E-05	1.5E-0S	
5-Me-deoxycytidine					
D-Ribose					
Thymidine			3.0E-06	1.5E-06	4.1E-05
Thymine					
TPN					
Uracil					
Uridine					4.1E-05
UTP					
Xanthine					

CMRL 1066	RPMI 1640	M199	L15	McCoy's 5A	Fischer	MB 752/1
		5.4E-05				
		5.8E-07				
		1.8E-05				
4.0E-05						
3.8E-05						
3.7E-05						
		3.7E-06				
9.5E-06						
1.2E-06						
1.9E-05						
		1.6E-06				
		2.2E-06				
4.1E-07						
		3.3E-06				
4.1E-05						
		2.4E-06				
1.3E-06						
		2.7E-06				
1.8E-06						
		2.0E-06				

(Continued)

TABLE 6.4 *(Continued)*

Component	MEM	DMEM	F12	DMEM/ F12	αMEM
Energy metabolism					
Cocarboxylase					
Coenzyme A					
D-galactose					
D-glucose	5.6E-03	2.5E-02	1.0E-02	1.8E-02	5.6E-03
Sodium acetate					
Sodium pyruvate		1.0E-03	1.0E-03	1.0E-03	1.0E-03
Lipids and precursors					
Cholesterol					
Ethanol (solvent)					
Linoleic acid		3.0E-07		1.5E-07	
Lipoic acid			1.0E-06	5.1E-07	9.7E-07
Tween 80					
Other components					
Peptone, mg/ml					
Phenol red	2.7E-05	4.0E-05	3.2E-05	3.6E-05	2.9E-05
Putrescine			1.0E-06	5.0E-07	
Gas Phase					
CO_2	5%	10%	2%	7%	5%

Mammalian Cell Lines and Characteristics

Mammalian cell lines are divided into two main types: primary cell lines and established cell lines.

Primary cultures are obtained directly from an animal and can be maintained in a differentiated state for the duration of a few days to weeks. When a culture is first set up using tissues, the cells derived from the same, begin to migrate almost immediately. The macrophages are the first cells to do so, followed by the fibroblasts. The subsequent fates of the different cells that migrate vary enormously. Some of them are short-lived, such as blood cells, while others survive for weeks or months, depending on the type of the cell. If the cells multiply repeatedly for a long time, they are often 'passaged'. This is achieved by suspending cells in a suitable media and then inoculating the suspension into a new culture vessel along with a fresh medium. Once the cells are 'passaged', the culture is known as the primary cell line.

CMRL 1066	RPMI 1640	M199	L15	McCoy's 5A	Fischer	MB 752/1
2.2E-06						
3.3E-06						
			5.0E-02			
5.6E-03	1.1E-02	5.6E-03		1.7E-02	5.6E-03	2.8E-02
6.1E-04		4.5E-04				
			5.0E-03			
5.2E-07		5.2E-07				
3.5E-04						
						8.9E-05
1.8E-05		1.8E-05				
			0.6			
5.3E-05	1.3E-05	4.5E-05	2.7E-05	2.9E-05	1.3E-05	2.7E-05
5%	5%	5%	Air	5%	2%	5%

Often, the primary cell lines continue to divide at a very rapid rate for a considerable period. In such cases, they are 'passaged' repeatedly. It is observed that a few cells from the cell population die during the subsequent passages, while some of them retain the ability to proliferate and are subcultured indefinitely in-vitro. Such cell lines are called the established cell lines.

Primary cell lines often stop growing before the medium is completely exhausted. This is due to a phenomenon known as contact inhibition. In certain types of cells, the fibroblasts become immobilized due to this phenomenon. The main cell activities that are affected by contact inhibition are: cell division, DNA synthesis, RNA synthesis and protein synthesis. Contact inhibition is absent during the initial stages of transformation of primary cell lines. As a result, the transformed cells grow to higher densities before eventually piling upon each other. This has a direct effect on the kinetics of the growth of primary cell lines. The kinetics is ideal only in case of dilute solutions. Here, once the small sheet of cells is formed, contact inhibition takes effect and growth slows down progressively.

Cell-Cell Interactions

One of the most fascinating features of cell culture is the ability of like cells to recognize and adhere to each other. It was demonstrated for the first time by Holtfreter (1955) that when amphibian cells were disaggregated in calcium- and magnesium-free salt solution, the cells first adhere in a random manner before eventually sorting themselves into groups of like cells. The entire embryo was reconstituted in the same manner. The different theories that attempt to explain this phenomenon range from the intercellular cement concept to the adhesion property that is an intrinsic property of the cell membrane themselves.

Kinetics of Growth

Cells demonstrate a type of growth pattern that can be studied across various time points divided into four different phases:

1. **Lag phase**: is when practically no growth occurs and may last for a few hours to a few days, depending on the type of cells before growth commences.
2. **Logarithmic (log) phase**: is when the growth proceeds with a steady pace with the population doubling in every 15–20 hours. This phase is the period of maximum growth when there is a linear relationship between the log of cell number (or the log of cell mass) and time.
3. **Stationary phase**: is when the maximum population number is reached and growth finally becomes stationary. This is a plateau stage during which the rate of production of new cells (the birth rate) equals the death rate.
4. **Death phase**: starts once the death rate is greater than the rate of production of new cells. This stage will continue until viability has dropped to zero and the culture is effectively sterile. The length of time this takes varies with species.

Figure 6.2 shows the progress of cell culture in week.

The growth rates of cultured cells are not necessarily uniform. It may vary to some extent from one stem cell strain to another, but in particular it varies considerably with environmental conditions such as pH, temperature and osmotic pressure.

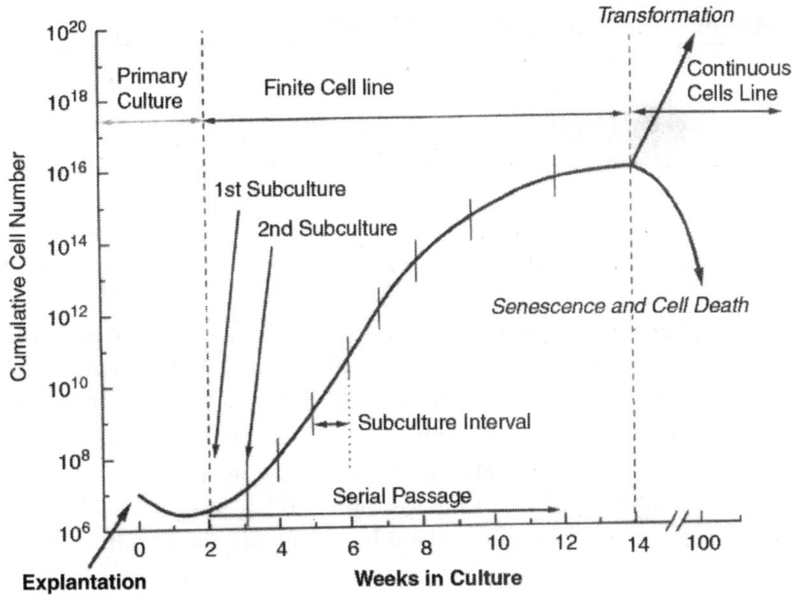

FIGURE 6.2 Weeks in Culture

Senescence

Normal cell can divide only for a limited number of times. For this reason, cell lines derived from normal tissues will die out after a fixed number of population doublings. This is a genetically determined event called senescence. This event is determined by the inability of the terminal sequences of the DNA in telomeres to replicate at each cell division. As a result the telomeres progressively shorten until finally the cell is unable to divide further. There are a few exceptions to this rule namely, the germ cells, stem cells, and transformed cells, which often express the enzyme telomerase and enable one to counter the limitation of chromosome shortening.

Transformed Cells

Transformation of cultured cells implies a spontaneous or induced permanent phenotypic change resulting from a heritable change in DNA and gene expression. It is basically associated with genetic instability and phenotypic change. Basically, transformation is seen as a series of events that may alter many of the cell lines' properties, such as growth rate, mode of growth, specialized product formation and longevity. Immortality of the cells is a major trait that is attributed to transformation.

MICROBIAL V/S ANIMAL CELL CULTURE

The basic premise of biotechnology has always been the manipulation of natural processes and phenomenon. Simulating the natural environment for propagating organisms of our choice in large numbers is another very important aspect of biotechnology. Cell culture is one such technique that provides the conditions optimum for the growth and proliferation of species. Plant, animal and microbial cells are cultured routinely. In this section, an effort is made to distinguish the microbial cell culture from animal cell culture.

Microbial cell culture exploits the inherent abilities of microorganisms to produce a wide range of economically viable products through metabolic processes. The process of microbial cell culture provides a nutrient source and optimum temperature. However, in case of animal cell culture, the requirements of growing cells are much more complicated.

Microbial cells divide at a very fast rate and can be harvested at regular intervals. Animal cell cultures, on the other hand, are sluggish in growth and take months to propagate. A major issue is that of contamination. While microbial cell cultures are equally prone to contamination, animal cell cultures are sensitive to necessitating regular subculturing and checks.

The infrastructure required to carry out microbial culture is relatively lesser as compared to animal cell culture laboratories, which come with additional requirements of low temperature storage and sophisticated incubators. Animal cell cultures have to be propagated to derive cell lines that have to be maintained at extremely low temperatures. Obtaining and maintaining cell lines require heavy infrastructure, are time consuming and expensive in nature.

Since microbes are dispersed throughout the environment, isolating them is not a major hassle. Besides this, different microbes present in varying environments have their own characteristics that can be exploited further. However, the case is not so simple in animal cell culture. Since many applications of animal cell culture are related to human therapies, they have to be isolated from human embryonic cells or from model animals used to simulate the same. In both the cases, obtaining cells in large numbers is a major task. In addition to this, very little is known about the nature and behaviour of certain types of animal cells, like the stem cells. This makes it difficult to work with animal cells.

While concerns about the use of genetically modified micro organisms exist, there have been no major issues regarding the use of naturally occurring life forms. Industries built around microbial cell cultures constitute a sizeable proportion of biotechnology-driven businesses. Fermentation technology and allied industries are some such examples.

As opposed to this, the animal cell culture technique is shrouded in debates about the ethical aspects and legislations to regulate the use of the same.

While it has shown considerable promise in numerous avenues, the technique is still in its nascent stage, as far as commercial application is concerned.

The near future may see an amalgamation of microbial and animal cell cultures, with newer antibiotics and drugs being isolated from microorganisms and stem cells emerging as candidates for drug delivery and testing. The integration of these technologies will boost the number of applications that pave the way for greater commercialization.

TECHNIQUES OF MAMMALIAN CELL CULTURE

Primary cell cultures may be obtained by various methods, either by allowing the cells to migrate from the tissue or by disaggregating the tissue for the production of suspension of cells. Figure 6.3 depicts a flowchart that shows the options for primary culture.

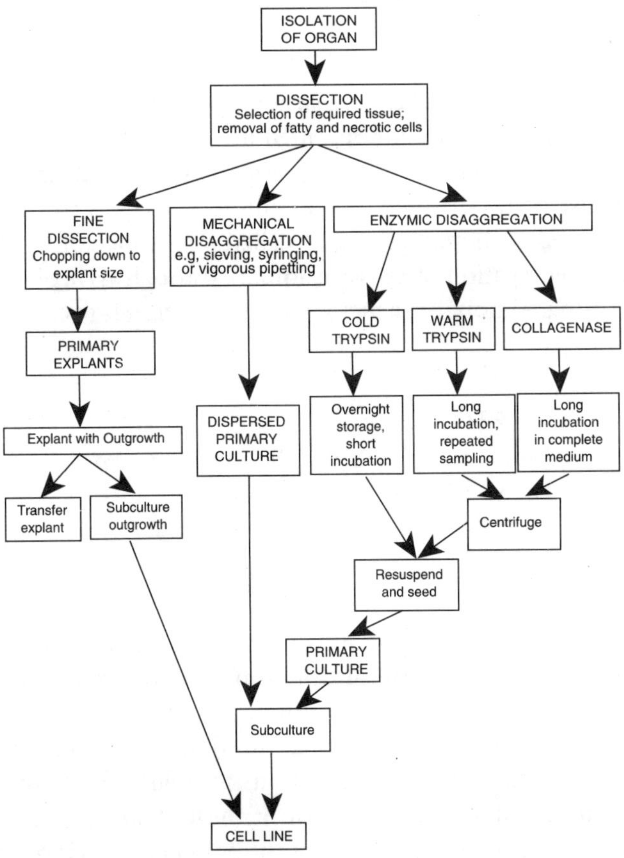

FIGURE 6.3 Options of Primary Culture

Isolation of Tissue

The first step in the cell culture is to isolate the tissue. Precautions should be taken to maintain sterile conditions while handling animal tissues. A balanced salt solution (BSS) containing 1,000 units of Penicillin and 0.5mg of Streptomycin or Neomycin per ml is used to wash the tissue to avoid contamination. The tissue is then washed with 70% Ethanol and retrieved surgically.

Disaggregation of Tissue

In some cases, the tissue consists of cells that are aggregated. This tissue has to be disaggregated either mechanically or using enzymes or chemicals to obtain cell suspensions. The different methods used for disaggregating the tissues are:

- Physical Disaggregation
- Enzymatic Disaggregation
- Treatment with chelating agents

Physical Disaggregation

After the tissue is isolated, it is kept in a sieve of 100 mm before being transferred to a petridish containing a buffered medium with BSS. The cells are then passed through sieves of decreasing pore size (50 mm and 20 mm mesh). The debris is discarded and the medium containing the cell is collected. The cell count is measured with an instrument called a haemocytometer. The other methods of disaggregation of cells are, forcing through syringe and needle or repeated pipetting. Although the method is quick, it results in extensive damage to live cells.

Enzymatic Disaggregation

Enzymes are used for dislodging cells. The use of enzymes results in a high number of cells. The two important enzymes in this method are collagenase and trypsin.

Collagenase

This enzyme is used for disaggregation of normal as well as malignant embryo tissues. The intracellular matrix of cells contains collagen, thus, the use of collagenase results in disaggregation of the same. In this method, the tissue is dissected into pieces in BSS containing antibiotics. It is then washed with sterile distilled water and transferred in complete medium containing collagenase. After about 4–5 days of treatment, the mixture is pipetted to disperse the medium. In some cases, clustering of the epithelial cells is observed at the bottom of the tubes. The clusters and collagenase are removed by centrifugation of the solution. The suspension consists of enriched fibroblast fraction that is plated on a medium. The cluster consists of the enriched epithelial fraction. Figure 6.4 shows the stages in disaggregation of tissue for primary culture by colllagenase.

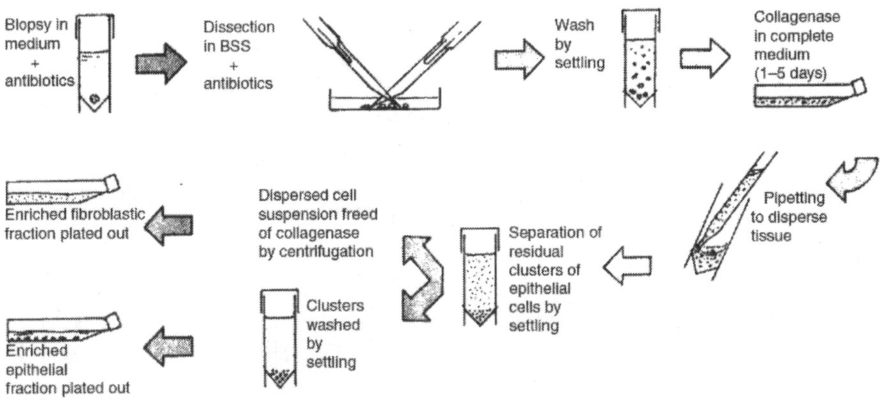

FIGURE 6.4 Stages in Disaggregation

Trypsin

The use of trypsin in the process of disaggregation is called trypsinization. Unlike collagenase, crude trypsin has to be selectively used depending on the type of cell that one deals with. Temperature plays a vital role in the activity of trypsin. Trypsin activity is of two main types: cold and warm trypsinization.

- **Cold trypsinization**: The tissue to be treated is first chopped into small pieces and then transferred to a solution containing BSS. The tissue, after dissection in the solution, is kept in a glass vial kept on ice and soaked in chilled trypsin for 4–6 hours. This aids the penetration of trypsin in the tissue. The tissue is then incubated at 36–37 degrees, for 30 minutes. Around 10–15ml of media containing serum is then added to the vial containing the tissue and the solution is pipetted repeatedly. The cells are counted using a haemocytometer.
- **Warm trypsinization**: The initial treatment for the tissue is similar to cold trypsinization. In this method, the chopped tissues are transferred to a 250ml flask containing around 100ml warm trypsin (36–37degrees). The content is stirred for 4–5 hours and the pieces are allowed to settle down. The dissociated cells are collected every 30 minutes, enabling minimum exposure of cells to warm enzyme. The trypsin is removed by centrifugation after 3–4 hours during which the complete tissue may be degraded. The vials containing the

disaggregated cells are then kept on ice and the cell count is measured using a haemocytometer.

Figure 6.5 shows Trypsin disaggregation.

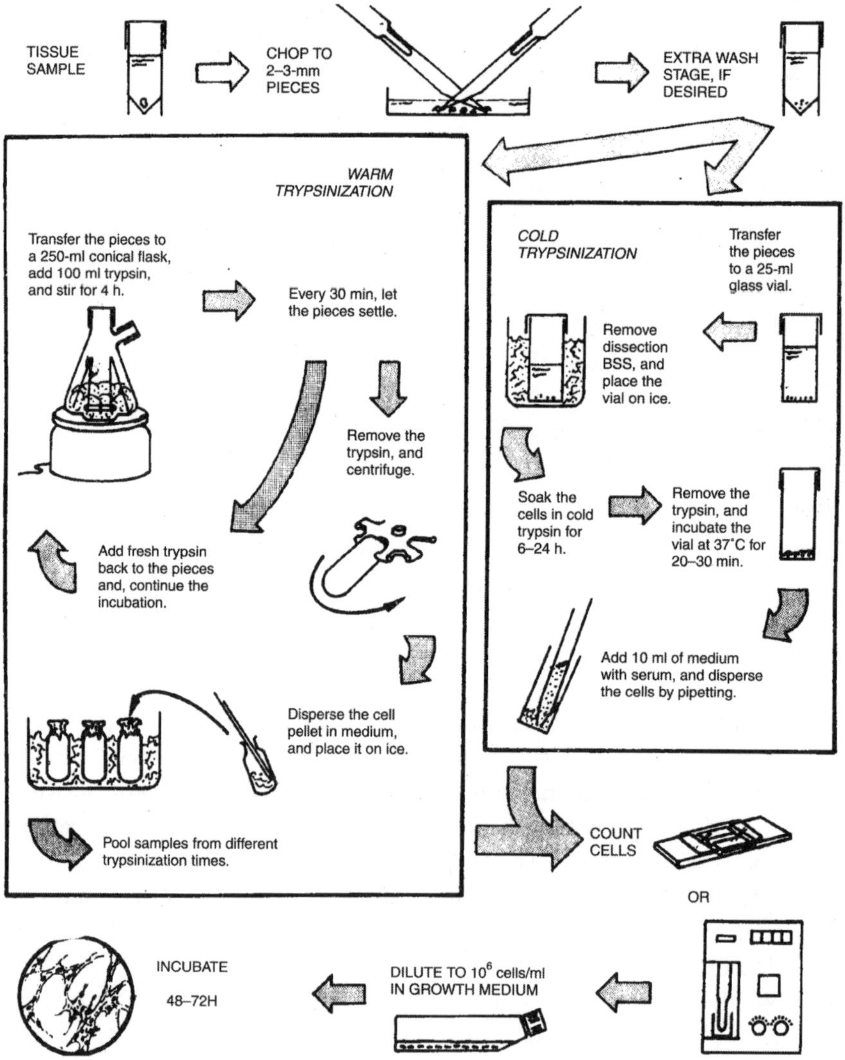

FIGURE 6.5 Trypsin Disaggregation

In each method, the cell density is maintained at around 100,000 cells/ml.

STEM CELL RESEARCH

Stem cells are unspecialized cells that renew themselves for long periods by repeated cell divisions. These stem cells can be induced under specialized physiological and experimental conditions to become cells with specialized functions. The two main types of stem cells that are used by scientists in research are:

- Embryonic stem cells
- Adult stem cells

Embryonic Stem Cells

As the name suggests, these stem cells are obtained from embryos. The embryonic stem cells are basically derived from the embryos that develop from eggs that have been fertilized in vitro (in an artificial environment). Naturally developing embryos are not used to derive the same. The embryos from which human stem cells are obtained are typically 4–5 days old and bear resemblance to hollow microscopic ball of cells called blastocyst. The blastocyst has three main layers:

- Trophoblast that surrounds the blastocyst.
- Blastocoel, a hollow cavity.
- Inner cell mass, a group of 30 cells concentrated at a pole.

Embryonic cells are grown in vitro in a process called cell culture. In the process, the inner cell mass of the blastocyst is transferred into a culture dish containing a nutrient broth called culture media. The embryonic cells grow and proliferate on the surface of the culture media. The inner surface of the culture dish is coated with an adherent material (chemically treated cells from mouse in most cases). This is to provide support for the growing embryonic cells and act as substratum during advanced growth. This layer is called the feeder layer. Feeder cells not only act as a substratum but also release nutrients into the media. This is the reason why feeder-less cultures are being preferred owing to risk of contamination and transmission of viruses from the feeder layer in case of mouse cells.

When the embryonic cells reach a certain stage of growth, they start crowding the plate and are transferred to a separate plate with fresh culture media. This process is repeated and is called subculturing. The subsequent cycles of subculturing is called a passage. After 6 months, the 30 odd cells of the inner cell mass divide into millions of embryonic cells. The cells that continue dividing without differentiating are called pleuripotent cells and form an embryonic cell line. These cells are then frozen in liquid nitrogen and stored.

The process of characterization, which is a series of tests performed on growing cells to check for signs of differentiation, is an important step. Some of the prominent tests include physical observation of cells to check for morphological changes and the use of surface markers found only on undifferentiated cells. Loss of the markers indicates differentiation of cells. As long as the cells are separated, they do not differentiate, but once allowed to clump or aggregate, they form embryoid bodies and differentiate further to form various other cells (muscle, nerve, etc.). The process of controlling the differentiation process by means of altering media, or introducing genes into cells to force embryonic cells to differentiate into specific organs is called directed differentiation. This is one of the processes used in stem cell therapy procedures.

Adult Stem Cells

These stem cells are undifferentiated cells found among differentiated cells, which yield specialized cells that produce various individual organs. Adult stem cells maintain and repair tissues and organs in which they are found. One of the striking aspects of these cells is that their origins are still unknown. Adult stem cells are also called somatic stem cells.

In the past decade, studies have revealed the presence of adult stem cells in many organs. There is also evidence about the ability of these cells to differentiate into various types of cells. Interest in adult stem cells began with the discovery that the bone marrow mainly contains two types of cells, the hematopoietic cells that form all the blood cells of the body and the bone marrow stromal cells. The stromal cells are responsible for production of bone, cartilage, fat and fibrous connective tissue.

Adult stem cells, distributed sparingly in the body and along with stem cells, remain quiescent for a number of years before being activated by disease or injuries. The major adult organs that contain stem cells are the brain, bone marrow, peripheral blood, blood vessels, skeletal muscle, skin and liver. Much of the research in the area is directed at ways of culturing cells for theraupeutic purposes. A very important property of the adult stem cells is transdifferentiation or platicity. It is the ability of the adult stem cell from one tissue to differentiate into another. Such kind of plasticity has been observed in case of bone marrow and brain cells where cells have transdifferentiated into muscle, bone and cardiac cells.

In spite of accumulating data on these cells, many questions regarding them are unanswered, such as: What is the source of adult stem cells? How do they transdifferentiate? What are the signaling events that are responsible for the process and can these cells be stimulated to grow in large enough numbers to produce fresh organs for transplantations?

Applications of Stem Cells

This section details the different uses of stem cells in practical fields of application.

Academic Interest

There are a number of areas that can benefit from the use of stem cells. These cells offer vital clues to the entire process of differentiation and lead us toward events that trigger the process. While signals that switch on and turn off the genes are being probed, not much is known about the actual mechanisms of these signals. Studies using stem cells are currently trying to unravel the mystery.

Therapeutics

Another important area that benefits from the use of stem cells is that of drug designing and testing. Potential drugs can be tested and screened for using differentiated stem cells. Cancer cell lines are already being used to test various anti-cancer drugs. The challenge scientists face in this regard is to maintain conditions and alter them sufficiently to stimulate the differentiation of cells.

The most important areas of application of stem cells are in the production of vital organs and tissues. Currently, transplantations from donors are the only means of therapy in certain cases of organ failure or tissue damage. This is, however, fraught with difficulties of transplant rejection and lack of sufficient number of donors. The ability to differentiate cells into particular organs or tissues might offer hope to many such patients suffering from diseases including Parkinson's and Alzheimer's diseases, spinal cord injury, stroke, burns, heart disease, diabetes, osteoarthritis and rheumatoid arthritis. Scinetists have successfully transplanted bone marrow cells in the heart of mice and obesreved the differentiation of cells into heart muscle cells while in some other quarters, scientists claim to have successfully differentiated adult stem cells into insulin-producing cells offering a treatment for diabetes.

Skin Transplantation

Burn victims can look forward to new hope since techniques involving stem cells are currently being used in skin grafts as transplantaion expriments. The skin keratinocyte stem cells from hair follicles are being used for this purpose. The patients' own epidermal layer can be rejuvenated using this approach without the risk of rejections.

Brain Cell Transplantations

Recent studies have negated the earlier concept that brain cells do not divide. Growing evidences have isolated neural stem cells from embryos and adults and have convinced scientists about the use of these in curing various ailments such as, stroke, spinal cord injury and neurodegenerative diseases, such as Parkinson's disease, by neural stem cell transplantation.

Clinical Applications of Haematopoietic Stem Cells (HSCs)

HSCs are used in treatment of cancers of the blood, such as leukemia and lymphoma that result from the uncontrolled proliferation of white blood cells. In these kinds of applications, the patient's own cancerous HSCs are destroyed by various therapies and the HSCs collected from a snatching donor are introduced in the body. The crucial element in this procedure is the compatibility of the donor, which in most cases are blood relatives. HSCs are also used in the treatment of hereditary blood disorders, such as different types of anaemia and inborn errors of metabolism.

A host of primary applications of HSCs are also used for the treatment of autoimmune diseases such as, diabetes and rheumatiod arthritis. However, the major hurdle that scientists face with such therapies are the life-threatening instances of graft rejection by the host's body. Research is currently on to outfox the host's immune system and in exploiting the property of plasticity to differentiate HSCs into specific organs and tissues.

Stem Cells and Gene Therapy

Non-embryonic stem cells have been used in cell-based gene therapy studies. In this approach, stem cells are removed from the body and genetically modified cells are introduced instead. Modified stem cells act as carriers for the desired gene that is to be replaced. The major advantage of such an approach is that stem cells are self-renewing population of cells and thus reduce or eliminate the need for repeated administrations of the gene therapies. The only types of stem cells that have been used in gene therapy are the HSCs.

Ethical Issues in Stem Cell Therapy

Stem cell therapy has raised some very important questions in ethics and legal practice. Some of these questions are discussed in this section.

Is it permissible to create life with an objective to destroy it?

This is one of the questions that have been raised by the academic and ethical communities across the world. Even as the potentials of stem cell based therapies emerge, there have been growing debates about the ethics on this issue. While religious heads across the world have deplored the use of embryos for stem cell culture branding it as "murder", scientists are divided on the issue of whether an embryo is animate or not.

Is it justified to seek human eggs for scientific research?

The biggest source of stem cells is human eggs that are raised in vitro. Since the average number of egg cells that a normal woman can produce is less, various chemical stimulants are used to stimulate the production of a large number of eggs. This is fraught with side effects and the morality behind such unnatural procedures is also being debated. In many cases, patients undergoing in vitro fertilization treatment are unsuspecting victims of such procedures, and the eggs that are produced are harvested without their consent. This trend is prevalent in Third World countries where the awareness levels are low and legislations are practically non-existent.

Protecting Donor Identity

The issue of protecting the donors' identity is an extremely sensitive one considering the fact that these donors could be targets for the media and anti-stem cell groups. Stem cell therapy depends heavily on the availability and willingness of healthy donors. Failure to safeguard their interests would only result in reticence on the part of the donors, which would ultimately stifle the entire field.

Terms to Know

Adult stem cell	An undifferentiated cell found in a differentiated tissue that can renew itself and (with certain limitations) differentiate to yield all the specialized cell types of the tissue from which it originated.
Antibody	A Y-shaped protein secreted by B cells in response to an antigen. An antibody binds specifically to the antigen that induced its production. Antibodies directed against antigens on the surface of infectious organisms help eliminate those organisms from the body.
Antigen	A substance (often a protein) that induces the formation of an antibody. Antigens are commonly found on the surface of infectious organisms, transfused blood cells, and transplanted organs.
B cells	Also known as B lymphocytes. Each B cell is capable of making one specific antibody. When stimulated by antigen and helper T cells, B cells mature into plasma cells that secrete large amounts of their specific antibody.
Blastocyst	A preimplantation embryo of 30–150 cells. The blastocyst consists of a sphere made up of an outer layer of cells (the trophectoderm), a fluid-filled cavity (the blastocoel), and a cluster of cells on the interior (the inner cell mass).
Bone marrow	The soft, living tissue that fills most bone cavities and contains hematopoietic stem cells, from which all red and white blood cells evolve. The bone marrow also contains mesenchymal stem cells that a number of cells types come from, including chondrocytes, which produce cartilage.

Chimera	An organism composed of cells derived from at least two genetically different zygotes. Theoretically, the zygote could be from separate species.
Differentiation	The process whereby an unspecialized early embryonic cell acquires the features of a specialized cell such as a heart, liver, or muscle cell.
Embryo	In humans the developing organism from the time of fertilization until the end of the eighth week of gestation is known as a fetus.
Embryonic germ (EG) cells	Cells found in a specific part of the embryo/fetus called the gonadal ridge that normally develop into mature gametes.
Embryonic stem (ES) cells	Primitive (undifferentiated) cells from the embryo that have the potential to become a wide variety of specialized cell types.
Feeder cell layer	Cells that are utilized in co-culture to maintain pluripotent stem cells. Cells usually consist of mouse embryonic fibroblasts.
Fibroblast	Cells that give rise to connective tissue.
Glial cells	Supporting cells of the nervous system, including oligodendrocytes and astrocytes in the vertebrate central nervous system and Schwann cells in the peripheral nervous system.
Green fluorescent protein (GFP)	Fluorescent-protein dye used to tag and trace particular genes and cells of interest.
In vitro fertilization (IVF)	An assisted reproduction technique in which fertilization is accomplished outside the body.
Leukocyte	A white blood cell or corpuscle. Leukocytes are formed from undifferentiated stem cells that give rise to all blood cells.

Melanocyte	A cell that produces the dark pigment melanin; responsible for the pigmentation of skin and hair.
Passage	A round of cell growth and proliferation in culture.
Pluripotent stem cell (PSC)	A single stem cell that has the capability of developing cells of all germ layers (endoderm, ectoderm, and mesoderm).
Somatic cells	Any cell of a plant or animal other than a germ cell or germ cell precursor.
T cells	A type of white blood cell that is of crucial importance to the immune system. Immature T cells migrate to the thymus gland in the upper chest cavity, where they mature and differentiate into various types of mature T cells and become active in the immune system in response to a hormone called thymosin and other factors. T-cells that are potentially activated against the body's own tissues are normally killed or changed ("down-regulated") during this maturation process.
Vimentin	The major polypeptide that joins with other subunits to form the intermediate filament cytoskeleton of mesenchymal cells. Vimentin may also have a role in maintaining the internal organization of certain cells.

7 Plant Biotechnology

Introduction

The last two decades have witnessed a remarkable progress in the fields of agriculture, horticulture, forestry and pharmaceutical industry. This has been largely possible due to the development of various prominent techniques, such as the tissue culture methods. In the earlier chapter, we have extensively dealt with the subject of animal tissue culture and its applications. The current chapter is designed to provide an insight in the area of plant tissue culture and enumerate the importance of the technique in the various spheres of biological sciences.

Cell and Tissue Culture

Plant cells can be grown in isolation from intact plants in tissue culture systems. The cells have the characteristics of callus cells, rather than other plant cell types. These are the cells that appear on cut surfaces when a plant is wounded and which gradually cover and seal the damaged area.

Pieces of plant tissue will slowly divide and grow into a colourless mass of cells if they are kept in special conditions. These are:

- initiated from the most appropriate plant tissue for the particular plant variety.
- presence of a high concentration of auxin and cytokinin growth regulators in the growth media.
- a growth medium containing organic and inorganic compounds to sustain the cells.
- aseptic conditions during culture to exclude competition from microorganisms.

The plant cells can grow on a solid surface as friable, pale-brown lumps (called callus), or as individual or small clusters of cells in a liquid medium called a suspension culture. These cells can be maintained indefinitely provided they are sub-cultured regularly into fresh growth medium.

Tissue culture cells generally lack the distinctive features of most plant cells. They have a small vacuole, lack chloroplasts and photosynthetic pathways and the structural or chemical features that distinguish so many cell types within the intact plant are absent. They are most similar to the undifferentiated cells found in meristematic regions, which become fated to develop into each cell type as the plant grows. Tissue cultured cells can also be induced to re-differentiate into whole plants by alterations to the growth media.

Plant tissue cultures can be initiated from almost any part of a plant. The physiological state of the plant does have an influence on its response to attempts to initiate tissue culture. The parent plant must be healthy and free from obvious signs of disease or decay. The source, termed explant, may be dictated by the reason for carrying out the tissue culture. Younger tissue contains a higher proportion of actively dividing cells and is more responsive to a callus initiation programme. The plants themselves must be actively growing, and not about to enter a period of dormancy.

The exact conditions required to initiate and sustain plant cells in culture, or to regenerate intact plants from cultured cells, are different for each plant species. Each variety of a species will often have a particular set of cultural requirements. Despite all the knowledge obtained about plant tissue culture during the twentieth century, these conditions have to be identified for each variety through experimentation.

Table 7.1 will enumerate the main milestones in the history of plant biotechnology:

The following table has been compiled from various sources on the Internet

TABLE 7.1 Timeline of Plant Tissue Culture

1838	MJ Schleiden hypothesized cell theory, suggesting totipotentiality of cells.
1902	G Haberlandt made first (but unsuccessful) attempt at tissue culture;
	Walter Suton coined the term "gene" and proposed that chromosomes carry genes (factors that Mendel said could be passed from generation to generation).
1904	B Hanning attempted embryo culture (crucifers).
1909	E Kuster achieved fusion of protoplasts although the products did not survive.
1921	M Molliard cultured fragments of plant embryos.

(Continued)

TABLE 7.1 *(Continued)*

1922	L Knudson germinated orchard seeds in vitro; WJ Robbins cultured root tips.
1924	Blumenthal and Meyer achieved callus formation on carrot root explants with use of lactic acid.
1925	Laibach used embryo culture to rescue interspecific hybrids of *Linum*.
1926	FW Went demonstrated that there were growth substances in coleoptiles from *Avena*.
1930–1940	Plant hybridization used in plant breeding.
1934	RJ Guatheret unsuccessfully cultured cambial tissues of trees and shrubs; PR White maintained long-term cultures of tomato roots; Kogl et al. identified the first plant growth regulator, IAA.
1936	CR LaRue cultured embryos of conifers.
1939	Successful continuously growing cambial cultures of carrot and tobacco.
1940	Gautheret cultured cambial tissues of *Ulmus* to study adventitious shoot formation.
1941	Van Overbeek used coconut milk to support growth and development of very young *Datura* embryos; Braun cultured crown gall tissues in vitro; George Beadle and Edward Tatum proposed the "one gene, one enzyme" hypothesis.
1942	Observation of secondary metabolites in callus cultures (Gautheret).
1943–1950	Braun elucidated the tumour-inducing principle of crown galls.
1944	Folke Skoog used tobacco cultures to study adventitious root formation.
1946	E Ball produced the first whole plants from shoot tip cultures of *Lupinus* and *Tropaeolum*.
1948	Formation of adventitious shoots and roots in tobacco culture (Skoog and Tsui).
1949	Culture of fruits in vitro (Nitch).
1950	Ball regenerated organs from sequoia; Morel reported the first successful cultures of a monocot using coconut milk.
1951	Nitch cultured excised ovaries in vitro; Skoog demonstrated chemical control of growth and organ formation in vitro.

(Continued)

TABLE 7.1 *(Continued)*

1952	First virus-free plants produced from meristem culture; First successful micrografts (Morel and Martin)
1954	Muir et al. regenerated calli from single cells using nurse cultures
1957	Discovery of the regulation of organ formation by changing the ratio of auxin: cytokinin (Skoog and Miller); Culture of excised anthers of *Allium* (Vasil); Francis Crick and George Gamov explain how DNA functions to make protein
1958	Regeneration of somatic embryos in vitro from the nucellus of *Citrus* ovules (Maheshwari and Rangaswamy); Proembryo formation in callus and cell suspensions of carrot (Reinert and Steward); Coenberg discovers DNA polymerase
1960	First successful test tube fertilization in *Papaver rhoeas* (Kanta); Use of the microculture method for growing single cells in hanging drops in a conditioned medium (Jones et al.); Enzymatic degradation of cell walls to obtain large numbers of protoplasts (Cocking).
1962	Development of Murashige and Skoog medium (Physiol Plant 15:473)
1964	Production of first haploid plants from pollen grains of *Datura* (Guha and Maheshwari)
1965	Regeneration of tobacco plants from single cells (Vasil and Hildebrant); Production of protocorms from orchids in vitro (Morel)
1967	Haploid plants produced from pollen grains of tobacco (Bourgin and Nitch)
1972	First report of interspecific hybridization through protoplast fusion in two species of *Nicotiana* (Carlson et al); Restriction fragments can be joined by DNA ligase regardless of their origin if they are cut with the same restriction enzyme (Mertz and Davis; Berg); Isolation of reverse transcriptase (Temin)
1973	Use of the Lobban and Kaiser technique to develop hybrid plasmid insertion of EcoRl fragment of DNA molecule into circular plasmid DNA of bacteria using DNA ligase. Gene from African clawed toad inserted into plasmid DNA of bacteria (Herbert Boyer and Stanley Cohen); Cytokinin found to be capable of breaking dormancy in Gerbera (Somatic hybridization of tomato and potato, resulting in pomato (Melchers et al.))

(Continued)

TABLE 7.1 *(Continued)*

1974	Regeneration of haploid petunia hybrid plants from protoplasts (Binding); Biotransformation in plant tissue cultures (Reinhard).
1975	Positive in vitro selection of maize callus cultures resistant to T toxin of *Helminthosporium maydis* (Gengenbach and Green).
1979	Cocultivation procedure developed for the genetic transformation of plant protoplasts with *Agrobacterium* (Marton et al.).
1981	Introduction of the term somaclonal variation (Larkin and Scowcroft).
1983	Transgenic plants were first created in the early 1980s by four groups working independently at Washington University in St. Louis, Missouri, the Rijksuniversiteit in Ghent, Belgium, Monsanto Company in St. Louis, Missouri, and the University of Wisconsin. The first three groups announced at a conference in Miami, Florida, that they had inserted bacterial genes into plants. The fourth group announced at a conference in Los Angeles, California, in April 1983 that they had inserted a plant gene from one species into another species.
1986	TMV virus-resistant tobacco and tomato (transgenic plants developed using cDNA of coat protein gene of TMV (Powell-Abel et al); Transformation of tobacco protoplasts by direct DNA microinjection (Crossway); First field trials of transgenic plants.
1987	Development of biolistic gene transfer method for plant transformation (Sanford et al; Klein et al.); Isolation of Bt gene from bacterium (*Bacillus thuringiensis*) (Barton et al.); First monocot (Asparagus) transformation by *Agrobacterium tumefaciens* (Bytebier et al.).
1988	Recovery of stable transformants through particle bombardment (Klein et al.); Automated mass propagation with organogenesis and embryogenesis (Levi R et al.).
1993	Flavr Savr tomatoes sold to public.

Basic Infrastructure

The basic requirements of a plant tissue culture laboratory are more or less similar to the animal tissue culture set-up. Care should be taken to maintain a sterile atmosphere since contamination is one of the most difficult things to tackle. Some of the vital requirements are:

- Media preparation area/room.
- Aseptic transfer chamber/area (laminar airflow).
- Environmentally controlled culture room (light/temperature/humidity).
- Analytical room.
- Acclimatization room (to be used in conditioning of plants).
- Hot air oven (for baking glassware).

Aseptic Techniques

One of the most difficult aspects of plant tissue culture is to maintain contamination-free conditions. This is due to the fact that in in vitro, both the plants and microorganisms have the same requirements for survival. This leads to growth of bacteria and other microbes in the media that are used for plant tissue culture. The microbes grow at a much faster rate and exhaust the media, eventually killing the plant tissue. This leads to the failure of the entire experiment. Therefore, a series of exhaustive procedures are followed to eliminate contaminants. Some of the methods are listed:

- **Autoclaving:** This is one of the most common and highly effective methods. In this, the media for growth along with all the glassware and other materials are sterilized under high pressure, using steam.
- **Use of chemicals:** The explant (plant part used for culture) may carry microbes with it, which are eliminated by surface-sterilizing using various chemicals, such as the chemicals listed in Table 7.2.

TABLE 7.2 Commonly Used Disinfectants for Plant Tissue Culture

Disinfectant	Concentration (%)	Exposure (mm)
Calcium hypochlorite	9–10	5–10
Sodium hypochlorite	0 5–5	5–30
Hydrogen peroxide	3–12	5–15
Ethyl alcohol	70–95	0.1–5.0
Silver nitrate	1	5–30
Mercuric chloride	0.1–1.0	2–10
Benzalkonium chloride	0.01–0.1	5–20

In addition to this, surfactants like Tween 20 or Triton X-100 are used in low concentrations to ensure proper surface sterilization without harming the plant tissue. Figure 7.1 shows a diagrammatic representation of the procedure for surface sterilization of plant material and inoculation of explant for culture.

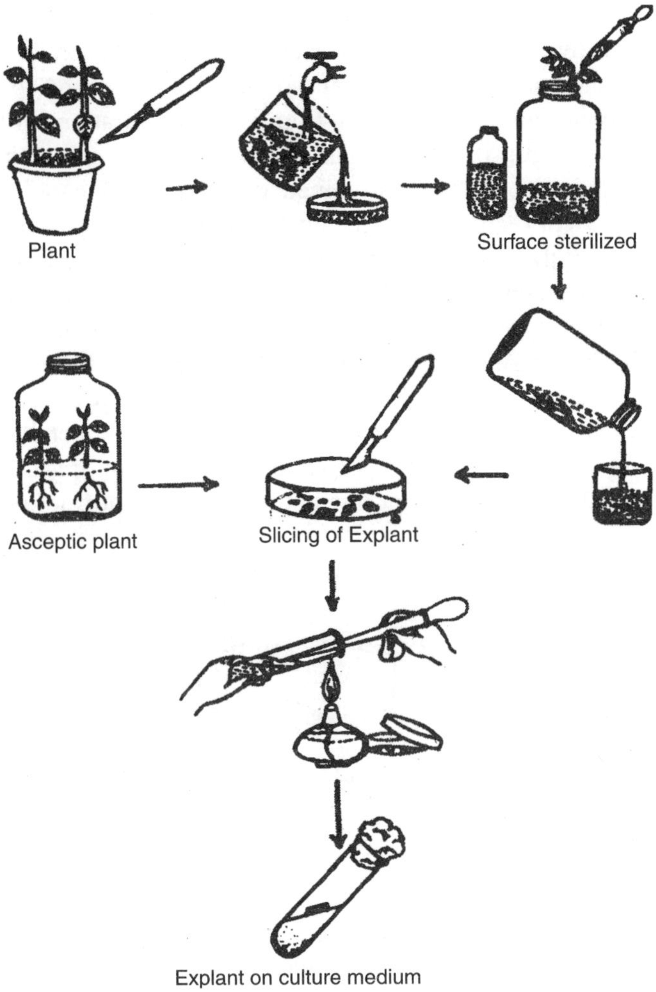

FIGURE 7.1 Surface Sterilization of Plant Material

Use of Antibiotics

Antibiotics are used to reduce, control or eliminate contaminants from the plant tissue cultures. Antibiotics are used only after autoclaving because most of them are heat-sensitive. They are filter sterilized after dissolving them in autoclaved solutions and then added to the media.

Nutrient Media

The nutrient media is one of the most vital components for the success of any tissue culture experiment. Some of the important constituents of the nutrient media are:

- **Carbon source:** The plant tissue used for culture is mostly photosynthetically inactive and they require an external source of carbon for the growth and proliferation of cells. The concentration of the carbon source varies from application to application, from low concentration for protoplast cultures, to higher concentrations in case of the embryo or anther culture. Sugars and hexitols are the prominent carbon sources.
- **Plant growth regulators:** Some hormones stimulate various biological activities in the nutrient media and are an integral part of plant growth and development. For this reason, these hormones are known as plant growth regulators. Some of the important plant growth regulators are Cytokinins, Auxins and Gibberellins. Cytokinins promote cell division and regulate growth and development, while Auxins resemble the indole acetic acids (IAA) and stimulate shoot elongation. Gibberellins are used rarely but play a very crucial role in the culture of apical meristems. Figure 7.2 shows the chemical structure of growth hormones.

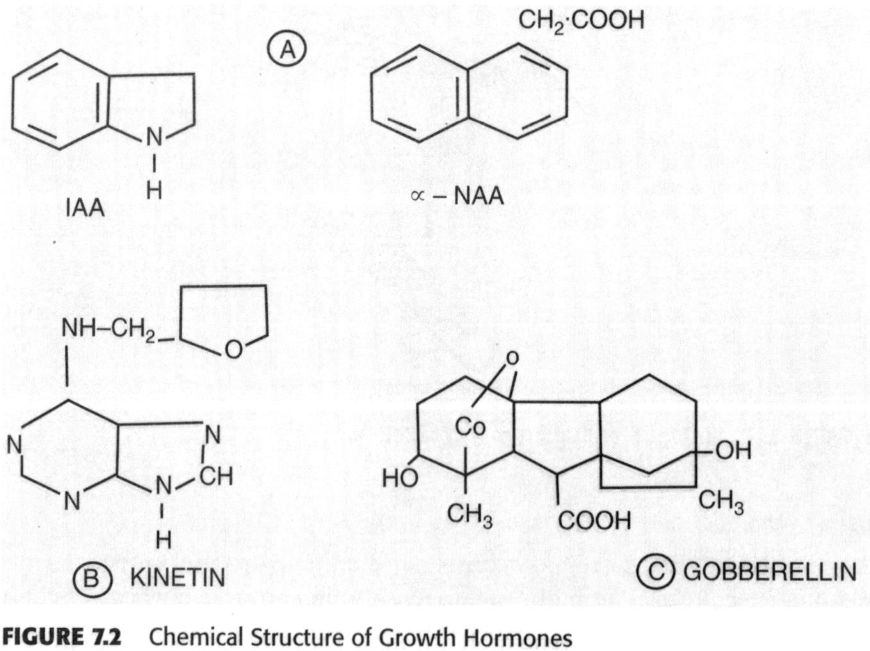

FIGURE 7.2 Chemical Structure of Growth Hormones

- **Vitamins:** Minute concentrations of vitamins are important in catalyzing various reactions during plant metabolism. Vitamin B1 (Thiamine) is the most commonly used vitamin for all plant tissue cultures. Other groups of vitamins that are used are niacine (Nicotinic acid), vitamin B2 (Riboflavin), vitamin B6 (Pyridoxine), vitamin C, vitamin H (Biotin) and vitamin B12 (Cyanocobalamine).
- **Inorganic materials:** These basically include macronutrients (such as Nitrogen, Phosphorous, Potassium, Calcium, Magnesium and Sulphur) and micronutrients (such as, Boron, Copper, Zinc) in their salt forms. These are used in varying concentrations and are mostly prepared as stock solutions and diluted prior to use.
- **Gelling agent:** These are used to provide a solid support medium for the growing plants. The gelling agents have to be chemically inert. The most commonly used gelling agent is agar, derived from seaweeds.

Apart from the above mentioned, many other components are also used, such as coconut milk, yeast extract, malt extract, tomato juice, malt extract, antibiotics and various antioxidants like citric acid.

Care has to be taken to ensure that all reagents and media are prepared in double distilled autoclaved/millipore grade water.

Plant Transformation Techniques

Plants have to respond to a wide range of environmental signals, in addition to the existing pervasiveness of networks within itself. The growth and productivity of plants are limited due to these constraints that consume a sizeable amount of the plant's resources and energy. These, in turn, directly affect the economic exploitation of these plants. It is against this background that various techniques are used to modify or transform the plants to bestow desirable traits on them.

Plant tissue culture is one of the most widely used and successful techniques in plant transformations. This technique is based on the concept of totipotency of cells. According to this concept, plant cells have the capacity to develop into a whole organism. Totipotency relies on using explants and stimulating growth and proliferation, ultimately resulting in the regeneration of complete plants. Figure 7.3 depicts the concept of totipotency of cells.

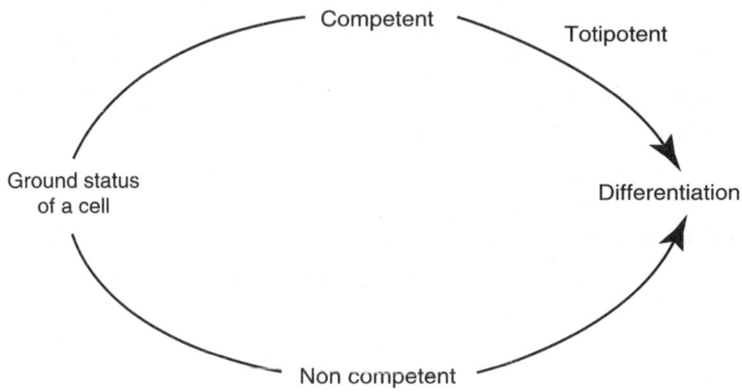

FIGURE 7.3 Totipotency of Cells

Callus Formation and Culture

The most common culture used in plant tissue culture is callus. A callus is wound tissue composed of undifferentiated and unorganized cells. In the natural environment, a callus develops by the infection of microorganisms from wounds due to the stimulation of various endogenous growth hormones, such as Auxins and Cytokinins.

In the tissue culture method, the explant is obtained and conditioned on media containing hormones and nutrients that stimulate callus formation. The callus, when formed, is an amorphous mass of loosely arranged thin-walled parenchyma cells. The most unique aspect of the callus formation is that it has the potential to develop normal roots, shoots and embryoids that ultimately form a plant.

Callus formation depends heavily on the source of the explant, nutritional composition of media and environmental factors. Callus is formed in three developmental stages: induction, cell division and differentiation. Once the callus grows on the media for some time, it is essential to transfer it into a fresh media to avoid nutrient depletion.

Cell Suspension Culture

This is prepared by transferring a fragment of callus (~300–500mg) to a liquid medium (~500–1000ml) and agitating them under sterile conditions to make the cells free in the medium. The chances of obtaining a suspension of a single cell are rare and generally there are a number of single cells, cell aggregates, residual innoculum and dead cells.

Cell separation is made possible by changing the medium to isolate different types of single cells. When such cells are transferred into a suitable medium, they divide after a lag phase and increase exponentially. This increases to a certain level based on the nutrient status of the medium before finally falling off, and becomes stationary at a point. To maintain the cells beyond this stage, one has to subculture them in a fresh medium on a regular basis. However, it is vital to determine the cell count before subculturing the medium. Growth essentially depends on a critical cell density below, which the culture does not grow. The main features of the cell suspensions are:

- They grow in sterile environments.
- They are homogenous in size.
- They have a doubling time longer than microorganisms but shorter than plant cells *in situ.*
- They can be grown on a large scale.

Organogenesis

The organs that are induced in plant tissue culture comprise root, shoots and leaves. Since the embryo is an independent structure and does not have a well-defined vascular supply, it is not considered to be a plant organ. Organogenesis (development of organs) starts with the stimulation caused by chemicals in the medium, substances that are carried over from the original explants and endogenous compounds produced by the culture. The concept of organogenesis was propounded by Skoog for the first time and extended by Torrey who stated that the genesis of organogenesis is the development of meristimatic cells that in turn induces development of various organs.

Micropropagation

Murashige was the first to establish micropropagation technique and its biotechnological applications in 1974. The technique is called micropropagation due to the minute size of the propagules used for the culture. The technique is divided into four main stages:

- **Stage I:** Selection of suitable explant and innoculation into the medium.
- **Stage II:** Multiplication and growth of culture that would take about 2 months, followed by a repeated subculture. Rootless cultures are derived at this stage.

- **Stage III:** Obtaining cultures by changing the medium. The medium is devised according to the application and type of plants to be cultured.
- **Stage IV:** Planting out stage where the plantlets are removed from the test tube environment to natural environment. Care should be taken to ensure that the roots are fully functional. During this procedure the plantlets routinely die due to desiccation, extreme climates and invasion of soil microorganisms. It is important to acclimatize (condition) the plantlets before planting them in the open. This can be done by inducing the development of leaves, roots, exposing the plantlets to the external environment in phases and recreating the external environment minus the extremities.

The advantages of micropropagation are numerous. Some of them are listed:

- Micropropagation is an alternative to vegetative propagation and is used to enhance the rate of multiplication. This is possible because a large number of plant propagules can be developed in a short time by using this technique.
- Micropropagated varieties can be introduced faster in the market due to the faster growth rates as compared to the wild varieties. This enables breeders to offer new lines and variants rapidly.
- The entire process of micropropagation can be completed in a very small space. Several thousands plants can be maintained inside a relatively small laboratory. This enables the growth and introduction of plants on a commercial scale without increased demand for space and resources.
- Since every precaution is taken to grow the plants in a sterile environment and fungal and microbial pathogens are eliminated upon detection the plants produced by the method are disease indexed, and represent clean, elite mother stock material for the future.
- Plant breeders normally maintain specific varieties of plants in green houses or fields for future propagation or testing. Micropropagation is emerging as an attractive alternative for this, considering the advantages this technique offers in terms of space, climate control and disease and pest control.

Protoplast Culture

Protoplasts are plant cells with a plasma membrane without a cell wall. This permits the fusion of similar or different species of protoplasts resulting in whole plants. This technique is known as somatic fertilization or fusion of two vegetative cells to form hybrids. Table 7.3 shows the various plant parts from which isolation of protoplast takes place.

TABLE 7.3 Isolation of Protoplast from Various Plant Parts

Plant Tissues	Young leaves, roots, stems
Flower parts	Petal, reproductive organs
Callus/cells	Friable callus tissues/cells
Suspension cultures	Fast growing cells
Aseptically grown plantlets	Leaves, stems, roots

Protoplasts are isolated from the cells by two main methods:

- Mechanical
- Enzymatic

Enzymatic Isolation

In case of the enzymatic method, as the name suggests, various enzymes are used for the purpose of isolation of protoplasts. They generate large number of protoplasts but are known to have a deleterious effect in some cases. The advantages of this method over the mechanical method are that the cells are intact and no physical damage is caused to the protoplasts. Once the protoplasts are derived, they are suspended in a culture medium. These protoplasts regenerate a cell wall, undergo cell division and form a callus. This callus can be subcultured. This is then followed by embryogenesis, which in turn develops into seedlings and finally become mature plants. Figure 7.4 illustrates the technique used for isolation, culture and regeneration of plants from leaf protoplasts.

Protoplast Fusion and Somatic Hybridization

Protoplasts from the same or different species can be fused together. The fusion of cytoplasm of two protoplasts results in coalescence of cytoplasms. The nuclei of the two protoplasts may or may not fuse together even after the fusion of the cytoplasm. The binucleate cells are called heterokaryons or heterocytes. When the nuclei are fused the cells are called hybrids or synkaryotes. In some cases, the cytoplasms fuse but the genetic material from one of the two nuclei is lost. Such a cell is called a cybrid. There are various methods

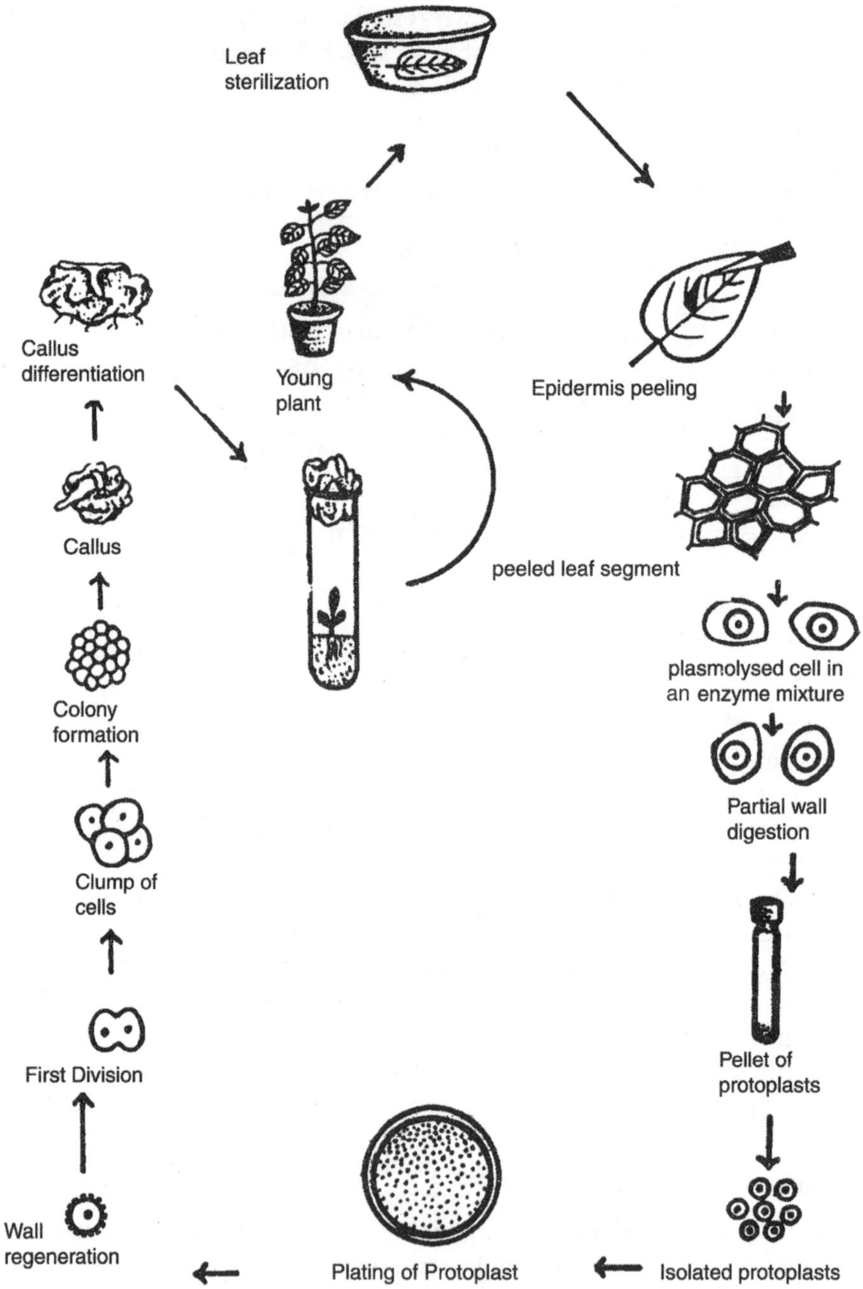

FIGURE 7.4 Technique for Isolation, Culture and Regeneration of Plants from Leaf Protoplasts

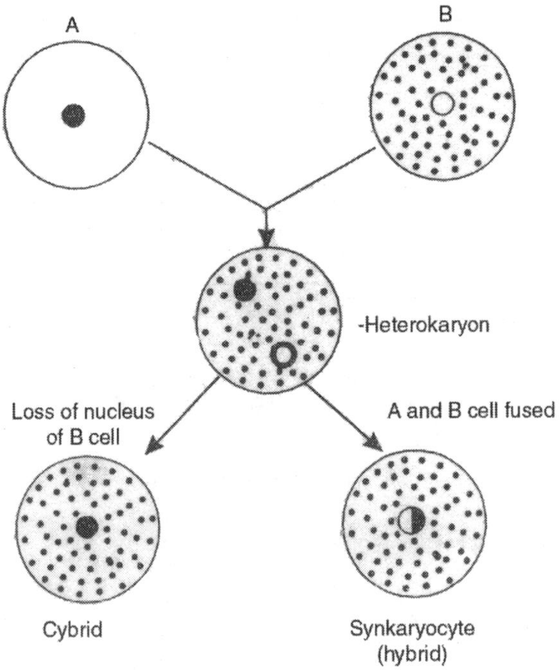

FIGURE 7.5 Hybrid Production Through Protoplast Fusion

of obtaining hybrids. Figure 7.5 shows hybrid production through protoplast fusion.

Anther and Pollen Culture

Anther is the male reproductive organ and is diploid in chromosome number. During the process of microsporogenesis, tetrads of microspores are formed from a single spore mother cell. The basic aim of anther/pollen culture is to obtain haploid plants by the process of induction or embryogenesis. Haploid plants have a single complete set of chromosomes that may in turn be useful for the improvement of many crop plants. In addition, by using various mutagens like colchicines one can obtain fertile homozygous diploids. The haploid plants are basically useful in the direct screening of recessive mutations and in the development of homozygous diploid plants. The most routine method of anther culture is depicted in Figure 7.6.

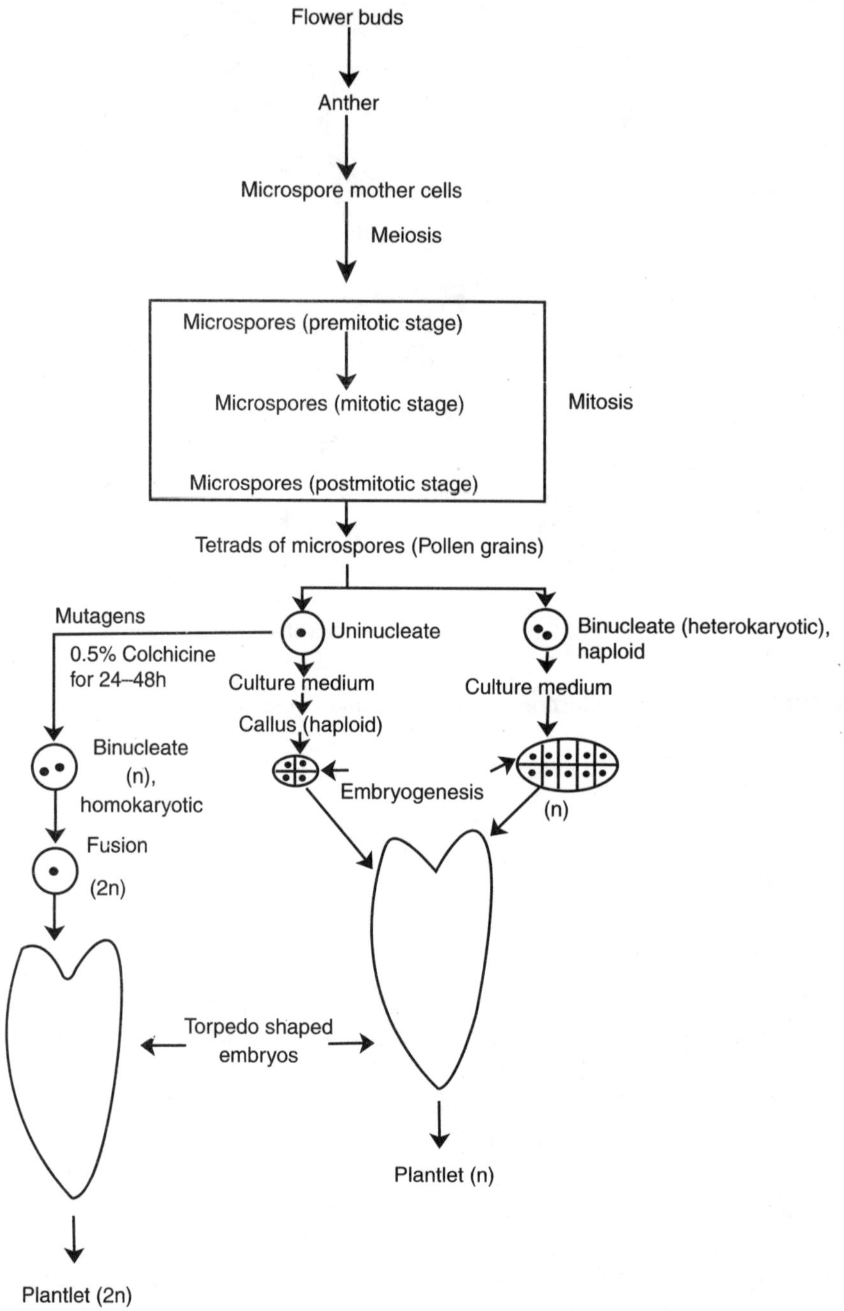

FIGURE 7.6 Anther Culture

Desirable traits can also be introduced in plants using genetic engineering. One of the methods mentioned in earlier chapters using Agrobacterium as a vector, is widely used in conjunction with plant tissue culture to introduce foreign genes in the plants.

Somaclonal Variation

Larkin and Snowcroft coined the term somaclonal variation in 1981. In case of plant cells, variability develops spontaneously during tissue culture. The plant variants that are obtained from tissue culture are called as somaclones. Somaclonal variation has yielded regenerated plants with altered growth habits, disease and pest resistance, fruit colour or size. Although somaclonal variation was limited to plants that were normally asexually propagated, it is now successfully applied to sexually propagated plants as well. Somaclonal variation can be extremely useful in establishing improved cultivars in different plants, with the advantage of avoiding the numerous crossings usually required to identify and establish a useful hybrid line by traditional breeding methods. Although it appears to be possible, to some extent, to induce variation by manipulation of the culture medium, it is not possible to predict what type of somaclones, if any, may be obtained by the development of a variant of a particular type, or whether the somaclones produced will have value. In other words, the scientist is limited by what randomly occurring variation might appear in his culture; the possibility always exists, therefore, that no agriculturally useful variants will appear. Table 7.4 shows somaclonal variation induced in morphological traits in some crop plants.

TABLE 7.4 Somaclonal Variation Induced in Morphological Traits in Crop Plants

Crop	Characters
Sugarcane	Cane diameter, stalk length and weight Cane yield, sugar yield, stalk number, length diameter, volume, density the weight.
Potato	Growth habit, maturity period, tuber uniformity and skin color. Photoperiod requirement, fruit production.
Tobacco	Yield, Days to flowering, plant height, stem diameter, leaf number, leaf length, leaf width and yield Leaf Shape, leaf number, plant height, type of inflorescence and yield.

(Continued)

TABLE 7.4 *(Continued)*

Crop	Characters
Rice	Number of tillers per plan, number of fertile fillers per plant, average panicle length, frequency of fertile seed, plant stature and flag leaf length. Flowering period, plant height, seed fertility and heading date.
Oats	Plant height, heading date, twin culms, yellow leaf stripe, awn morphology and fertility.
Maize	Twin stalks from a single node, reduced pollen fertility and male sterility.
Brassica	Altered leaf way, multiple branching of the stem, precocious flowering from the apex, stem or leaf, abnormal leaves, reduced lamina in leaves, spontaneously absorbing vegetative buds, slow growth, failure to flower, large pollen grains. Delayed flowering altered growth habit and gross morphology.
Pelargonium	Leaf shape, size and form, flower morphology, plant height, fasciation, pubescence and anthocyanin pigmentation.
Pineapple	Leaf colour, foliage density, leaf width and leaf spine formation.
Tomato	Male sterility, jointless pedicel, tangerine vircscent leaf, flower and fruit colour, lethal chlorophyll deficiency mottled leaf appearance, fruit ripening and growth habit.

There are various theories that attempt to offer an explanation for these variations. The most prominent among them is the theory that somaclones occur due to:

- Preexisting genetic variations in explant tissue.
- Spontaneous mutation, which tends to occur and accumulate during innumerable cell divisions that cells of the explant undergo before finally differentiating into a whole plant.
- Numerical and structural changes in the chromosomes during growth.
- Intracellular mutagens that are produced during growth.
- Activation of transposable elements or jumping genes that are genetic elements capable of moving across the genome and integrate at various loci.

Forms of Somaclonal Variations

There are various forms of somaclonal variations. The most common forms include point mutations, chromosomal aberrations, and an increase or decrease in the number of nuclear chromosomes. A number of approaches are available to detect somaclones. Some of them are:

- Visual detection due to accumulation of pigments produced due to mutations.
- Isolation of herbicide and antibiotic resistant varieties by growing them on the appropriate media.
- Detection of morphological variations by looking at individual traits and comparing them with the wild-type varieties.

Applications in Breeding

Variations that occur from the gametophytic or gametic cells are called gametoclonal variation. Along with somaclonal variation, there are important sources of introducing genetic variations, such as disease resistance and stability, which are valuable to plant breeders. Some of the examples are shown in Table 7.5.

TABLE 7.5 Applications of breeding

Structure	Cell wall, Plasma membrane, Organelle–structure and biosynthesis
Function	Biochemistry, Photosynthesis, Absorption
Genetics	Fusion, Organelle and DNA uptake
Plant Pathology	Virus, RNA uptake, Pathotoxin
Differentiation	Morphogenesis, Embryogenesis

Particle Bombardment

The method of particle bombardment was developed by Professor Stanford and the team at Cornell University in 1987. The method shoots foreign DNA into plant tissues or cells at high speed. This technique is also known by many names, such as particle gun method, biolistic process, microprojectile bombardment and particle acceleration. This method is used when plants do not respond to the Agrobacterium-mediated transfer of foreign DNA. The apparatus consists of a chamber connected to an outlet to create vacuum. At the top is a cylinder that is sealed from the rest of the chamber into which helium gas flows. A plastic microcarrier containing DNA coated with tungsten particles (micropellet) is close to the rupture disk that is close to the gas chamber. Helium is introduced into the cylinder at high velocity rupturing the disk and the consequent shock wave propels the micropellets into the plant cell/tissue. Table 7.6 depicts the problems and possible solutions in plant tissue culture work.

TABLE 7.6 Problems and Possible Solutions in Plant Tissue Culture Work

Symptoms	Possible Causes	Possible Solution
Culture contamination	Source heavily infested/ infected with microorganisms Poor sterilization.	Improve sterilization method (70% Ethanol, drop of detergent, antimicrobial agents) clean plant material/select unexposed tissue.
Explant dies	Strong disinfectants. Media too strong Wrong stage of growth.	Use weaker disinfectant/ change sterilizing agents. Use 1/2 or 1/4 strength. Obtain explants at different stage of growth.
Culture blackens and dies	Contaminated by microorganisms Bleeding Agar problem Water problem Wrong formula media constituents.	Discard with care. Review sterile technique and sanitation. Subculture immediately. Transfer more frequently. Try different agar concentrations. Check water purity. Try different formula.
Explant live but no growth	Dormant Media too harsh Wrong formula.	Chill for a month Use explant at different stage of growth Lower salts and hormones try different formula.
Culture live but no growth	Too cold. Wrong medium.	Increase temperature. Try different medium.
Shoots tool long (leggy), and poormultiplication leaves are yellow, watery	Too little cytokinin.	Increase cytokinin Change cytokinin/ auxin ratio.
Leafy shoots too short	Hormones too strong.	Decrease or omit hormones.
No multiplication	Too little cytokinin Needs chilling Too cold Requires dormacy period.	Increase cytokinin Cold store 4–8 weeks Run cytokinin/auxin grid Increase temperature Cold treat 3–8 weeks.

APPLICATIONS

Plant tissue culture has found an important place in various applications today. This section will provide the current perspective on the application of plants and their products. The plant products are fast emerging not only as an economical source but also as a more eco-friendly alternative.

Green Chemistry

The use of plants rather than petroleum to produce various chemicals is proving to be an attractive alternative. This has many advantages over the existing methods. First, global warming, which is largely attributed to burning of fossil fuels, will be reduced. This is possible owing to the growth of plants that fix atmospheric carbon dioxide, which can also be used for producing alternative fuel sources. Another major advantage is that the plants products will be biodegradable, which is crucial for sustaining the environment.

A major success story in this regard is the use of maize to produce ethanol. The production of millions of tons of ethanol using this method has established the credibility of this method beyond doubt. According to the current estimates, approximately 7% of the total maize produced in the USA is currently being utilized to produce 3.5 billion liters of ethanol.

A plastic with the trade name NatureWorks, which is produced as an alternative to synthetic plastics, is based on the fact that bacteria that are fed glucose derived from starch can produce a high concentration of lactic acids. The process is scaled up in a bio fermentor. After recovery, the lactic acid is polymerized to polylactic acid (PLA) that is used as an alternative for synthetic plastic. The biggest advantage is the biodegradable nature of the product.

PLA is used as plastic films that would serve as a replacement for polyethylene films that are currently being used to solarize vegetable production. In this method, the soil is first covered with this film to minimize evaporation. At the same time, it also protects the seeds from weeds and kills harmful nematodes. Since these films are biodegradable, they can be ploughed back into the soil at the end of the growing season.

Edible Vaccines

In humans and other higher organisms, the immune system performs a vital task of defending the body against pathogenic life forms. Pathogens are engulfed using various proteins produced by the immune system. These proteins are known as antibodies. Vaccination makes people immune by triggering the immune system to produce antibodies by simulating with greatly attenuated strain of bacteria or viruses.

In the early 1990s, Charles J. Arntzen from Texas A&M University conceived a way to introduce selected genes into plants and inducing the altered or transgenic plants to manufacture these antibodies. These edible plants used for this purpose, were locally grown using standard conditions. In addition to this, the logistics pertaining to the transport and storage of vaccines were also settled. By using this method, the plants do not require special conditions.

Edible vaccines contain DNA fragments from the original pathogen, such as the measles virus. These DNA fragments code for proteins that are normally located outside the virus and provoke the body's immune response. Various edible vaccines are being developed all over the world. The GM tomatoes are being developed in Cornell University as a vaccine against the Norwalk virus that causes severe diarrhoea. In Australia, scientists have developed a vaccine against measles in tobacco plants. Current efforts are concentrated at developing the vaccine in a more palatable form in rice. Table 7.7 shows a list of production of vaccines in plants:

TABLE 7.7 Production of Vaccines in Plants

Potential Application	Protein Expressed in Plants
Dental caries	*Streptococcus mutans* surface proteins
Cholera and *E. coli*	*E. coli* heat-labile enterotoxin
Hepatitis B	Hepatitis B antigen
Hoof-and-mouth disease	Hoof-and-mouth viral antigen
Malaria	Malaria B-cell epitope
Influenza	Haemagglutin
Rabies	Rabies virus glucoprotein

Production of both the chemicals as well as vaccines in plants is in its stage of infancy but is expected to grow as scientists unravel the complex metabolic networks in plant models. Using transgenic plants as edible vaccines will greatly reduce the cost of vaccines enabling the Third World countries to grapple some of the major diseases prevalent there.

Applications in Industries

The compounds derived from cultured plant cells are of special interests in various biotechnological programmes. However, in the recent past, most of the attention has centred on the production of secondary compounds, such as essential oils, food flavourings and colours that are used for industrial applications.

Agriculture and Biotechnology

It is hardly surprising in a country like ours that is predominantly agrarian, that biotechnology has been looked upon to offer solutions to various problems. In a very broad sense, there are two main areas where biotechnology has played a major role as far as agriculture is concerned:

- Improvement of genetic engineering techniques due to which the time for generating and evaluating new varieties has drastically reduced.
- Chemical and biotechnological processes have opened new avenues to the products of improved value from agricultural raw materials.

Needless to mention, plant tissue culture has played a major role in the development of agriculture and allied fields. Tissue culture technologies have been employed in order to bring about an improvement in the existing varieties and also create new improved plant lines. Some of the applications of plant tissue culture in agriculture are discussed in this section.

Improvements of Hybrids

The development of cell fusion and hybridization techniques has enabled plant breeders to produce improved varieties of plants in a short time span. Besides this technique, anther and pollen culture have also resulted in the development of hitherto impossible varieties of plants. The most famous example is that of the anther culture-derived rice and wheat in China. About 50 varieties of rice and 20 of wheat have been developed so far.

Production of Disease-Resistant Varieties

There are many plants that are routinely affected by virus, bacteria, fungi and nematodes. The infection is not confined to a single generation and results in periodic attacks and subsequent huge losses. Thus, disease-free crops are of paramount importance to improve the quality and productivity. Tissue culture techniques have solved this problem considerably. One example of the use of tissue culture in eliminating viruses was demonstrated in 1983 by Sood and Palni, for the plant Easter Lily (*Lilium longoflorum*).

The advantages of using the technique are:

- Host cells can be isolated without wounding.
- Control of pathogen and number of host cells.
- Ability to test the effect of change in medium on the pathogen.
- Rapid isolation and detection of the pathogens.

In-Vitro Selection of Cell Lines for Disease Resistance

To study the disease resistance in vitro, one of the most important considerations is the selection of suitable type of culture, that is, callus tissue, suspension culture, isolated cells or protoplasts. Callus cultures have been preferred over the other methods because:

- Of the ease to instantiate and subsequently maintain.
- Of the ability to add innoculum (spores) directly on the callus to avoid interference of the culture media.
- Of the ability to follow the progress of infection and colonization.

In 1980, Shepherd demonstrated an increased resistance to the leaf blight disease in *bajra* This marked a new era in the entry of genetic engineering into plant breeding.

Transgenic Plants and Applications

Plants in which foreign genes with desirable traits are introduced are called transgenic plants. The methods for introduction of foreign genes have been covered in detail in the earlier chapters.

One of the earliest examples of a commercially genetic engineered plant was the FLAVR SAVR tomato developed by Calgene Inc. of Davis, California. It was claimed that the FLAVR SAVR tomato remains on the vine longer to ripen to full flavour before harvest. To reach markets without being crushed, other tomato varieties are harvested while they are green and ripening is induced after shipment with ethylene gas, the natural ripening agent in tomatoes. The FLAVR SAVR tomato stays firm after harvest because, through gene manipulation, the production of the enzyme *polygalacturonase* (PG) is suppressed. PG breaks down the pectin found in cell walls, causing ripe tomatoes to soften. PG is suppressed by introducing a reverse copy of the PG gene called an antisense gene into the plant. As a result, FLAVR SAVR tomatoes have lower levels of PG allowing them to soften more slowly when ripe.

Apart from other applications, such as herbicide resistance and insect resistance, transgenic plants have attracted considerable attention due to their potential to act as bioreactors or reservoirs for the production of immunotherapeutic molecules. In 1989, Hiatt demonstrated that plants could be used to produce antibodies that can produce positive immunization.

In 1992, Arntzen expressed the hepatitis B surface antigen in tobacco to produce immunologically active ingredients using transgenic plants. Subsequently, various molecules including serum albumin, human interferon, human erythropoietin and murine IgG and IgA immunoglobins have been successfully expressed in plants.

The advantages of using plants for production of antibodies are:

- Transgenic material in the form of seed or fruit can be easily stored and transported.
- Several products can be created using various transgenic plants and by crossing them.
- Transgenic material from plants also has a greater shelf life, in some cases.

Terms to Know

adventitious	Developing from unusual points of origin, such as shoots or roots from callus or from leaf explants, or embryos from other sources than a zygote.
anther culture	Culture of anthers in tissue culture.
antioxidant	Product which is preferably oxidized; mainly used in case of browning problems, to avoid or to minimize the formation of polyphenolics.
bacteria free	The use of this term should be avoided and be replaced by "tested for a particular bacteria" (we do not have the adequate diagnostic tools at our disposal to prove cultures are completely bacteria free).
binary vectors	The use of two compatible plasmids for genetic engineering; one of them is a broad host-range vector containing the T-DNA border regions flanking selective markers and other foreign genes, the second plasmid carries the virulence genes (often called the helper plasmid) but has a large deletion in the T-DNA region.
callus	Unorganized growth of cells, does occur in nature for healing wounds, does occur in tissue culture as a proliferative mass of differentiated cells.
cell suspension	Culture of cells in a suspension; most often it is not really isolated cells but aggregates of up to 10 cells.
clonal propagation	Asexual propagation of plants that are considered to be Genetically uniform and originated from a single individual or explant; a synonym for clonal propagation in vitro is micropropagation.
contaminant	Unwanted organism invading cultures; differs from a pathogen in the sense that a contaminant is not necessarily responsible for diseases.
differentiation	Acquisition of specialized cell functions; the differentiated status is usually stable and in plant tissue culture the term is often taken to indicate the formation of shoots, roots or even other organs.
electroporation	Creation of transient pores in the plasmalemma, by means of an electrical current, for the purpose of introducing exogenous material, specially DNA, from the medium.
epigenetic	Selective gene expression, any change in phenotype which does not result from an alteration in DNA sequence; is inheritable in the sense that there is no

	transmission of the change to meiotic progeny but is heritable in the sense that it can be passed from one cell generation to the next; an example is switching off of genes by base methylation.
explant	Tissue or organ taken from its original site and transferred to an artificial medium for growth or maintenance.
hardening	Application of mild environmental stress to prepare the plant for more rigorous growing conditions.
heterokaryon	A cell possessing two or more genetically different nuclei in a common cytoplasm, usually as the result from cell to cell fusion.
inoculum	Tissues or organs taken from an established tissue culture and transferred to fresh medium for subculture.
meristem culture	Culture of a shoot apical meristem in view of avoiding contaminations, especially specific viruses and bacteria.
pectinase	An enzyme that hydrolyses pectin, a heterogeneous polysaccharide mainly found in the middle lamella between adjacent cell walls. Incubation of tissue with pectinase can produce individual, walled cells.
regeneration	Term used in plant cultures to indicate a morphogenic response that results in the neoformation of organs (roots, buds, embryos, . .).
primary metabolite	Essential metabolites (e.g., carbohydrates, proteins, lipids), components of the essential metabolic pathways.
somaclonal variation	Genotypic or phenotypic variations occurring in tissue culture, the latter being either genetic or epigenetic in origin.
secondary metabolite	A metabolite that is not part of the primary metabolic. Network of the cells. Primary metabolites are components of fundamental biochemical pathways (e.g., glycolysis) that are present in all cells. Secondary metabolites only appear in cells that are specialized in some way e.g., in defence.
suspension culture	A type of culture in which cells, or aggregates of cells, multiply while suspended in liquid medium.
totipotentiality	A cell characteristic in which the potential for forming all the cell types in the adult organism is retained, theoretically it means that each cell can give rise to a complete plant (vs competence).
vitrification	Formation of amorphous ice during the process of cryopreservation; formerly also used to describe the process of hyperhydricity.

8 Medical and Industrial Biotechnology: The Fine Blend

- Introduction
- Gene Therapy
- Bioproducts
- Biomaterials
- Nanotechnology
- Words of Caution

Introduction

Biotechnology has generated a lot of interest among academicians as well as the industrial fraternity. This has been largely due to developments in the fields of genetics and molecular biology. The ability to manipulate DNA has changed our perception of medicine, agriculture and environment management and control. A number of scientific breakthroughs in gene expression studies, protein engineering and cell fusion are currently being translated by a growing biotechnology industry into revolutionary new products and services.

Although new products and processes can be conceived and partially developed in laboratories bringing modern biotechnology into a tangible form, expanding it at the industrial level requires advanced engineering skills and know-how. The principle behind this is that biological molecules follow the laws of chemistry and physics, and therefore are easily amenable to engineering analysis and applications. This has resulted in the evolution of a new allied branch called bioprocess or bioengineering that has resulted in an avalanche of biomaterials and bioproducts.

Currently, bioprocessing is an essential part of food, chemical and pharmaceutical industries. Bioprocess operations employ microbial, animal and plant cells and components of cells, such as enzymes, to manufacture new products. There are a number of exhaustive steps that require close coordination and inputs from specialists of diverse fields to develop a complete bioprocess.

From stumbling upon plants that have healing properties, to mending broken bones and to developing vaccines and advanced procedures of surgery, mankind has come a long way with medicine and treatment. The promise of being able to manipulate human genetic material to treat incurable diseases has rendered new hope to scientists and patients alike. Today, many of us know that the origin of many diseases is defective genes.

The goal of applications like gene therapy is to produce normal gene products, or alternatively switch on or turn off the defective genes to regulate gene products and in turn control diseases that may arise as a consequence. The field of gene therapy like bioprocess is extremely multi disciplinary, consisting of researchers in life sciences, molecular biology, biochemists, physiologists and many others who track the origin of diseases and determine the functionality of various proteins.

The entire process is akin to a relay race where scientists involved in genetic engineering are constructing and manufacturing plasmids and virologists are modifying viruses. Chemists, on the other hand, are synthesizing newer better polymers and lipids, and the physiologists are studying diseases and their symptoms. The toxicologists then devise the correct therapeutic strategies by clinical trials and tests.

This chapter would try to put the two fields of bioprocess and medical applications in the right perspective for the reader in order to grasp the fine blend between the two. These have truly revolutionized science as we know it, by making it a more cohesive and concerted approach.

Gene Therapy

Gene therapy is a technique for correcting defective genes responsible for disease development. The various approaches that are being adopted as part of gene therapy combine one or many techniques. Most of the methods used are based either on the swapping of an abnormal gene for a normal gene through homologous recombination, repairing the abnormal gene(s) or, more recently, inserting a correct gene within the genome in a non-specific location.

Mode of Action

Most of the gene therapy protocols revolve around the use of a normal functional gene to replace the defective one. This is made possible by delivering the gene into the host. As has been elaborately described in the earlier chapter, this involves the use of a carrier called the vector. The vector molecule is a virus loaded with the normal gene and is targeted within the host. Delivery of the gene takes place as described earlier. Target cells such as the liver or lung cells are infected with the viral vector. This vector unloads the genetic material containing therapeutic gene into the target cell. A functional protein product is generated from the therapeutic gene that restores normal function of the gene. Figure 8.1 shows a flow chart depicting gene therapy.

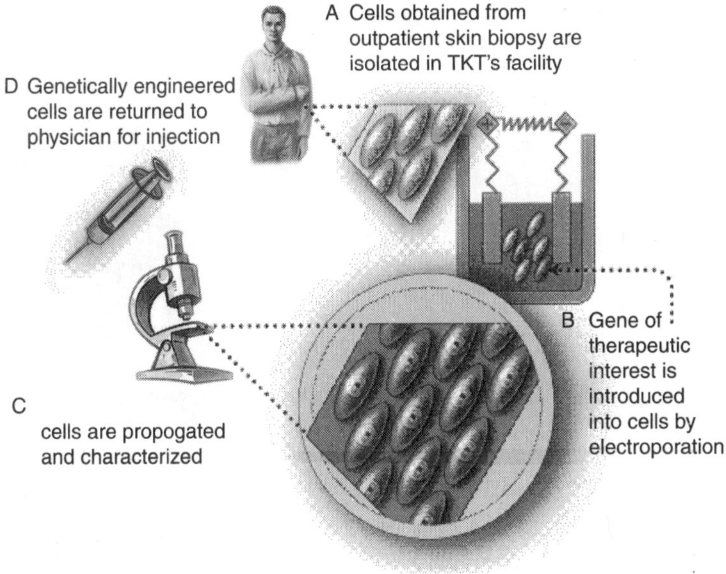

FIGURE 8.1 Gene Therapy

A variety of viral vectors are used for gene therapy. A notable few are shown in Table 8.1:

TABLE 8.1 Viral Vectors used for Gene Therapy

Virus	Description	Example
Retroviruses	Create double-stranded DNA copies of their RNA genomes that can be integrated into chromosomes of host cells.	Human immunodeficiency virus (HIV)
Adenoviruses	Double-stranded DNA genomes that cause respiratory, intestinal and eye infections in humans.	Common cold influenza virus
Herpes simplex viruses	Double-stranded DNA viruses that infect a particular cell type, neurons.	Herpes simplex virus type 1
Adeno-associated viruses	Small, single-stranded DNA viruses that can insert their genetic material at a specific site on chromosome 19.	

Apart from viral vectors, there are other options for gene delivery. The simplest method is the direct introduction of therapeutic DNA into target cells. This is, however, limited in its application because the technique is possible only for selected tissues and cells and requires large amounts of DNA. Figure 8.2 shows an example of a viral vector used for introducing therapeutic gene.

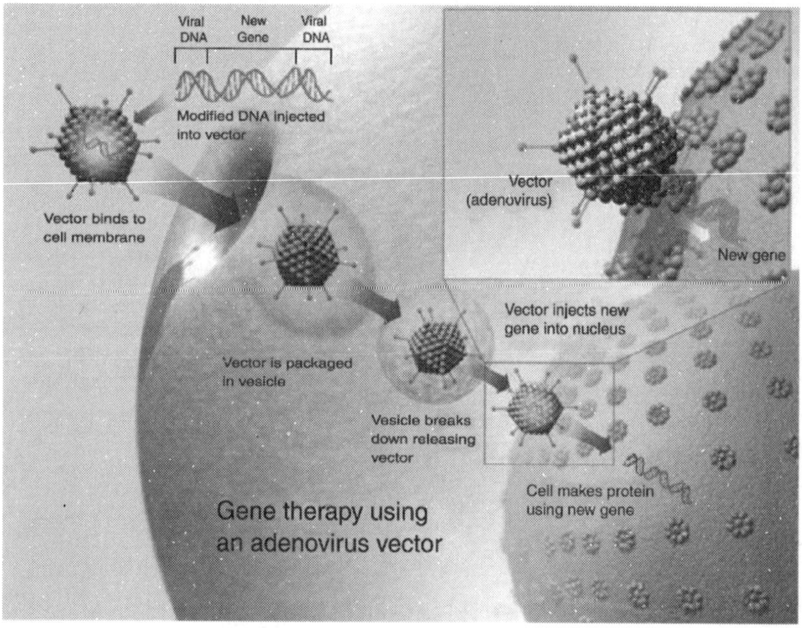

FIGURE 8.2 Viral Vector for Introducing Therapeutic Gene
Source: U.S. National Library of Medicine

Researchers also use an artificial lipid sphere with an aqueous core, called a liposome, which carries the therapeutic DNA and delivers it by passing through the target cell's membrane. This DNA can also be introduced into target cells by chemically linking the DNA to a molecule that will bind to special cell receptors. Once bound to these receptors, the therapeutic DNA constructs are engulfed by the cell membrane and passed into the interior of the target cell. This delivery system tends to be less effective than other options considering the complexity involved in the receptor-ligand interactions.

Synthetic Vectors

An inorganic particle, to which is bonded a cell binding component and a nucleic acid, is provided for delivery of a nucleic acid to a cell. The disclosed particle acts as a synthetic vector for achieving efficient transfection of associated nucleic acid into a cell.

Gene Expression Systems

Successful gene therapies depend on some important factors. It is important to choose a relevant therapeutic gene and the apt gene delivery system with the ability to express (turn on) the delivered gene. There are many promoters used for gene therapy applications. Some promoters are cell-specific and are used to target gene activity to specific cells, while others are expressed in a wide variety of cell types and may be used when cell-specific expression is not required. The required level of gene expression is crucial in promoter selection. Inducible promoter systems that can be turned on or off, in response to certain agonists are being investigated. These systems are touted to be self-regulating expression systems providing very high levels of expression. The variety of gene expression systems that are available and under development, expand the potential applications of gene therapy to new indications.

Routes of Administration

The therapeutic effect of the drug is largely affected by the route through which the desired gene is introduced. This is being currently done in cells that have been removed from the patient's body and the gene is introduced into the cell and subsequently transferred back into the patient. This is called the *ex vivo* gene therapy.

Another approach that is routinely being used is that of delivering the gene into the body and target specific tissues for delivery. The approach is called the *in vivo* gene therapy and is finding applications in the treatment of tumours.

Potential Benefits of Gene Therapy

The biggest advantage with this therapy is that it corrects the defect that causes the disease rather than merely alleviating its symptoms. This offers a long term and permanent cure for the disease, which is seen as a major advantage.

The fact that DNA is more stable as compared to proteins and other therapeutics used in treating diseases is exploited in case of gene therapy. The fact that a functional DNA, which replaces the defective one, is able to synthesize the desired product indefinitely does away with the need for recurring therapies and medications. This is especially beneficial for patients suffering from diseases like haemophilia and diabetes where the therapeutic proteins have to be taken daily or weekly.

Gene therapy can eliminate cancer cells without damaging normal, healthy tissue. This is advantageous over cancer therapies that are highly invasive and painful, resulting in the disruption of the patient's day-to-day life. Additionally, these strategies are not failure-free, which result in frequent remissions. As the genomics revolution identifies new genes and as gene delivery technologies continue to evolve, the role of gene therapy in the treatment of a wide variety of diseases is highly critical.

Current Status of Gene Therapy Research

Gene therapy products are yet to be cleared by the Food and Drug Administration (FDA), which has laid down certain basic criteria to be met for the same. A proposed gene therapy trial, or protocol, must be approved by at least two review boards at the scientists' institution from where the protocol is first reported. In the U.S. Gene therapy protocols must also be approved by the FDA, which regulates all gene therapy products. In addition, trials that are funded by the National Institutes of Health (NIH) must be registered with the NIH Recombinant DNA Advisory Committee (RAC). The NIH, which includes more than 20 institutes and offices, is the Federal focal point for biomedical research in the United States.

FDA regulations were further tightened due to the death of 18-year-old Jesse Gelsinger in 1999, who was a participant in a gene therapy trial for ornithine transcarboxylase deficiency (OTCD), and died from multiple organ failures 4 days after starting the treatment. His death is believed to have been triggered by a severe immune response to the adenovirus carrier.

FDA placed a temporary halt on all gene therapy trials using retroviral vectors in blood stem cells. After it learned that a second child treated in a French gene therapy trial had developed a leukemia-like condition. Both this child and another who had developed a similar condition in August 2002 had been successfully treated by gene therapy for X-linked severe combined immunodeficiency disease (X-SCID), also known as "bubble baby syndrome".

Ethical Considerations

Like any new technology, the use of gene therapy has also triggered a worldwide debate about the pros and cons of the same. Gene therapy is currently focused on correcting genetic flaws and curing life-threatening diseases. This technology is tightly regulated and its applications are limited to clinical trials involving volunteers. However, once the technology becomes mainstream, it will raise several questions that the society and the scientific fraternity will be forced to tackle.

One such question is related to the possibility of genetically altering human reproductive cells that pass genes on to future generations. Another question is related to the potential for enhancing human capabilities, be it stamina levels, or intelligence by means of genetic intrusion. The cost of the technology would ensure that only the rich and powerful have access to it. This would lead to the creation of a new super class within the society, laying the foundation for a new era of eugenics, where discrimination will be genetic in nature. These and many such issues continue to bog down gene therapy at present. This technology is certainly on the path to providing recourse to many of the health-related issues that plague the society. Table 8.2 shows the diseases that are currently treated using gene therapy.

TABLE 8.2 Diseases Treated Using Gene Therapy

Developmental and metabolic disorders	Cystic fibrosis, retinoblastoma
Neurological disorders	Parkinson's, Huntington's chorea and Alzheimer's
Infectious disease	AIDS, Malaria, other viral infections (including human papilloma virus)
Immune-mediated diseases and immunodeficiency	Severe Combined Immunodeficiency Disease, Rheumatoid Arthritis, Type 1 Diabetes and Multiple sclerosis
Acquired disorders	Heart disease, Cancer (including brain cancer)

Bioproducts

Bioproducts can be defined as any commercial or industrial product composed, in whole or in significant part, of biological products or renewable domestic plant, animal, marine, or forestry-based material, that generally does not contain synthetics, toxins, or environmentally damaging substances. Biomaterials, on the other hand, can be defined as materials with novel chemical, physical, mechanical, or "intelligent" properties, produced through processes that employ or mimic biological phenomena.

The production of bioproducts relies largely on the various instrumentations that are associated with the process. While isolating these on a small scale is relatively straightforward, scaling up the same at an industrial level is a difficult task that requires standardization.

Growing microbes on a laboratory scale in Petri dishes, test tubes and flasks using a suitable nutrient medium and the right conditions is both routine and easy. Producing substances such as antibiotics from microbes on an industrial scale is a more complicated exercise considering the massive numbers of organisms have to be grown for commercial use. This has been facilitated largely due to the design and manufacture of huge containers called fermenters that are capable of supporting the growth and propagation of bacterial cultures on a large scale. Figure 8.3 shows the schematic of a typical fermenter.

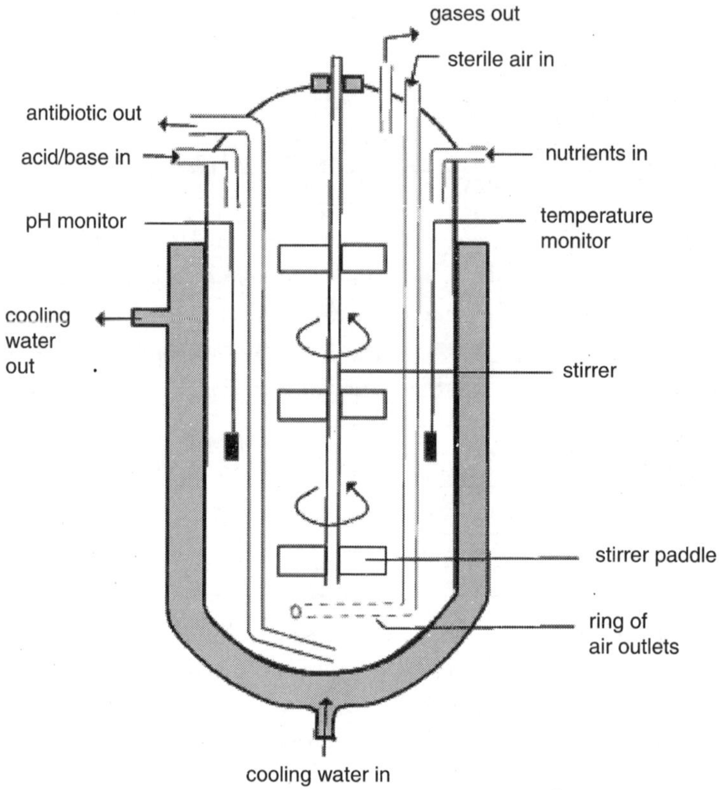

FIGURE 8.3 A Typical Fermenter

A fermenter consists of a large stainless steel cavity filled with a sterile nutrient solution that is inoculated with a pure culture of selected microorganisms. Paddles called agitators rotate the mixture to mix and homogenize the contents. Nutrients are added periodically to replenish the microorganism's requirements Probes monitor the mixture and changes in various parameters, such as pH, oxygen concentration and temperature, which are all controlled by highly advanced software that is integrated with the set-up. Since fermentation is a heat-generating process, a water jacket surrounding the fermenter containing fast flowing cold water is used to cool the fermenter. Air and residual gases generated during the process, produced by cell metabolism, leave the fermenter by an exhaust pipe.

There are two main types of cultures used in industrial processes: **batch cultures** and **continuous cultures.** These two essentially differ on various counts and are used depending on the ultimate application and product desired. The differences between the batch and continuous cultures are displayed in Table 8.3.

TABLE 8.3 Batch Culture and Continuous Culture

Batch Culture	Continuous Culture
Cells are grown in a fixed volume of liquid medium in a closed vessel.	Volume of suspension is kept constant.
No microorganisms, fluid or nutrients are added or removed from the culture during the incubation period.	Nutrients are added and cells harvested at a constant rate. Fermenter does not have to be emptied, cleaned and refilled very often.
These are primarily used for producing secondary metabolites, such as penicillin and other antibiotics, which are relatively unstable and not essential for the growth of the culture.	Production is almost continuous.
Secondary metabolites can be extracted economically only when they reach a high concentration in the culture.	Continuous cultivation needs sophisticated equipment to maintain constant conditions. Highly trained staff needs to operate the equipment. Therefore this process can be expensive.

The critical requirements for production of microbes in fermenters are:

- **Oxygen** for aerobic respiration of some microorganisms; others are strict anaerobes and oxygen must be excluded.
- **Carbohydrate** as an energy source for respiration to release energy needed for growth.
- **Nitrogen** for protein synthesis. **Ammonia** (NH_3) and **urea** ($(NH_2)_2CO$) are both widely used as sources of useable nitrogen.

Downstream Processing

After fermentation is completed for the necessary time, recovery and purification of the desired product is performed in a process called downstream processing. In the first stage, cells are separated from the liquid part of the suspension. This is done with techniques, such as sedimentation, centrifugation or filtration. In cases where the cells themselves are the desired product (e.g., for single-cell protein production for animal feed), they have to be first sterilized, washed, dried and packaged. If the desired product is a chemical within the cells, these cells are broken apart to release the chemical and the cellular components removed. The desired chemical is then extracted and purified by a number of techniques, such as precipitation and chromatography. In the last stage, the purified chemical has to be dried and packaged in a suitable form.

Brewing

Brewing is a natural process in which yeast (*Saccharomyces cerevisiae*) reacts with the sugars in fruit or malted barley to produce ethanol and carbon dioxide. In the natural environment, yeast competes with bacteria for the available sugar in wild fruits. This is achieved by the production of ethanol during fermentation. Ethanol is both toxic to the bacteria and a byproduct of the yeast metabolism. Fermentation takes several days or weeks and results in a product with maximum alcohol content—12%—above which the yeast itself is killed. Most alcoholic beverages are made by distilling raw brew.

Yeast is also mixed with flour and water to make bread. Hence, brewing and baking are closely related. In baking, the carbon dioxide is the important product, since it makes the dough rise. The ethanol evaporates in the baking.

Manufacture of Beer

The main ingredients of beer have remained constant over centuries. It is the methodology of mixing and matching these components in different proportions for varying time intervals that makes one beer different from the other. The process at the industrial level takes anywhere between 15 days to a month. Brewing takes nine and a half hours, while fermentation and aging combined take between 21 and 35 days for ales and lagers, respectively.

The major ingredients and processes while making beer are:

- Water: is one of the most crucial ingredients in good beer and a lot of attention is paid toward the source and purification of the brewing water. The water used in brewing is purified with proper calcium and acidic content, for the maximum activity of enzymes in the mash.
- Malt: Barley is used to make brewers' malt. Barley is soaked, germinated, then dried and/or kilned/roasted to arrest further growth. During the period of controlled growth in the malting plant, specific barley enzymes are released to break down the membranes of the starch cells that make up most of the kernel. These changes are largely intrinsic and apart from a slight change in colour, the external characteristics

remain essentially unchanged. In order to keep the husks as whole as possible, the malt is screened and crushed rather than ground to flour in the brewery. This process not only prevents the extraction of undesirable materials from the husks but also allows them to act as a filter bed for separation of the liquid extract formed during mashing.

- Mashing: Malt is added to heated, purified water and, through a carefully controlled time and temperature process, the malt enzymes break down the starch to sugar and the complex proteins of the malt to simpler nitrogen compounds. Mashing takes place in a large, round tank called a "mash mixer" or "mash tun" which is temperature controlled. At this point, depending on the type of beer desired, the malt is supplemented by starch from other cereals such as corn, wheat or rice.
- Lautering: The mash is transferred to a straining (or lautering) vessel, which is usually cylindrical with a slotted false bottom two to five centimetres above the true bottom. The liquid extract drains through the false bottom and is run off to the brew kettle. This extract, a sugar solution, is called "wort" but it is not yet beer. Water is "sparged" (or sprayed) though the grains to wash out as much of the extract as possible.
- Boiling and Hopping: The brew kettle, which is a huge vessel, is usually made of shiny copper or stainless steel. It is fitted with coils or a jacketed bottom for steam heating and is designed to boil the wort under carefully controlled conditions. Boiling, which usually lasts about two hours, serves to concentrate the wort to a desired specific gravity, to sterilize it and to obtain the desired extract from the hops. The hop resins contribute flavour, aroma and bitterness to the brew. Once the hops have flavoured the brew, they are removed. Undesirable protein substances that have survived the journey from the mash mixer are coagulated, leaving the wort clear.
- Hop Separation and Cooling: After the beer has taken on the flavour of the hops, the wort then proceeds to the "hot wort tank". It is then cooled, usually in a simple-looking apparatus called a "plate cooler". As the wort and a coolant flow past each other on opposite sides of stainless steel plates, the temperature of the wort drops from boiling to about 10 to 15.5°C, a drop of more than 65.6°C, in a few seconds.
- Fermentation: The wort is then moved to the fermenting vessels and yeast (usually a highly purified and engineered strain) is added to it. Apart from breaking down the sugar in the wort to carbon dioxide and alcohol it also adds many beer-flavouring components. Two species of yeast belonging to the genus *saccharomyces* are used in the process. One yeast type, which rises to the top of the liquid at the completion of the fermentation process, is used in brewing ale and stout. The other, which drops to the bottom of the brewing vessel, is used in brewing lager.

During fermentation, which lasts about 7 to 10 days, the yeast may multiply six-fold and in the open-tank fermenters used for brewing ale, a creamy, frothy head may be seen on top of the brew. When the fermentation is complete, the yeast is removed. For one to three weeks, the beer is stored cold and then filtered once or twice before it is ready for bottling or "racking" into kegs. The draught beer is placed in sterilized kegs ready for shipment. Figure 8.4 depicts, with the help of a flow chart, the different stages in beer manufacture.

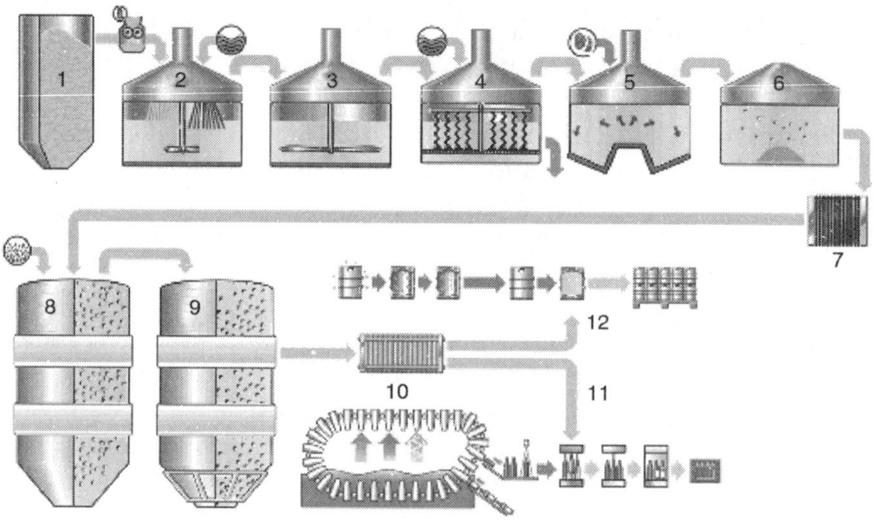

1. malt-silo
2. mash vat
3. mash pan
4. clarification vat
5. wort pan
6. whirlpool
7. wort cooler
8. fermentation tank
9. lagering cellar
10. beer filtration
11. filling in bottles
12. filling in kegs

FIGURE 8.4 Different Stages in Beer Manufacture

Types of Beer

This section will detail on some of the different types of beers that are prepared.

Lager

Derived from the German verb "lagern", which means "to store", this is one of the most popular beers among Europeans. It originated in the late middle ages, when fermentation was a matter of chance rather than technology. To ensure a supply of beer for the summer when fermentation was dicey, brewers in the Bavarian Alps stored kegs of spring brew in icy mountain caves. As the beer slowly aged, the yeast settled, creating a drink that was dark but clear and sparkling with a crisper, more delicate flavour. In 1842, lager acquired its familiar golden colour when a brewery in Pilsen, Czechoslovakia perfected a pale, bottom-fermented version of the beer. Lagers typically take more time to brew and are aged longer than ales.

Bock Beer

The other bottom-fermented beer is bock, named for the famous medieval German brewing town of Einbeck. Heavier than lager and darkened by high-coloured malts, bock is traditionally brewed in the winter for drinking during the spring.

Ale

The term covers a variety of styles retaining some hallmark characteristics. Top-fermentation and the inclusion of more hops in the wort give these beers a distinctive fruitiness, acidity and a pleasantly bitter seasoning. All ales typically take less time to brew.

Dry

"Dry" refers to the amount of residual sugar left in a beer following fermentation. This type of beer is fermented for longer than normal brews so that practically all of the residual sugar is converted into alcohol.

Winemaking

The process of winemaking involves a number of steps:

1. **Determining the pick date:** The decision to harvest grapes depends many factors; weather being the most important. Extreme heat or cold can have a detrimental effect on the quality of grapes that would determine the quality of the wine. The ripeness of grapes is analysed and judged. It is then checked for sugar level, acidity levels, pH, colour, health, seed colour, aromas and flavours.
2. **Picking ripe grapes:** Once the decision to pick has been made, containers are placed into the vineyard and picking begins. It is important that the grapes arrive freshly picked. The grapes are then visually checked and weighed.
3. **Destemming:** The fresh grapes are tipped out of the bins and placed into the destemmer. As the name suggests, this machine removes stems because there are astringent, bitter and tannic compounds in the stems that can transfer to the juice or wines if left in contact too long. Once destemmed, the grapes are pumped to one of three vessels (including the press) depending on the grape variety and style of wine to be made.

4. **Skin Contact:** White aromatic grape varieties receive up to 24 hours of skin contact. This affects a greater release of phenolics (secondary metabolites) and other compounds from the skin of the grape. (It is in the skin that most of these aromas and flavours are stored.) With red wine varieties, the skin also contains the colour pigments. To release the colour, the red wine grape varieties are actually fermented in their skins. A combination of heat, alcohol and natural acids extract colour from the grape skins.
5. **Pressing:** At some point, all grapes pass through the press. The function of the press is to extract the grape juice or wine from the skins and seeds of the grape. In the case of non-aromatic varieties or champagne-base wines, the grapes are either destemmed first or placed directly into the press. In the case of aromatic varieties, they are pumped directly into the press after 24 hours of skin contact. Fermented or fermenting red wine grapes are either pumped or dumped from bins into the press. The first juice or red wine to leave the press is classified as the free-run because it leaves the press without any mechanical pressure being placed on it. This juice or wine is always separated and represents the best portion of the pressing. Depending upon the wine style and winery, the second pressed juice might be kept separated and treated for high phenolics or blended away.
6. **Cold Settling:** White wine "must" is cold-settled and racked fresh from the press. This removes any grape skin particles that might have traveled with the juice from the press. The result is a clean grape must which is ready for fermentation.
7. **Fermentation:** The process of fermentation involves the metamorphosis of grape juice to wine. Yeast creates the wine by metabolizing the grape sugars. It digests and breaks down the sugars, which then form into carbon dioxide and alcohol. Like in the case of beer pure cultured yeasts are selected for their influences on each individual grape variety.
8. **Racking:** Once fermentation is complete, racking begins. Racking is the process in which the yeast that has settled to the bottom of the barrel is removed. The wine is pumped out of the barrel or tank, leaving the yeast behind in the bottom of the container. This process may take place several times before the wine becomes clear. Racking can take place a week or years after fermentation, depending on the style of wine that is being made.
9. **Fining:** In this process, unwanted natural substances, like excess protein or tannins, are removed. These substances are removed because they may cause the wine to be cloudy or may give it an unbalanced taste.

10. **Cold Stabilization:** This process makes the wines "tartrate stable". In wine, there are two main types of acid; malic and tartric. These acids are in abundance in the grape juice, but become less soluble in the presence of alcohol. The higher the alcohol, the less soluble the acid. Potassium and calcium combine with the malic and tartric acids, producing a white crystal. If left uncontrolled, these crystals can form and precipitate in the wine bottle instead of the tank.
11. **Filtering:** This process takes place in many different ways. Some wines are ready to bottle right from racking and aging. Others require diatomaceous earth (DE) filtering, pad filtering or micron filtering. DE filters are very gentle and work on a sieve principle. Filter pads work on absorption and micron filters, which filter out extremely small particles, also work on the sieve principle. Micron filters are also known as police filters because they screen bacteria and yeast cells.
12. **Bottling:** Filling the bottles, corking, capsuling, labeling and palletizing are the last steps to finishing the wine. Red wines often mature for years before being sold. White wines tend to come to market earlier. Wine is made up of over a thousand different minerals, vitamins and other components and once in the bottle, the wine continues to mature and improve.

Vinegar Production

It was Louis Pasteur, who in 1841, at the behest of emperor Napoleon started investigations into the conversion or "spoilage" of wine. He showed that bacteria act on the wine and convert it into a product called vinegar. To make vinegar, wine is slowly pored over oak chips in a tall tower, open to the air. Bacteria (*Acetobacter*) on the wood oxidize the ethanol in the wine and turn it into ethanoic acid or vinegar, releasing energy in the form of heat. If the vinegar is made from fermented raisins and stored in oak vats then the sweet, highly prized *Balsamic vinegar* is produced that is a famous product of Italy.

Yoghurt Production

Yoghurt is the result of controlled fermentation of milk by two species of bacteria (Lactobacillus sp. and Streptococcus sp.). The sugar in milk (called lactose) is fermented to lactic acid and causes the curd to form. The acid also restricts the growth of food poisoning bacteria and some spoilage bacteria. Milk is a potential source of food poisoning and only has a shelf life of a few days. Yoghurt, however, is safer and can be kept for up to ten days, under proper storage conditions. Yoghurt can be either stirred or set. Stirred yoghurt is prepared by fermentation in bulk, stirred and then dispensed into pots or poured into customers' containers. Set yoghurt is made by pouring the inoculated milk into pots and fermenting it. Fruit and nuts can be added to each type but care is needed to ensure that they are thoroughly cleaned and blanched to avoid contamination. In some countries, a layer of fruit syrup on the top of set yoghurt is a popular alternative. In other places, thicker, stirred yoghurt is preferred. This can be made by adding dried skimmed milk to the milk before pasteurising. The steps for the commercial production of stirred yoghurt are:

1. Heat milk to 85–95°C for 15–30 minutes;
2. Homogenize milk;
3. Cool milk to incubation temperature;
4. Add bacterial starter;
5. Incubate mixture at 40–46°C for 4–5 hours (until pH has fallen to around 4.3);
6. Stir thickened yoghurt and cool it to 5°C;
7. Add fruit, flavourings and colorants (if required);
8. Pack and despatch yoghurt, keeping it refrigerated.

Leather

The first step while using leather for manufacturing is called softening. It has to be bated, which means that some of the protein fibres are removed. By the early 1900s, it was known that the animal waste was rich in bacteria that produced proteases, which degraded part of the leather. It was a highly skilled job to prevent the enzymes damaging the leather, which is largely made up of protein. The German scientist Ršhm, developed a standardized *bate* in 1908, based on an extract from the pancreases of slaughtered animals. This contained trypsin, of a mixture of enzymes found in the digestive system. Since then, all bates have been based on enzyme preparations, with bacterial and fungal origin.

The Roman writer Pliny reported the use of pigeon droppings for this process, over 2,000 years ago. Later, leather was bated by smearing it with dog excreta.

Washing Powders

Ršhm was the first to examine the chemical composition of dirt on laundry and he thought of using pancreatic extracts to wash clothes. He patented his idea and in 1914, developed the first enzymatic washing agent. It was so effective that only a small quantity was required, and was sold as a spot remover. However, the product was greeted with widespread speculation, since washing powders produced lots of lather. In 1915, some people even thought it was a hoax. The product was investigated by scientists who found that it really did work. However, it was only in the 1960s that enzymatic detergents gained widespread acceptance. Today, the detergent industry is huge and various advanced techniques, such as enzyme immobilization are being used to come up with even better and more effective enzymes.

Cheese Manufacture

To prepare cheese, warm milk is mixed with enzyme rennin (*rennet*) (in a ratio of 2000:1). This mixture is allowed to react for several hours. The *caesinogen* in the milk clots to *casein* and solidifies the milk into curd. The whey (liquid) is drained and the curd is chopped, salted and placed in a mould before squeezing to remove trapped air. This process is known as 'cheddaring', hence the name "cheddar cheese".

The enzyme rennin used to be extracted from (dead) calves' stomachs. However, today, this enzyme is produced from bacteria. Rennin, in young mammals, clots mother's milk in the baby's stomach. This tricks the stomach into keeping the contents there for several hours, allowing protein digestion. Most mammals stop producing rennin after weaning, but Caucasians, who drink milk throughout life continue to produce the enzyme even as adults!

Cheeses that undergo a ripening process are treated with rennet. This ripening process is also known as a curing, maturing or ageing. However, before this procedure, the cheese goes through a number of steps. The general sequence is as follows:

1. Warm the milk;
2. Add a starter culture and rennet;
3. Milk coagulates into single huge curd;
4. Stir the curd and cut to drain off whey;
5. Heat the curd, and press to remove more whey;
6. Mould and shape the curd into a cheese;
7. Salt the curd before and after step 6;
8. Allow the cheese to mature under controlled conditions.

Sometimes, the cheese is dipped in brine or a solution of fungal spores to inoculate it and produce a surface 'rind'. These rinds of cheese are formed during the ripening process. Rinds may be brushed, washed, oiled, treated with a paraffin wax coating or simply not touched at all. Traditional Cheddars are wrapped with a cotton bandage. The rind protects the interior of the cheese and allows it to ripen. The rind affects the final flavour of the cheese. Heavily salted cheeses develop thick, tough outer rinds, characterized by the Swiss range of cheeses. Cheddar, another natural rind cheese, is less salted than the Swiss varieties and consequently has a much thinner rind.

There are many types of cheese available today. Some of them are:

- **Fresh, unripened cheeses:** such as cottage, cream and ricotta. These cheeses do not undergo a maturing process. They are the mildest, least flavoursome of all cheeses, and are usually unsalted.
- **Swiss-style cheeses:** usually have tough hard rinds and interiors dotted with holes. The most famous is Emmentaler with its distinctive 'eyes' and characteristic mild, sweetish, nutty flavour. These holes are caused by expansion of gas within the cheese curd during the ripening period. Relatives of Emmentaler include Appenzell, Gruyere, Royalp, Sbrinz and Raclette.
- **Cheddar cheeses:** such as Gloucester, Cheshire, Leicester, Lancashire, Derby, Wensleydale and Caerphilly are some of the most popular and widely copied cheeses in the world. Cheeses that are 'cheddared' are cut into pieces and stacked and turned at the bottom of the cheese vat for a period. The archetypal Cheddar cheese is firm textured, yellow in colour with a clean, mellow taste that develops a sharp and tangy bite, the further it matures. Apart from the aforementioned English cheeses, America, Australia, Canada and New Zealand make varieties of Cheddar in varying degrees of strength. Vermont is one of the best-known American types and the Canadian Cheddars, made from unpasteurised milk, are highly rated. Cantal is a French Cheddar-style cheese and Kashkaval is a distant Greek relative made from sheep's milk.
- **Blue veins:** cheeses are characterized by their internal veining of blue, blue-black or green and their pungent aromas and tangy flavours. All blue-vein cheeses are internally ripened after being inoculated with a Penicillium spore. They are usually classified as soft cheeses but can be crumbly in texture. Some, however, are exceptionally soft, even to the point of spreadability. Roquefort, Stilton and Gorgonzola are considered the three best blue veins in the world.
- **Whey cheeses:** such as Gjetost, are made from the whey. It has milk sugar (lactose) and cream added to it during the making process. Ricotta is also made from whey.
- **Strong-smelling cheeses:** such as, the (in)famous Limburger, possess a powerful aroma usually of the washed-rind variety. Limburger is Belgian in origin but often thought of as German due to its great popularity in that country. Maroilles and Livarot are other strong smelling and strong tasting cheese varieties.

- **Processed cheese:** has its ripening process arrested at a given point by heat treatment. It is usually made from one or two cheese types blended together and can never develop individual flavour of natural cheese because the microorganisms that create them are effectively killed off. Processed cheeses are enormously popular partly because of their keeping properties, their economy and their blandness. Today, there is an enormous range of processed cheese on the market, flavoured by a variety of ingredients, ranging from salmon to pineapple.

BIOMATERIALS

In the last century, man-made materials and devices have been used successfully to replace body parts and systems in human bodies. These materials function as part of the body without being recognized as foreign and are called biomaterials. Some of the salient features of a biomaterial are:

- It displays traits of biocompatibility and must not elicit negative response from the immune system.
- It must display biodurability, resisting wear or corrosion.
- It should enhance the healing process.

Figure 8.5 shows the application of biomaterials in the human body.

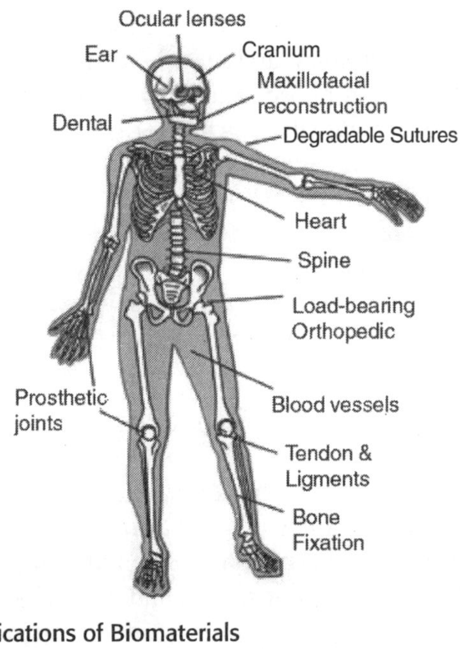

FIGURE 8.5 Applications of Biomaterials

Biomaterials have been used in various applications, from healing fractures to filling and replacing dental cavities. The main issues regarding the use of the biomaterials have been the materials that have been used to manufacture them. The three areas of concern are design, material selection and biocompatibility. With major advances taking place in the medical field, the average life expectancy of individuals has been increasing steadily. This has, however, resulted in an increase in degenerative diseases, particularly among the aging population. Today, organs, joints, and other critical body parts that are worn out must be replaced if people are to maintain a good quality of life. Biomaterials play a major role in replacing or improving the function of every major body system.

Bioactive Materials

In cases where biomaterials are introduced into the body, there are various ways in which the immune system can react based on the tissue and the material. Depending on the tissue response, they elicit materials that can be classified into inert, bioresorbable and bioactive. Inert materials, such as Titanium and Alumina (Al_2O_3) are nearly chemically inert in the body and exhibit minimal chemical interaction with adjacent tissues. A fibrous tissue capsule will normally form around inert implants. This morphological fixation is not ideal for the long-term stability of permanent implants and often becomes a problem with orthopedic and dental implant applications. Bioresorbable materials, such as tricalcium phosphate and polylactic-polyglycolic acid copolymers, are designed to be slowly replaced by tissue (such as bone) or for use in drug-delivery applications. Certain glasses, ceramics, and glass-ceramics that contain oxides of silicon, sodium, calcium, and phosphorus (SiO_2, Na_2O, CaO, and P_2O_5) have been shown to be the only materials known to form a chemical bond with bone, resulting in a strong mechanical implant/bone bond. These materials are referred to as bioactive because they bond to bone (and in some cases to soft tissue) through a time-dependent, kinetic modification of the surface triggered by their implantation within living bone. Biomaterials can be broadly classified on the basis of the areas where they are used most often.

Orthopaedic Biomaterials

In cases where physical therapy, nonsurgical treatments, or surgical repairs, fail to achieve the desired results, orthopaedic surgeons often advise joint replacement surgery in which the deteriorated joint is removed and replaced with a man-made device. These devices are called as Orthopedic Biomaterials. Artificial joints consisting of a plastic cup made of ultrahigh molecular weight polyethylene are placed in the joint socket, and a metal (titanium or cobalt chromium alloy) or ceramic (aluminum oxide or zirconium oxide) ball affixed to a metal stem. This type of artificial joint is used to replace hip, knee, shoulder, wrist, finger, or toe joints to restore function that has been impaired as a result of arthritis or other degenerative joint diseases or trauma from sports injuries or other accidents.

Heart Valve Materials

The most widely used and successful implants consisting of biomaterials have been in case of the artificial heart valves. Although the product suffered early setbacks due to the design and choice of materials, it has emerged as a major success story. Figure 8.6 shows a model of an artificial heart valve.

FIGURE 8.6 Model of Artificial Heart Valve

Currently, two main types of materials are used for artificial heart valves: soft bioprosthetic materials, such as denatured porcine aortic valves or bovine pericardium, and hard man-made materials used in mechanical heart valves. While both have had potential complications, the latter have had a relatively higher degree of success. Bioprosthetic valves have been reported to show high failure rates due to calcification (calcium from the blood stream forms deposits on the implant), which results in mechanical dysfunction. These are also susceptible to mechanical fatigue. The principal problem with mechanical heart valves is thrombosis, or haemorrhage.

NANOTECHNOLOGY

Many scientific and technological advances have been made recently, that depend on the properties of materials at a very small scale. Such technological advances are known as 'nanotechnology', the prefix 'nano-' being derived from the Greek word '*nanos*', which means dwarf. Nanotechnology involves lots of complex concepts, which are not easily seen or understood, making the consequences of developments disputable. Nanotechnology is the area of science and technology where extremely small structures play a critical role. Nanotechnology basically occurs at the order of a nanometer (nm), i.e., from 0.1nm to around 100nm. One nanometer equals one billionth of a metre, which is thousandth of a millionth of a metre. In biological terms a nanometer is about:

- width of a DNA molecule
- size to which an average humans fingernails grow in a second
- 1/80,000 the size of a human hair

Table 8.4 shows the milestones in the field of nanotechnology. This table has been compiled from *http://www.ptei.org/ Pittsburgh Tissue Engineering Initiative*

TABLE 8.4 Milestones in Nanotechnology

Year	Event	Description
1857	Michael Faraday discovers colloid gold	Faraday introduced 'colloidal gold' samples to the Royal Society. This suspension of gold nanoparticles in solution was transparent in some lighting, but in other lighting conditions could produce differently coloured solutions of 'ruby, green, violet or blue' (*Philosophical transactions of the Royal Society*, 1857, 147, 145).
1905	Albert Einstein explains existence of colloids	Einstein provided a thoroughly quantitative theory for the state of a colloid dispersion. He considered colloids to behave as 'big atoms' and explained their movement in terms of Brownian motion.
1932	Langmuir discovers layers of atoms one molecule thick	Langmuir established the existence of monolayers (layers of atoms or molecules one atom thick). These monolayers have peculiar two-dimensional qualities, and led to the development of a totally transparent glass produced by forming a thin film of fluorine compound on the surface. He was awarded the Nobel prize in 1932 for this work on thin films.

(Continued)

TABLE 8.4 *(Continued)*

Year	Event	Description
1981	IBM invents a machine that can move single atoms around	Gerd Binning and Heinrich Rohrer invented the Scanning Tunneling Microscope (STM) at IBM. This microscope allows atomic-scale three-dimensional profiles of surfaces to be obtained. The microscope relies on a tip that is positioned within 2nm of the surface and measures the electron density of the surface. They were awarded the Nobel prize in 1986 for this work.
1985	A new form of carbon is discovered: C60	Richard Smalley, Robert Curl and Harold Kroto discovered C60 while investigating the outer atmosphere of stars, for which they were awarded the Nobel Prize in 1996. Officially known as buckminsterfullerene, C60 is more commonly known as a buckyball as the 60 carbon atoms are arranged into a sphere made of 12 pentagons and 20 hexagons (like a football).
1990	IBM demonstrates ability to control position of atoms	IBM research scientist Don Eigler showed that the position of atoms could be controlled precisely.
1991	Carbon nanotubes discovered	Sumino Iijima discovered a process to make 'graphitic carbon needles ranging from 4nm to 30nm in diameter and 1 micron in length'. The needle-like tubes he described consisted of multiple sheets of graphite rolled into hollow tubes, which have now become known as carbon nanotubes. In 1993 the first single-walled nanotubes (SWNT) were produced (*Nature* 354, 1991, 56).
1993	First high-quality quantum dots prepared	Murray, Norris and Bawendi synthesize the first high quality quantum dots of nearly monodisperse CdS CdSe and CdTe. Quantum dots are very small particles with interesting optical properties: they absorb normal white light and, depending on their size, emit a range of bright colours. This property arises directly from the very small size of the particle (*Journal of the American Chemical Society*, 1993, 115).
1997	Nanotransistor built	Lucent Technologies fabricated the 'nanotransistor' –a complete metal oxide semiconductor transistor.

(Continued)

TABLE 8.4 *(Continued)*

Year	Event	Description
2000	DNA motor made	The first DNA motor was created by Lucent Technologies with Oxford University. These devices are similar to motorized tweezers and have the potential to make computers 1,000 more powerful than today's machines. The hope is that DNA motors can be attached to electrically conducting molecules to assemble rudimentary circuits by acting as switches (*Nature* 406 (6796), 2000, 605–608).
2001	Prototype fuel cell made using nanotubes	Nanohorns, irregularly shaped nanotubes, were developed as fuel cell for hydrogen-based fuel such as methanol. They group together creating a high surface area ideal for catalysts.
2002	Stain-repellent trousers reach the high street	Clothing embedded with nanoparticles that produce a stain-repellent coating has been developed. Nano-care™ khakis have the fabric fibres coated with nanowhiskers 10–100nm in length. This new stain-repellent fabric is available from a number of high street retailers and is available in trousers, shirts and ties.
2003	Prototype nano-solar cells produced	Prototype solar cells have been made by Nanosolar Inc. in California. They use conducting polymers and nano-based particles. This technology has great advantages, compared to that for traditional silicon-based solar cells, including making the products much cheaper and easier to make. These cells are also produced in flexible sheets, making them suitable for many applications.
2004	Research and development continues to advance	Research and development in many nanotechnology fields continues apace; some recent developments include: Nanospectra Bioscience has used gold-coated nanoshells to destroy cancer tumours in mice (*Cancer Letters*, 209, 171). NanoScale Materials Inc. have developed a family of non-toxic nano-engineered products that counteract a variety of chemical warfare agents and toxic chemicals.

Nanotechnology-Derived Products

This section details various products that are derived from nanotechnology.

Colour Displays

Organic Light Emitting Diodes (OLED) are nanostructured polymer films, less than 500nm thick which can be used as lighter, thinner, more flexible, and less power-consuming colour displays. Companies like Cambridge Display Technologies, Universal Display Corporation and Kodak are developing the films for use in a whole range of flat panel displays (e.g., cameras).

Solar Panels

Solar cells using conducting polymers and nano-based particles are being produced by Nanosolar Inc. in California. The polymer and nanoparticles are mixed and then printed on to a polymer surface a 200nm thick layer. It produces cheap, flexible and easier-to-make solar cells. Figure 8.7 depicts a solar panel.

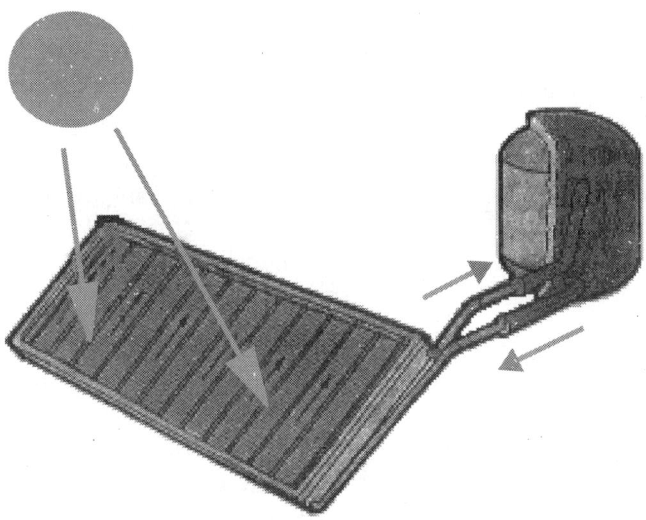

FIGURE 8.7 Solar Panel

Sunscreen

65nm particles of titanium oxide are being used in new sunscreens. These particles, made by companies like Oxonica, absorb UV light for longer with significantly less free radical formation (which leads to cell damage and skin ageing) than existing sunscreens.

Nanomagnetic Particles

An example of how nanotechnology is offering a new means of diagnosing and actively treating cancers is through the introduction of nanomagnetic particles into tumours. Under the influence of a magnetic field, these particles heat up and dissolve the tumour cells, which are resorbed into the body.

Quantum Dots

Quantum dots are nanoparticles which, being smaller than the wavelength of visible light cannot be seen. They can impart new properties while remaining invisible themselves. The quantum dots luminesce under ultraviolet light, with the size of the dots controlling its colour, e.g. 2nm quantum dots luminesce bright green, while 5nm quantum dots luminesce red.

They fluoresce or stay lit much longer than dyes conventionally used for tagging cells. Dots are tagged to proteins and their glow enables the identification of specific proteins or DNA making it possible to diagnose various diseases. Researchers can tell how much protein is on each cell by the amount of light transmitted in a particular colour. A change in the concentration of proteins on each cell is an early indication of cancer. This is a great leap forward as using current techniques, by the time the doctor can see the physical changes in cells that indicate cancer, it is often too late and the cells may be damaged and cancer may have spread. When DNA in a test sample binds with a specific DNA on a quantum dot probe, the sample fluoresces under UV radiation and this signal can then be analysed. This technique can be designed to detect genes of harmful pathogens ranging from staphylococcus to anthrax.

Drug Delivery Systems

Drugs can be encapsulated in a variety of carriers. The carrier can be a carbon nanotube, made of materials other than carbon, having a structure like a silicon wafer with antibodies or some other molecule that bind to the drug.

Nanoparticles enter damaged cells and release enzymes that initiate "apoptosis". Alternatively they can release enzymes to try to repair the cell and return it to normal functioning. Drugs in nanocrystalline form can be administered in smaller doses because they can be delivered directly to the tissue and in controlled doses related to the patient's requirements.

Nanoparticles as Drug Carriers

Encapsulated drugs can be protected from degradation. The drugs may be in particles from microns to millimeters in size, with coatings only a few nanometers in thickness. The drugs are normally ingested or implanted and designed to deliver a controlled release of drug, which may last for many months and can be activated at different sites in the body.

Nanoporous Membranes

Nanopores can act as tiny turnstiles for releasing drugs. By making the nanopores only slightly larger than the molecules of drugs, they can control the rate of diffusion of the molecules keeping it constant, regardless of the amount of drug remaining inside a capsule. Nanoporous membranes can also be used on implantable devices for the treatment of chronic diseases.

Words of Caution

The introduction of any new technology has always sparked debates that have covered diverse areas from economics to ethics and spirituality. While many of the arguments raised against emerging technologies result due to the lack of information and awareness, those that are justified have to be sorted out by the industry and those associated with the technology.

Biotechnology is an emerging technology and the manufacture of bioproducts and biomaterials has provided a major thrust to its acceptance and development. In case of bioproducts, issues that have been raised are concerning the processes that are used to obtain the end products. Plants are a major source of most bioproducts and the lure of quick money and large returns are driving farmers into cultivating these crops. This results in the deficiency of many important food crops. This is a disturbing trend considering that worldwide production of food crops has dwindled alarmingly in the past decade in spite of the increase in population. It is imperative that while reaping the harvests of biotechnology-driven bioproducts, one also has to take care that the basic demands of the populace are met. In addition to this, the unabated cultivation of such crops whose growth is assisted by fertilizers and pesticides will result in further pollution and reduce the land cover. Hence, some amount of caution has to be exercised vis-à-vis the use and production of bioproducts.

Biomaterials are a relatively new entrant in the biotechnology arena. In such a short span, biomaterials have shown a lot of promises that have projected it as the technology of the future. However, since it depends heavily on a number of allied technologies, one has to ensure that the amalgamation of these are closely watched and regulated. Ethical and spiritual questions have already been raised in argument. While the use of biomaterials is increasing in medicine, it is a fact that a number of issues regarding the compatibility and durability of these products are yet to be assessed. Gene therapy has shown immense potential in curing various, hitherto incurable ailments. However, the regulation and approval of such techniques is still being debated in many quarters and it will be a while before a consensus is reached on the same.

It is, therefore, up to the scientific community to adopt a measured approach and sensitize the common man about the pros and cons of new technologies. A lot will depend on it considering the fact that the success of any technology revolves around mass acceptance and not merely academic recognition.

Terms to Know

(*Adapted from http://www.ptei.org/* Pittsburgh Tissue Engineering Initiative)

Angiogenic	Which promotes or develops blood vessels; which promotes an increase in vascularization.
Biodegradable Polymer	A polymer that breaks down when placed in a biologic environment.
Bone Substitute	Synthetic or natural materials for the replacement of bones or bone tissue. They include hard tissue replacement polymers, natural coral, hydroxyapatite, beta-tricalcium phosphate and various other biomaterials. The bone substitutes as inert materials can be incorporated into surrounding tissue or gradually replaced by original tissue.
Dental amalgam	Amalgam is the most common material used for fillings. Also known as silver filling it is a mixture of mercury and an alloy of silver, tin and copper. Mercury makes up about 45–50 percent of the compound and is used to bind the metals together and to provide a strong, hard, durable filling. After years of research, mercury has been found to be the only element that will bind these metals together in such a way that can be easily manipulated into a tooth cavity.
Dental implant	A dental implant is an artificial tooth root surgically placed directly into the jawbone where a tooth is missing. Once the implant has healed, a bridge or replacement tooth will be attached to the artificial root. Implants provide a good alternative to crowns and bridges due to the fact that they are in no way attached to neighboring teeth and thus look and feel like natural teeth. Dental implants are generally made of titanium, a surgical metal that bonds well to bone. The cylindrical implant is designed to integrate into the structure of your jawbone, forming an anchor for your replacement tooth. Implanting the artificial root requires surgery, usually under a local anesthetic. During surgery, an incision is

	made in the gum to expose the jawbone, the bone is prepared, and the implant is inserted into the prepared site. Usually this area is left undisturbed for 4–6 months to ensure proper healing. Once the site has healed, a second procedure is required to clear the gum away from the implant and fasten a post, which will attach to the replacement tooth.
Fixed prosthodontics	The three main types of prostheses are **removable dentures, tooth supported fixed bridge** and **dental**.
Implant supported bridge	Removable dentures are the least expensive of the three options and initially are esthetically pleasing. Due to physical health and/or economic conditions, removable dentures are the only options available for patients. However, in most cases they are simply a good interim solution for patients undergoing osseointegration (dental implant treatment.) Of the other main types of prosthesis, **tooth supported fixed bridges** are a better option than removable dentures. The main benefit is that they are removable only by a dental specialist and not by the patient. They are also generally more esthetically pleasing than removable dentures and also offer better chewing ability. The best option for prosthesis is the **dental implant support bridge**. Dental implants provide a solid anchor for teeth and actually stimulate bone growth. Additionally, adjacent teeth are not impacted. They are also the most aesthetically pleasing. However, dental implants are the most initially expensive of the three main options of prostheses. They also require the presence of adequate bone in order for it to be placed. If adequate bone is absent, a bone graft procedure may be required.
Guided Bone Regeneration (GBR)	Refers to procedures that attempt to regenerate bone prior to the placement of bridges or, more commonly, **implants**. This is accomplished using bone grafts and biocompatible membranes that keep out tissue and allow the bone to grow.

Guided Tissue Regeneration (GTR)	Guided Tissue Regeneration (GTR) is a procedure that enables bone and tissue to re-grow around an endangered tooth or if the tooth is lost, to increase the amount of bone for implant placement.
Orthodontics	Orthodontics is the branch of dentistry devoted to correcting tooth alignment and bite problems. Orthodontists, dentists trained in orthodontics, straighten teeth and correct over or underbite. A perfect smile should consist of straight teeth, be gap and overlap-free, and align top teeth with the bottom. Mouth injuries, prolonged pacifier use in infancy, thumb-sucking in early childhood, and supernumary or impacted teeth can all cause misalignment. In some cases, a gap left by a knocked-out tooth creates a space for other teeth to shift into. Most often, though, orthodontic problems occur simply because the teeth and mouth are growing at different speeds.
Osseoconduction	A physical effect by which the matrix of the graft forms a scaffold on which cells in the recipient are able to form new bone.
Rapid prototyping	A host of related technologies that are used to fabricate physical objects directly from CAD data sources. These methods are unique in that they add and bond materials in layers to form objects. Such systems are also known by the general names solid freeform fabrication and layered manufacturing and offer advantages in many applications compared to classical subtractive fabrication methods such as milling or turning: • Objects can be formed with any geometric complexity or intricacy without the need for elaborate machine set-up or final assembly; • Objects can be made from multiple materials, or as composites, or materials can even be varied in a controlled fashion at any location in an object; • Solid freeform fabrication systems reduce the construction of complex objects to a

	manageable, straightforward and relatively fast process.
Scaffold	A three-dimensional biocompatible construct (may be seeded with cells) that serves as a temporary implantable tissue; generally fated to biodegrade and be replaced by natural tissue.
Tissue Engineering	The application of the principles of life sciences and engineering to develop biological substitutes for the restoration or replacement of tissue or organ function.

9 Environmental Biotechnology

- INTRODUCTION
- MICROBIAL ECOLOGY
- BIOTECHNOLOGY: A WEAPON AGAINST POLLUTION

INTRODUCTION

Environmental biotechnology is the application of biotechnology to solve problems associated with natural, agricultural and man-made environments. Although environmental biotechnology is normally associated merely with bioremediation, it includes a variety of other applications such as the use of microorganisms in metal recovery by ore leaching and the development of biofertilizers and biopesticides for agricultural purpose. Along with micro-organisms, application of higher plants in environmental biotechnology is gradually increasing, notably in bioremediation. As a result, environmental biotechnology can no longer be considered merely as an extension of applied microbiology.

MICROBIAL ECOLOGY

Microbial ecology deals with the study of the interactions of living micro-organisms with one another and with their nonliving environment. The study of microbial ecology is essential for the best possible application of micro-organisms in environmental biotechnology. The immediate surroundings of any living organism comprise physical, chemical and biological aspects that together make up the environment. The physical aspects include soil type and structure, water temperature and microclimate. Chemical aspects are the minerals and gases dissolved in water that in turn, influence pH (acidity/basicity) and the mineral composition of the soil. The biological aspect of a microorganism's environment includes other living organisms residing in the same environment. The relationship that various microorganisms living in an environment share with each other is known as the ecosystem of that microorganism.

Ecosystems are made of communities of similar organisms. A microbial community is an integrated assembly of microbial populations occurring and interacting within a given location. These communities, in turn, are made up of related microorganisms called guilds. Guilds are formed by populations of one type of microorganism. The number of different populations of microorganisms in any guild depends upon the environment. Figure 9.1 shows the hierarchical structure of a microbial ecosystem.

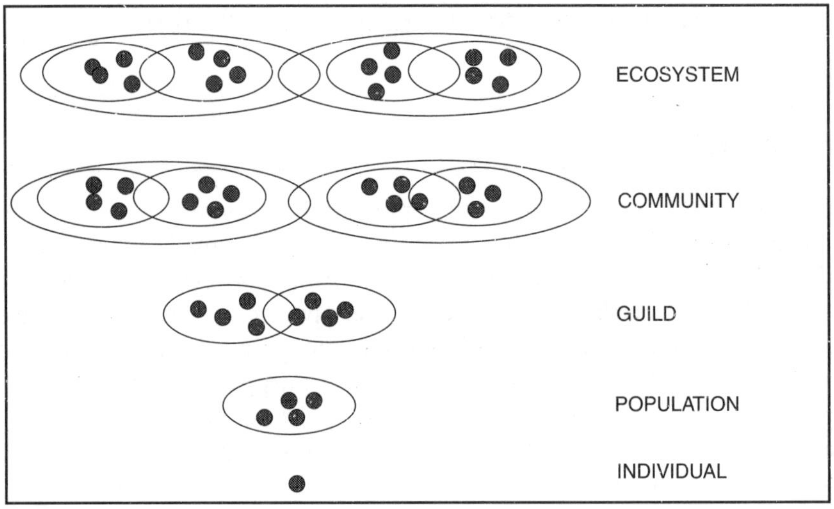

FIGURE 9.1 Microbial Ecosystem

Interaction and relationship among various microorganisms residing in an environment can be categorized on the basis of effect, as follows:

- Neutralism: is defined as a lack of interaction. A complete lack of interaction however is rarely seen in nature. The term can thus signify a relationship in which each species derives neither benefit nor harm to any measurable degree from the other species.
- Commensalism: is a relationship in which one organism gets benefited while the other organism is neither benefited nor harmed.
- Synergism: is an interaction in which the effect of the two microbes acting collectively is greater than the effect of either one of them acting alone.
- Competition: is an association between two species in which both struggle for the same limited environmental factor for growth.
- Predation or parasitism: is an interaction between organisms in which one organism captures biomass from another. It is generally thought of

as a synonym for carnivory but formally it includes herbivory, parasitism and parasitoidism also.

- Symbiosis: is an obligatory relationship between two organisms in which the partners are constantly in contact with each other and most often one lives inside the other. A symbiotic relationship generally implies mutualism, it also includes other types of relationships like parasitism and commensalism.
- Mutualism: is an interaction where both populations are benefited from each other. It is mostly a non-obligatory or temporary association.
- Ammensalism: is detrimental to one species while neutral to the other.
- Synnecrosis: is detrimental to both species in an association. The interaction is rare and essentially short-lived as evolution selects against it.

Classification of Microorganisms

Microorganisms are generally classified as follows:

Bacteria

Bacteria (singular bacterium) are a group of very small relatively simple, single-celled (unicellular) organisms. The genetic material of these organisms is not enclosed in a special nuclear membrane and hence, they are called prokaryotes (primitive type of nucleus. Bacteria are further classified on the basis of their cell shapes as Bacillus (rod like), coccus (spherical or ovoid) and spiral (curved) being the most commonly observed shapes. These types of bacteria are shown in Figure 9.2:

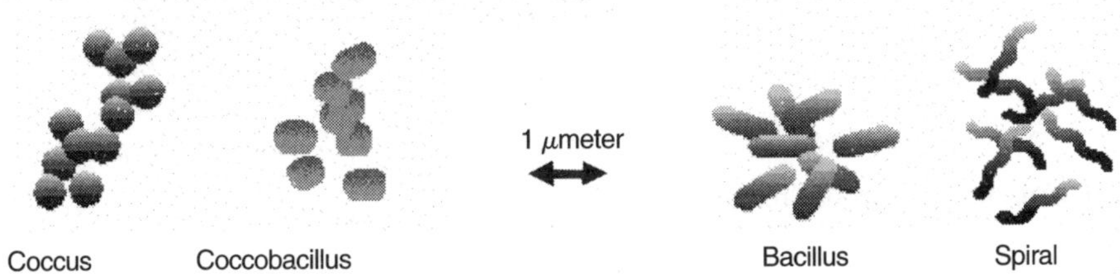

FIGURE 9.2 Types of Bacteria

Fungi

Fungi (singular fungus) are eukaryotic organisms whose cells have a distinct nucleus that contains the cells genetic material, surrounded by a nuclear membrane. Both unicellular and multicellular organisms are found in the kingdom fungi. True fungi have their cell walls composed primarily of a substance called chitin. Figure 9.3 shows some of the fungi found in nature:

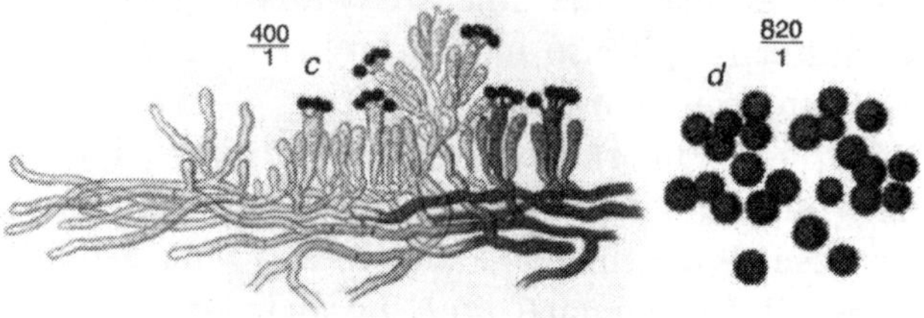

FIGURE 9.3 Fungi

Protozoa

Protozoa (singular protozoan) are unicellular, eukaryotic microbes belonging to the kingdom protista. They are further classified on the basis of their means of locomotion as pseudopods (false feet), flagellates and ciliates. Figure 9.4 shows some of the Protozoa:

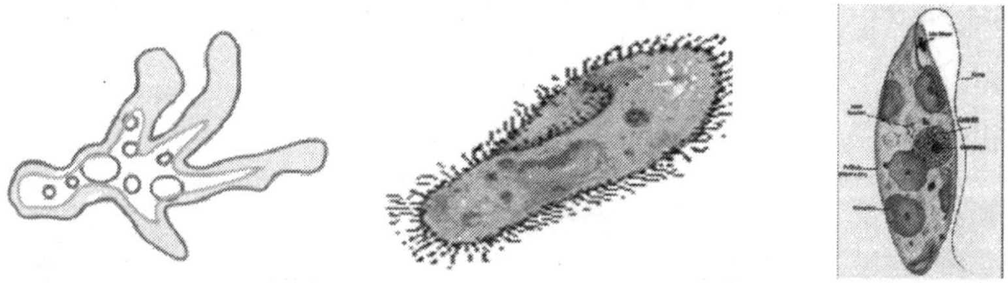

Amoeba (pseudopod) Paramecium (ciliate) Paranema (flagellate)

FIGURE 9.4 Types of Protozoa

Algae

Algae (singular alga) are photosynthetic eukaryotes found in wide variety of shapes and have both sexual and asexual reproductive forms. Their cell walls are composed of cellulose like higher plants. Algae are abundant in fresh and salt water. They are also found in soil and in association with plants. Figure 9.5 shows some algae:

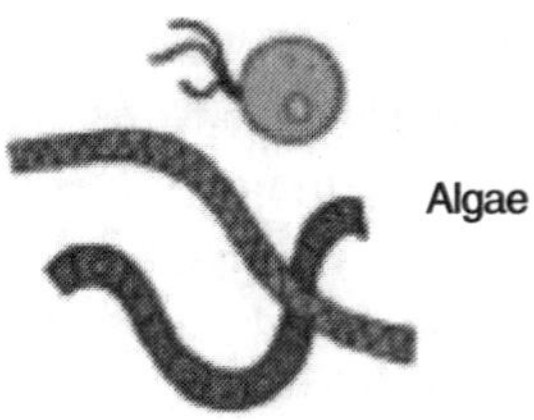

FIGURE 9.5 Algae

Virus

Viruses are very different from other microorganisms as they are acellular (not cellular). They are so small that most of them can be seen only with an electron microscope. Viruses have a very simple structure. A virus particle typically contains a core made up of only one type of nucleic acid, either DNA or RNA. This core is, in turn, surrounded by a protein coat. Sometimes, the coat is encased by an additional lipid membrane called envelope. Viruses can multiply only by using the cellular machinery of other organisms and thus all viruses act as parasites for other forms of life. Figure 9.6 depicts a virus:

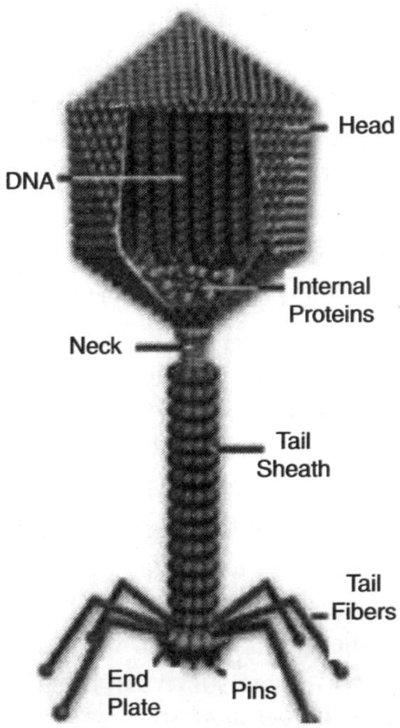

FIGURE 9.6 A Virus

Microbes, especially bacteria, are found in a wide variety of habitats on earth. This diversity of microbial populations is an indication of the fact that they are capable of taking advantage of any ecological niche that they inhabit. An ecological niche can be defined as an organism's place and role in the ecosystem: where it lives, what it consumes, what consumes it, and how it interacts with all associating biotic and abiotic factors. Microbes which have adapted to various extreme environmental conditions of temperature, acidity, alkalinity or salinity (such as polar, deep sea, hydrothermal vent and hot sub-vent) through various modifications in their metabolism are called extremophiles and play a crucial role in bioremediation and need especial mention in environmental biotechnology. Most of these microorganisms are members of the archaea. Extremophiles can be broadly categorized as:

- Anaerobes: can live and are active even in the absence of oxygen. Such organisms lack mitochondria. The role of the mitochondrion in energy generation through respiration has been replaced by the hydrogenosome.
- Thermophiles: can survive even at relatively high temperatures (up to about 60°C and above). For example, the archaebacteria *Pyrococcus* is known to thrive in a temperature above 100°C. The best-studied eukaryote found at high-temperature is the acidophilic phototroph *Cyanidium caldarium* reported to grow in a temperature range of 45°C to 80°C.
- Psychrophiles: extremophilic organisms found in cold temperatures. For example, the bacteria *Polaromonas vacuolata* has an optimum growth temperature of 4°C. The ciliate *Holosticha* sp. was reported to be unable to divide and breed above -2°C. Photosynthetic eukaryotes, such as *Chlamydomonas nivalis, Chloromonas (Scotiella), Ankistrodesmus, Raphionema, Mycanthococcus* and certain dinoflagellates are also Psychrophiles.
- Acidophiles: can survive in acidic conditions. Until recently, the acidophiles identified so far included only four organisms—all eukaryotes, *Cyanidium caldarium* and three fungi, *Acontium cymatium, Cephalosporium spp.*, and *Trichosporon cerebriae* which were found growing in pH values close to 0. Recently, two new prokaryotic members have been added to the list, *Picrophilus oshimae* and *P. torridus*. Both are thermophilic archaeans from Japanese solfataras.
- Alkalophiles: can tolerate alkaline conditions. For example, *Natronobacterium magadii*, which can survive in the pH, range 8.5 to 12 and has an optimum growth pH value of 10.

- Halpohiles: are found in highly saline conditions. For example, *Halobacterium salinarum* shows optimum growth at 25% salt concentration. Minimum and maximum salt concentrations that it can bear are 15% and 32%, respectively.
- Barophiles: can stay alive even at high-pressure conditions. For example, optimum growth conditions for the strain MT41 (Mariana Trench) are 700 atm and 4°C. This strain has not received a formal genus and species name up till now.
- Xerophiles: grow in extremely dry conditions.
- Others: Heavy metals are known to be toxic to living organisms but still there are organisms that can tolerate high concentrations of such metals. For example, the bacteria *Ferroplasma acidarmanus* can survive at high concentration of metals such as Copper, Arsenic, Cadmium and Zinc. Another microorganism *Ralstonia* sp. CH34 can grow even in the presence of high concentrations of Zinc, Cobalt, Cadmium, Mercury and lead.

Microbial Metabolism

Microbial populations bring about significant changes in their environment as a result of their metabolic activity, thereby playing an important role in material recycling and stabilization of natural ecosystems. The term "metabolism" refers to the sum of all chemical reactions taking place within a living organism. Metabolism is a combination of various catabolic and anabolic reactions. Catabolic reactions are those in which energy is released and complex organic compounds are broken down to simpler ones. Anabolic reactions involve utilization of energy and construction of more complex organic compounds from simpler compounds. A proper balance of these catabolic and anabolic reactions is required for proper functioning of a living organism.

All living organisms, including microorganisms, can be classified metabolically on the basis of their nutritional pattern, which includes source of energy and source of carbon for an organism. Firstly, on the basis of energy source, organisms can be classified as phototrophs or chemotrophs. Phototrophs use light as the primary source of energy while chemotrophs depend upon the oxidation-reduction reactions of inorganic or organic compounds for energy. Autotrophs (self-feeders) use carbon dioxide as their principal carbon source, while heterotrophs (feeders on others) require an organic carbon source. Autotrophs are sometimes also referred to as lithotrophs (rock eating), and heterotrophs are referred to as organotrophs. Figure 9.7 shows a flow chart depicting the classification of organisms on the basis of nutritional pattern:

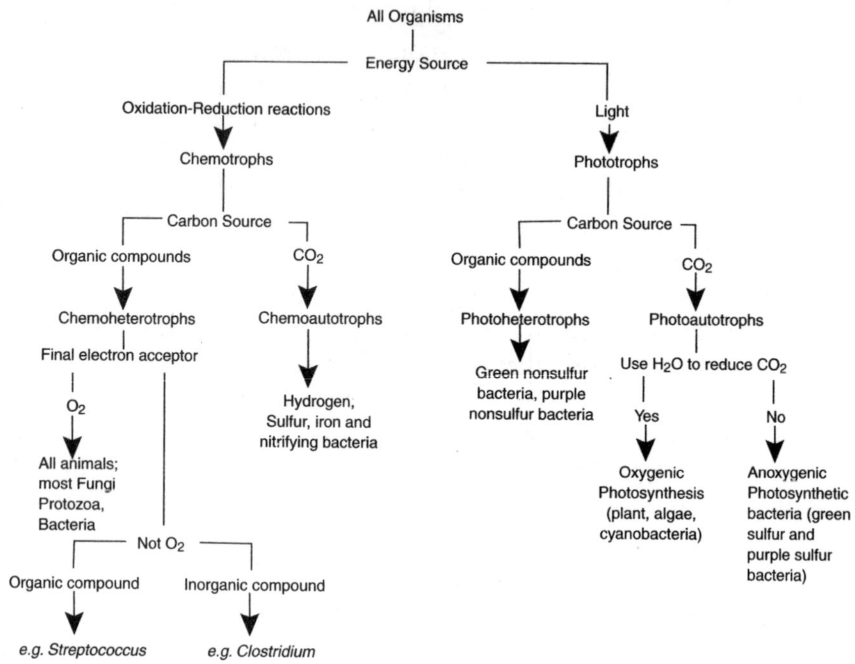

FIGURE 9.7 Classification of Organisms Based on Nutritional Pattern

Most of the microorganisms, especially bacteria and fungi, are heterotrophic in nature. They consume and not produce organic matter. However, a small group of autotrophic bacteria are also known which are photosynthetic and chemoautotrophic in nature, and fix carbon dioxide. These organisms are of rare occurrence in nature, and their share in the total production of organic matter is also small. The photosynthetic bacteria are known to have two types of photosynthesis. In the cyanobacteria, the photosynthesis is similar to that of plants, and oxygen is produced in the process. In bacteria other than cyanobacteria, oxygen is not evolved during photosynthesis and they are able to use a wide range of electron donors for reducing power, such as H_2S (hydrogen sulfide) or organic acids. Such bacteria are obligate anaerobes or microaerophiles. Another group of autotrophic microorganisms consists of chemosynthetic bacteria, which satisfy their energy requirements by oxidizing reduced inorganic substrates, such as compounds of inorganic nitrogen, sulphur and iron or elements, such as sulphur and hydrogen. Most chemosynthetic bacteria require free oxygen as the electron acceptor but some other bacteria such as *Thiobacillus denitrificans*, *Desulfovibrio desulfuricans*, etc. have adapted themselves to use bound oxygen derived from nitrate or sulphate. The reduced substrates of these bacteria are mostly derived from the decay of organic matter. Therefore, chemosynthetic bacteria are not considered as the primary producers in the sense of a food or energy in a food chain. The transformations taking place as a result of the metabolic activity of the chemosynthetic bacteria are very important in element cycling and involve large amounts of material.

Among all the heterotrophic microorganisms, bacteria are much more important than fungi or algae since they are more abundant and diverse in nature. Fungi are present in low numbers especially in aquatic habitats but they are much plentiful in streams or marshes where large amounts of water from terrestrial plants are present. Heterotrophic algae have also been identified but their importance is not very clear.

Microbes and Minerals

Understanding the interaction of microbes and minerals is important in context of bioremediation, an important application of microorganism in environmental biotechnology. The impact of microbes on the cycling of important minerals or elements such as carbon, nitrogen, sulfur and other dissolved biogeochemical species is well recognized. A biogeochemical cycle can be defined as a natural process in which nitrogen, carbon, and other inorganic elements of the soil, atmosphere, etc. of a region are converted into the organic substances of animals and plants of the region and released back into the environment. Microbes are key players in these cycles. In biogeochemical cycles, elements are oxidized and reduced by microorganisms to meet metabolic needs. Two significant biogeochemical cycles (the carbon and nitrogen cycle) are shown in Figure 9.8 and Figure 9.9, respectively, to make the function of microorganisms in biogeochemical cycle understandable.

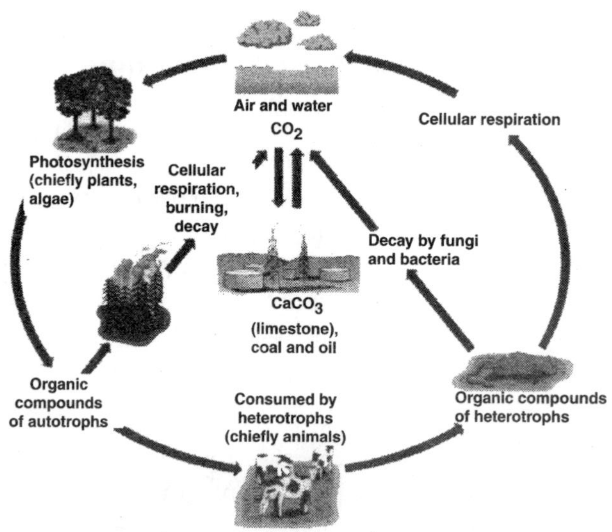

FIGURE 9.8 Carbon Cycle

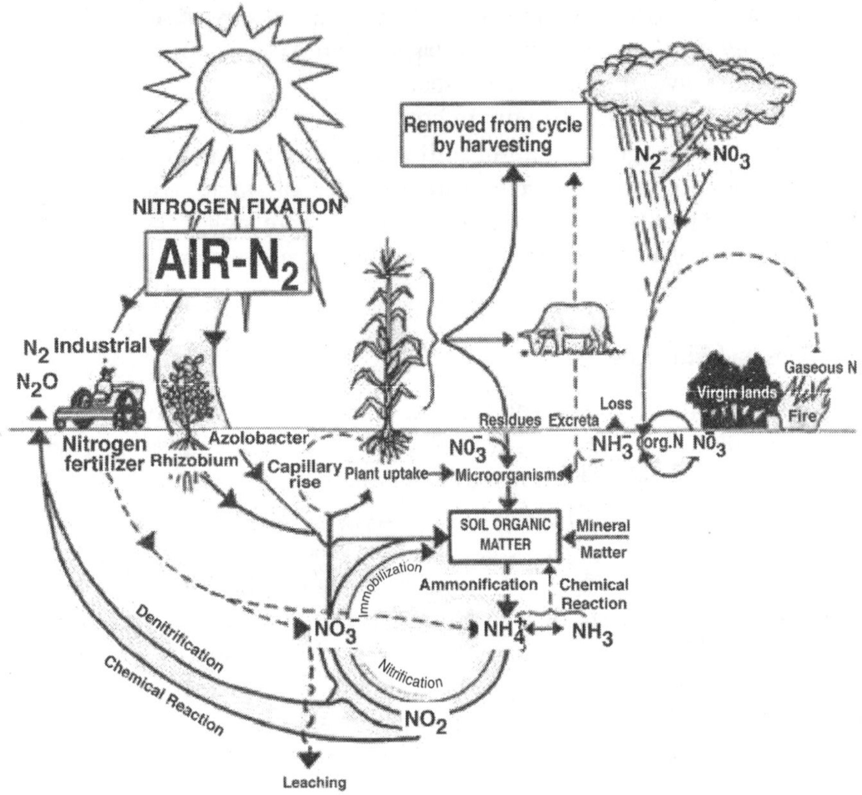

FIGURE 9.9 Nitrogen Cycle

BIOTECHNOLOGY: A WEAPON AGAINST POLLUTION

An important application of environmental biotechnology in combating the problem of environmental pollution is "bioremediation" which employs the application of natural and genetically engineered organisms, mostly microbes, but also higher plants to improve the efficiency of pollution control and reduce the cost of process. The graph shown in Figure 9.10 depicts the worldwide Sales of Environmental Biotechnology Products for waste treatment, 2001 and 2006 ($ Millions) by U.S. manufacturers.

Bioremediation

Bioremediation is a general term covering a whole range of environmental de-polluting techniques used to degrade harmful or hazardous contamination into less harmful or benign components. It can also be defined as the use of biological organisms such as plants or microbes to aid in degradation of

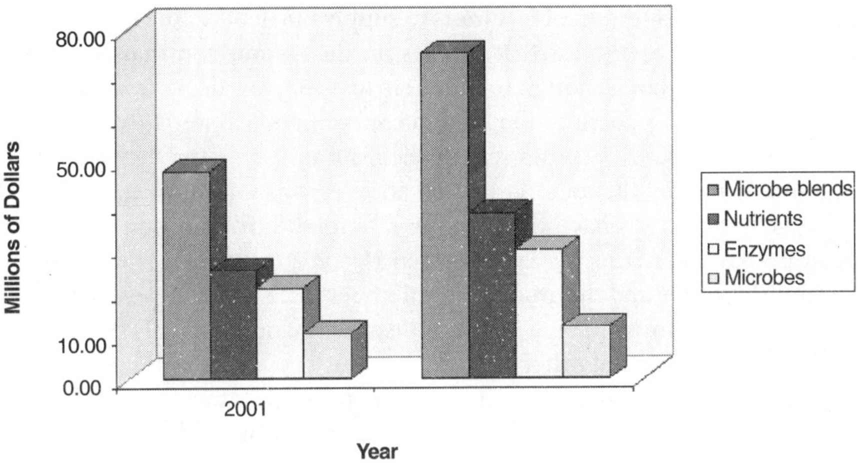

FIGURE 9.10 Sales of Environmental Biotechnology Products for Waste Treatment, 2001 and 2006
(Ref: Business Communications company, Inc., www.bccresearch.com)

environmental pollutants from an area. It can be thought of as a technology for removing pollutants from the environment and restoring the contaminated sites using natural process, thereby preventing future pollution. The process of bioremediation is based on the enormous natural capacity of microorganisms to degrade organic compounds. In addition to degradation of complex organic compounds into simpler compounds, microorganisms can also influence the nature of inorganic compounds in two important ways that are of potential use in bioremediation. These include:

- Concentrating metals or radionuclides on or within cells
- Bringing about change in the mobility of metals and radionuclides in the environment through modification in their physicochemical conditions

The effectiveness of microorganisms to degrade environmental pollutants can be further improved through genetic manipulation and introduction of GMMs (Genetically Modified Microorganisms) at polluted site.

A wide variety of microbes are found naturally existing in all types of soils and waters that take care of degrading the waste usually found in its surrounding ecosystem. These are called *indigenous* microbes. The process of using the naturally occurring microbial populations for remediation of the contamination entering an area is called *natural attenuation.* However, more often than not, very little remediation takes place by this process. To improve the growth and efficiency of indigenous microbial population in bioremediation, *biostimulation* is required that involves aeration of the polluted site and addition of fertilizers to supply nutrients to the microbes. The drawback of the method is that it relies on the assumption that the existing microbes will be large enough in numbers to effectively degrade the contaminant and to readily adapt to the foreign contaminants to initiate degradation. However, the process becomes very time consuming. To improve the efficiency of bioremediation, the method usually adopted by biotechnologists is to add microbes that already have an aggressive affinity toward common organic and inorganic contamination and is described as *bioaugmentation*. The approach is generally effective and the amount of microbes to be added depends upon the volume of contaminant, type and volatility of contaminants, pH of the soil or water, temperature, and other factors.

An alternative to the bioaugmentation approach is the microbial ecology approach. This involves adjustment of environmental conditions to increase microbial metabolism and activity in the environmental samples undergoing treatment. The microbial ecology approach is relatively simple and often involves very low technology. For example, land farming, which as its name suggests, is merely cultivating the land in order to produce soil conditions that are favourable to microbial activity. Land farming involves simple procedures such as ploughing (which helps in aeration of the site) and addition of lime (to increase the soil pH). No addition of microorganisms from outside is done in land farming, instead it is assumed that the ideal conditions provided to indigenous microbial flora will increase microbial efficiency to break down the pollutant. The microbial augmentation approach is more useful when the site has been recently contaminated or where high concentration of a particular pollutant is present and thus rapid cleanup is required. The ecological approach, on the other hand, is more useful for cleaning up sites that were exposed to a multiplicity of uses and contains a cocktail of pollutants.

Another factor that greatly affects bioremediation is bioavailability of material to be degraded together with the availability of microbial populations and may be the limiting factor in breakdown. Therefore, pollutants from smaller soil particles, which have a large surface area and can thus be readily mixed with nutrients and easily aerated, are more easily degraded than the pollutants from large clumps of soil. The binding of contaminants to clay or organic matter may also significantly influence its rate of degradation.

There are two methods for bioremediation on the basis of removal and transportation of wastes for treatment: *in situ* bioremediation and *ex situ* bioremediation.

In situ *Bioremediation*

In this approach to bioremediation, the organisms known to be capable of degrading the pollutant under consideration are usually added to the environment together with the required amounts of nitrogen, phosphorus and potassium and micronutrients. Degradation may be limited by temperature and occur only, or at least at a significant rate, during the summer months. In addition, the pH of the environment may need to be adjusted by the addition of lime. The *in situ* degradation of many pollutants, including hydrocarbons is limited by oxygen supply. Oxygen gas can be readily supplied to waters, but oxygen supply to soils is generally achieved by adding hydrogen peroxide. Soil peroxidases break this down to water and oxygen, the water itself being useful in dry environments. There are two types of *in situ* bioremediation:

- Intrinsic bioremediation: Conversion of environmental pollutants into harmless forms through the innate capabilities of naturally occurring microbial population is called intrinsic bioremediation. The conditions of site that favour intrinsic bioremediation are groundwater flow throughout the year, carbonate minerals to buffer acidity produced during biodegradation, supply of electron acceptors and nutrients for microbial growth and absence of toxic compounds. The other environmental factors such as pH, concentration, temperature and nutrient availability determine whether or not biotransformation takes place.
- Engineered bioremediation: When site conditions are unsuitable, bioremediation requires construction of engineered systems to supply materials that stimulate microorganisms. Intrinsic bioremediation is ineffective when there are poorly adapted microorganisms, limited ability of electron acceptor and nutrients, cold temperature and high concentration of contaminants. Engineered bioremediation encourages growth of more microorganisms via optimizing physico-chemical conditions. When the contamination is deeper, amended water is injected through wells. But in *in situ* bioremediation systems, both extraction and injection wells are used in combination to control the flow of contaminated groundwater along with the above ground bioreactor treatment and subsequent reinjection of nutrients spiked effluent are done.

Figure 9.11 shows the physical configuration of an engineered *in situ* bioremediation.

Ex situ *Bioremediation*

In this approach, the soil is excavated and treated on site. It incurs costs associated with solid handling process, for instance, excavation, screening and fractionation, mixing, homogenizing and final disposal. On the basis of

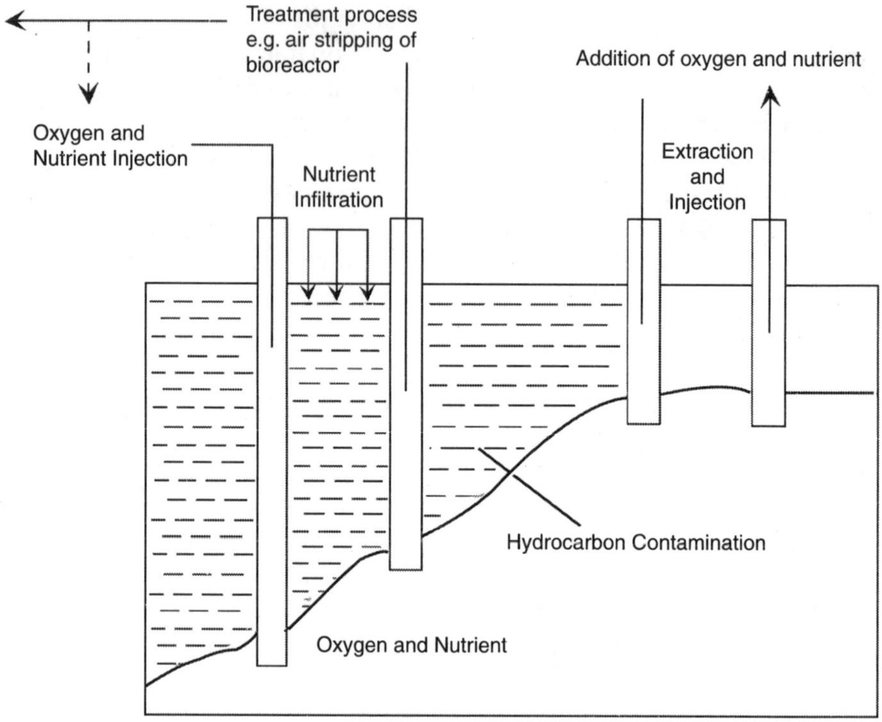

FIGURE 9.11 Physical Configuration of *in situ* Bioremediation

phases of contaminated materials under treatment, *ex situ* bioremediation is classified into two: solid and slurry phase systems.

Solid Phase Treatment

It includes organic wastes (leaves, animal manures and agriculture wastes) and problematic wastes (domestic and industrial wastes, sewage sludge and municipal solid wastes). The traditional cleanup practice involves production of composts that may be used as soil amendment.

Composting is a self-heating, substrate dense, managed microbial system, and one solid phase biological treatment technology which is suitable for the treatment of large amount of contaminated solid materials. As composting is a solid phase biological treatment, target compounds must be either solid or liquid associated with a solid matrix. The hazardous compounds reported to disappear through composting include aliphatic and aromatic hydrocarbons and certain halogenated compounds. The possible routes leading to disappearance of hazardous compounds include volatilization, assimilation, adsorption, polymerization and leaching.

Composting can be done in open system or in closed system. The open land system can be an inexpensive treatment method, but the temperature fluctuates from summer to winter. Therefore, the rate of biodegradation of waste materials declines. Secondly, land treatment system may become oxygen limited, depending on the amount of substrate, depth of waste, application, etc. However, efficiency of open treatment system can be increased by passing air. This approach is referred to as engineered soil piles and forced aeration treatment. Figure 9.12 shows the physical configuration of an open composting.

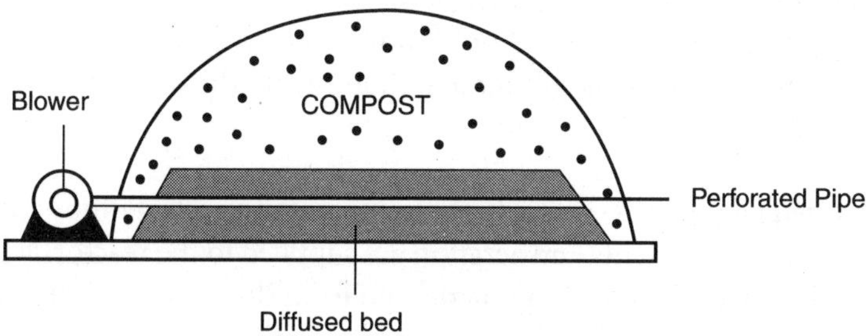

FIGURE 9.12 Physical Configuration of Open Composting

The closed treatment system is preferred over the open land system because controlled air is supplied to maintain the microbial activity. Use of blowers for air circulation and exhaust for removal of toxic volatiles are set up in closed treatment system. Ventilators supply oxygen and remove heat through evaporation of water. Figure 9.13 shows the physical configuration of a closed composting system.

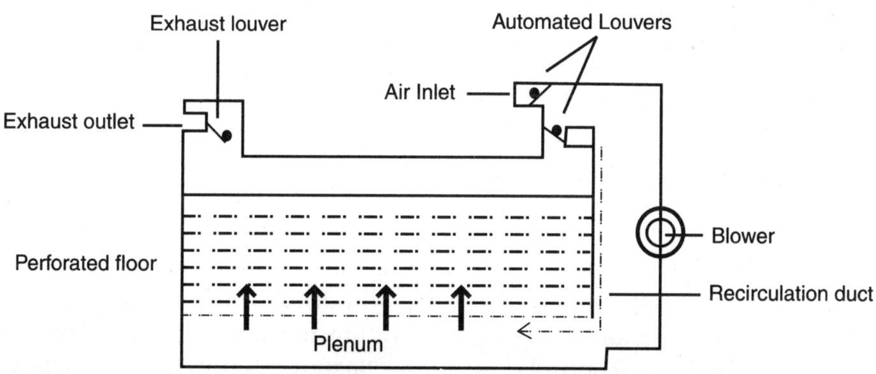

FIGURE 9.13 Physical Configuration of Closed Composting System

Slurry Phase Treatment

The contaminated solid materials (soil, degraded sediments), microorganisms and water formulated into slurry are brought within a bioreactor i.e. fermenter. Thus, slurry phase treatment is a triphasic system involving three major components: water, suspended particulate matter and air. Here, water serves as suspending medium where nutrients, trace elements, pH adjustment chemicals and desorbed contaminants are dissolved. Suspended particulate matter includes a biologically insert substratum consisting of contaminants (soil particles) and biomass attached to soil matrix or free in suspending medium.

Biologically, there are three types of slurry phase bioreactors: aerated lagoons, low shear airlift reactor, and fluidized bed soil reactor. The first two types are in use of full-scale bioremediation, while third one is in developmental stage.

- **Aerated Lagoons:** is similar to the treatment of small common municipal wastewater. Nutrients and aeration are supplied to the reactor. Mixers are fitted to mix different components and form slurry, whereas surface aerators provide air required for microbial growth. The process may be used as single stage or multistage operation. If the waste contains volatiles, this reactor is not appropriate. Figure 9.14 shows a slurry phase lagoon.

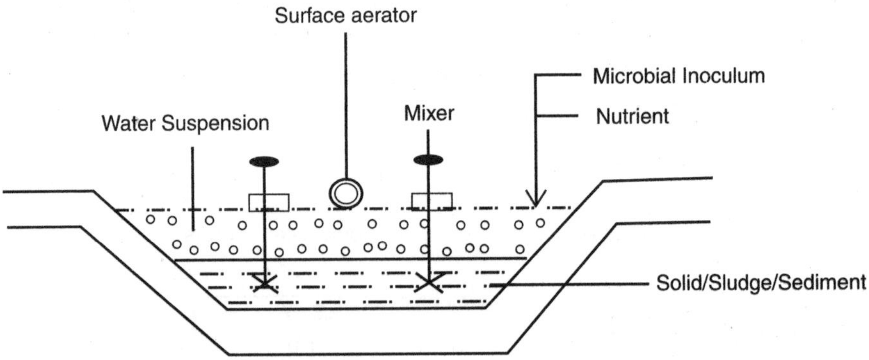

FIGURE 9.14 Slurry Phase Lagoon

- **Low Shear Airlift Reactors** (LSARs): are useful when waste contains volatile components. LSARs are cylindrical tanks made of stainless steel. In this bioreactor, pH, temperature, nutrient addition, mixing and oxygen can be controlled as desired. Shaft is equipped with impellers. It is driven by motor set-up at the top. The rake arms are connected with blades used for resuspension of coarse materials that tend to settle on the bottom of the bioreactor. Airlift provides to bottom circulation of contents in reactor. Pre-treatment processes include size fractionation solids, soil washing, and milling to reduce particle size and slurry preparation. Certain surfactants such as anthracene, pyrene, perylene, etc. are added to enhance the rate of biodegradation. These act as co-substrate and utilized as carbon source. Figure 9.15 shows a low shear airlift slurry phase bioreactor.

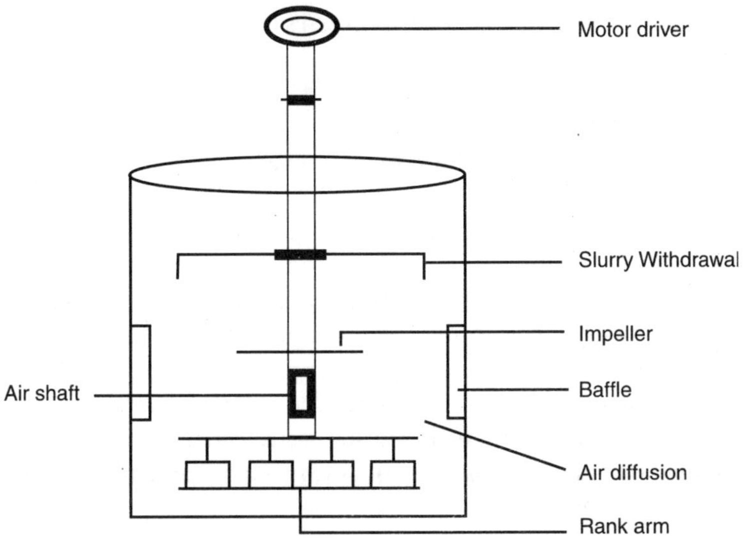

FIGURE 9.15 Low Shear Airlift Slurry Phase Bioreactor

- Fluidized bed soil reactors are a hybrid of column and batch reactors in that aqueous chemical conditions can be kept constant while maintaining a well-mixed system and minimizing physical control on reaction kinetics. Unlike batch reactors, however, used, fluidized-bed reactors have been used rarely for sediment reactivity experiments.

Advantages of Bioremediation

Some of the advantages of bioremediation are:

- It is a natural process.
- It can completely destroy a wide variety of contaminants. This eliminates the chance of future liability associated with treatment or disposal of contaminated materials.

- It involves lower human risk than alternative remediation technologies.
- Biological treatment of soils can frequently be conducted under difficult site conditions. Because large volumes of soil can be treated at one time in an above ground engineered system, and because *in situ* applications avoid the need for excavation and subsequent disposal or treatment, bioremediation has the major benefit of being a cost-competitive remediation technology.
- Post-cleanup costs can be substantially reduced because bioremediation methods minimize site disturbance compared with conventional cleanup technologies.
- It is one of the few treatment processes that are currently acceptable in the minds of the general public.

Disadvantages of Bioremediation

Some of the disadvantages of bioremediation are:

- Bioremediation is highly specific and not all xenobiotic compounds are susceptible to rapid and complete biodegradation.
- Concerns have also been raised that, in some cases, metabolic end products of biodegradation may be more persistent or hazardous than the initial parent compound(s). For example, the biotransformation of the Chlorinated Aliphatic Trichloroethylene (TCE) can result in formation and accumulation of vinyl chloride, a suspected carcinogen.
- Reintroduced organisms do not always survive and are outdone by the indigenous populations.

Success Stories of Bioremediation

Bioremediation has emerged as an effective tool in controlling spill sites of organic contaminants with relatively lesser infrastructure and manpower. Another major advantage is that the method is extremely useful in cases of space constraints. In addition, the method also does not come without any long-term liabilities since the contaminants are completely degraded and turned into harmless byproducts of CO_2, water and trace inorganic salts. This is of tremendous importance considering the illicit waste trafficking that some countries indulge in, by dumping their harmful toxic wastes into the poorer third world countries. Apart from the ethical considerations, the practice is disturbing considering the fact that these poor countries are hubs of biological diversity, which is threatened by such indiscriminate acts.

Bioremediation came into the picture for the first time in a big way in 1975 when about 8,000 tons of jet fuel leaked from a military installation in the U.S. Though immediate steps were taken to contain the leak, it did manage to seep into the groundwater. Within a decade, the groundwater reached the residential areas spewing toxic wastes and creating an environmental catastrophe.

The solution was offered by the U.S. Geological Survey (USGS), which had shown that microorganisms naturally present in the soils were actively consuming fuel-derived toxic compounds and transforming them into harmless carbon dioxide. Addition of nutrients was found to accelerate the rate of decomposition manifold. In the next two years, the agency completed the pumping and treating of groundwater with microorganisms and monitored the detectable contamination. By the end of the second year, the limits of the pollutants had dropped to less than 75% of the original.

Bioremediation of Oil Spills

Oil spills at sea are relatively frequent and pose a considerable environmental problem. At this point the oil can be removed mechanically using oil skimmers. The most common of these employ sorbent surface skimmers which use adhesion to metal or plastic surfaces to lift oil off the water surface. Natural sorbents such as peat and wood products can also be used. It should be emphasized that natural recovery of oil-polluted sites is often more effective than is generally thought and that even the most distressing contamination will eventually be cleared up by self purification by the indigenous microbial populations. However, if such natural remediation is too slow then it can be speeded up bioremediation. For example, microbial degradation of light and medium crude oils on shorelines generally takes a year but can be speeded up by enhanced by using fertilizers and possibly inoculants. Genetically engineered microorganisms have also been developed for oil degradation, but due to governmental restrictions have yet to be used in the environment. Fertilizers are effective at speeding up oil degradation, but they need to be in place in oil-contaminated environments for as long as possible. However, because of tides and wave action, long exposure periods are not always achievable. To avoid removal and dilution by the sea slow release fertilizers have been developed. Oleophilic fertilizers have also been employed.

An excellent example of using novel approaches to environmental problems is application of human hair in oil bioremediation. Hair has an affinity for oil and grease. Oil tends to be adsorbed in layers on hairs and does not chemically bond with keratin molecules of which hair is composed. The hair is placed in nylon bags and added to oil-polluted water course; the oil that is soaked up can then be removed by shampooing. The hair can then be returned and go through a large number of adsorption cycles. One and a quarter pounds of hair adsorb a gallon of oil in less then two minutes.

Degradation of petroleum sludge occurred within two weeks when the bacterial culture (*Bacillus circulans* C1) was used. Moreover, significant degradation of petroleum sludge was noticed in 10 days when the fungus + *B.*

circulans and a prepared surfactant were exogenously added to petroleum sludge.

Crude Oil Spill, Bemidji, Minnesota

In 1979, a pipeline carrying crude oil burst and contaminated the underlying aquifer. It was observed that natural microbial populations rapidly degraded toxic chemicals leaching from the crude oil. This was the first documented instance of natural bioremediation without any human intervention.

Bioremediation of Contaminated Groundwater

Compounds contaminating groundwater include solvents, such as benzene and toluene, soil fumigants, nematocides and herbicides such as 2,4-D. In the United States, a total of 39 pesticides have been detected in groundwater from 34 states, with concentrations ranging from 0.1–5µg per litre. It might be expected that the problem could be readily solved by flushing well with fresh water. However, because these contaminants generally sorb to soils they need to be broken down by bioremediation. The treatment methods used depend on the type and degree of contamination and its location. The most frequently used approach is to "pump and treat", that is, remove the contaminated water and treat it on the surface. However, *in situ* treatment may prove a more effective alternative. At greater depth, *in situ* bioremediation can be achieved by infiltrating or injecting nutrients and oxygen into the subsurface, to stimulate aerobic degradation.

Chlorinated Solvents, New Jersey

Chlorinated solvents are a particularly common contaminant in heavily industrialized areas. Microorganisms, with their highly adaptable metabolic processes, use chlorinated compounds as oxidants when other oxidants are not available. Many areas of New Jersey have readily benefited from such microorganisms that have depleted many contaminants like chlorine and nitrate over a period of time.

Bioremediation of Xenobiotics

Xenobiotics are compounds foreign to biological systems, and are often referred to human-made compounds that are resistant or recalcitrant to biodegradation and decomposition. Many xenobiotics persist in environment and do not undergo biological transformation. Microorganisms play an important role in degradation of xenobiotics, and maintaining of steady state concentrations of chemicals in the environment. The complete degradation of a pesticide molecule to its inorganic components that can be eventually used in an oxidative cycle removes its potential toxicity from the environment. Biodegradation of pesticides occurs by aerobic soil microbes.

Pentachlorophenol (PCP) is a biocide that has been used as fungicide, insecticide, herbicide, algicide, disinfectant and antifouling agent. Bioreactors containing immobilized PCP degrading *Flavobacterium* (ATCC39723) cells have been used to remove PCP from contaminated water. DDT [1,1,1–trichloro–2,2–bis (p-chlorophenyl) ethane] is an insecticide that persists in soil for four years. Microorganisms associated with DDT degradation are *Aspergillus flavus, Fusarium oxysporum, Mucor alternans, Pseudomonas chrysosporium, Trichoderma viride*, etc.

The number of xenobiotic degrading microorganisms is increasing but only a few pesticide-degrading genes have been characterized. Most of the genes responsible for catabolic degradation are located on the chromosomes, and rarely on transposons or plasmids. Chakrabarty and Gunsalus (1971) found that camphor-degrading genes of Pseudomonas putida are located on plasmid, soon after naphthalene (NAH) degrading plasmid was isolated. Genes responsible for degradation of environmental pollutants, such as toluene, chlorobenzene acids and other halogenated pesticides and toxic wastes have been identified and grouped into four categories:

- OCT plasmid: degrades octane, hexane and decane
- XYL plasmid: degrades xylenes and toluenes
- CAM plasmid: decomposes camphor
- NAH plasmid: degrades naphthalene

Dr. Anand Mohan Chakrabarty produced a new product of genetic engineering called superbug (oil-eating bug) by introducing plasmids from different strains into a single cell of *P. putida*. This superbug can degrade all four types of substrates for which four separate plasmids were required.

Bioremediation of Dyes

There is limited study on microbial degradation of azo and reactive dyes. A number of dyes undergo degradation through reduction. Kulla (1981) discussed azo-reductase of Pseudomonas strains in the chemostat culture. This enzyme catalyzes azo-linkage of the dye. During degradation process of azo, NAD (P) acts as electron donor. Later, in 1995, it was observed that the degradation of black liquor pulp mill effluents by the strains of *Pseudomonas putida*. Some anaerobic bacteria, streptomyces and fungi (e.g., *Phanerochaete chrysosporium*) have been characterized for decolouration of chromogenic dyes. The enzymes involved in dye degradation are lignases (lignin peroxidase), Mn (II) dependent peroxidase and glyoxal oxidase. These enzymes are well associated with lignin degrading system.

Bioremediation of Hydrocarbons

Petroleum and its products are hydrocarbons. Oils constitute a variety of hydrocarbons viz., xylanes, napthalenes, octanes, camphor, etc. If present in the environment these cause pollution. In addition, oil and petrol leakage in marine environment is of usual phenomenon. For example, 11 million gallon oil spill from the supertanker *Exxon Valdez* ran around near Prince William Sound, Alaska in March 1989. It is interesting to note that in Pennsylvania about 27,000 litres (6,000 gallon) of petrol leakage occurred. It contaminated the underground water supplies. The input of oil to the environment can be ecologically devastating and the cost of cleaning can go up to several million dollars.

Oil is insoluble in water and is less dense. It floats on water surface and forms slicks. The microorganisms, which are capable of degrading petroleum, include pseudomonads, various corynebacteria, mycobacteria and some yeast. The method can be improved by using a mixture of bacteria and using genetically engineered microbial stains.

Bioremediation of Heavy Metals

Bacteria, algae, fungi, actinomycetes and higher plants accumulate high amount of heavy metals in their cells.

Algae

Several species of *Chlorella*, *Anabaena*, and marine algae have been used for the removal of heavy metals. Rai et al., (1998) studied biosorption(both adsorption and absorption) of Cd^{2+} by a capsulated cyanobacterium, *Microcystis* both from field and laboratory. The naturally occurring cells showed higher efficiency for Cd^{2+} and Ni^{2+} as compared to laboratory cells.

Fungi

They are also capable of accumulating heavy metals in their cells using various mechanisms, such as:

- **Metabolism independent accumulation:** The positively charged ions in the solution are attracted to negatively charged ligands in cell materials resulting in biosorption of metal ion on microbial cell surface. But composition of biomass and other factors affect biosorption. For example, in *Rhizopus arrhizus* adsorption depends on ionic radius of Li^{3+}, Mn^{2+}, Cu^{2+}, Zn^{2+}, Cd^{2+}, Ba^{2+}, Hg^{2+} and Pb^{2+}. However, binding of Hg^{2+}, Ag^{2+}, Cd^{2+}, Al^{3+}, Ni^{2+}, Cu^{2+} and Pb^{2+} strongly depends on concentration of yeast cells.
- **Metabolism dependent accumulation:** In fungi and yeast, heavy metal ions are transported into the cells through cell membrane.

However, as a result of metabolic processes ions are precipitated around the cells, and synthesized intracellularly as metal-binding proteins. Moreover, intracellular uptake is influenced by certain external factors such as pH, anions, cations and organic materials growth phase, etc. Metal uptake by growing batch culture was found maximum during lag phase in *Aspergillus niger*, *Penicillium spinulosum* and *Trichoderma viride.*

- **Extracellular precipitation and complexation:** Fungi produce several extracellular products that can complex or precipitate heavy metals. For example, many fungi and yeast release high affinity Fe binding compounds that chelate iron and are called siderophores. The Fe^{3+} chelates that are found outside the cell wall are taken up into the cell. In *Saccharomyces cerevisiae* removal of metals is done by their precipitation as sulphides e.g., Cu^{2+} is precipitated as CuS.

Phytoremediation

Phytoremediation is the use of higher plants for bioremediation. It is also referred to as botanical bioremediation or green remediation. The potential use of plants to metabolize toxic organic products in soils also appears promising. Categories of phytoremediation include:

- **Phytoextraction:** is the use of plants to remove contaminants from soil.
- **Phytovolatilization:** is the use of plants to make volatile chemical species of soil elements. A number of heavy metals can be volatilized. These include selenium that is converted to dimethylselenide and mercury that is converted to its methyl derivatives. However, volatile derivatives are usually far more toxic than the element itself and present an environmental problem in themselves.
- **Rhizofiltration:** is the use of plant roots to remove contaminants from running water.
- **Phytostabilization:** is the use of plants to convert toxic metals to less toxic forms, while not removing them from soils. This approach has use with two toxic elements, chromium and lead. For example, Cr^{6+}, which poses a considerable environmental risk, can be reduced to highly insoluble Cr^{3+}, which presents less of a risk. Chemical species of lead are biologically available and toxic, whereas the lead phosphate mineral, chloropyromorphite, is both extremely insoluble and non-bioavailable. This fact can be put to good use since, the roots of the grass *Agrostis capillaries* are known to cause the formation of pyromorphite from soil lead and phosphate.

Figure 9.16 shows a diagrammatic representation of various methods adopted by plants for bioremediation of soil pollutants.

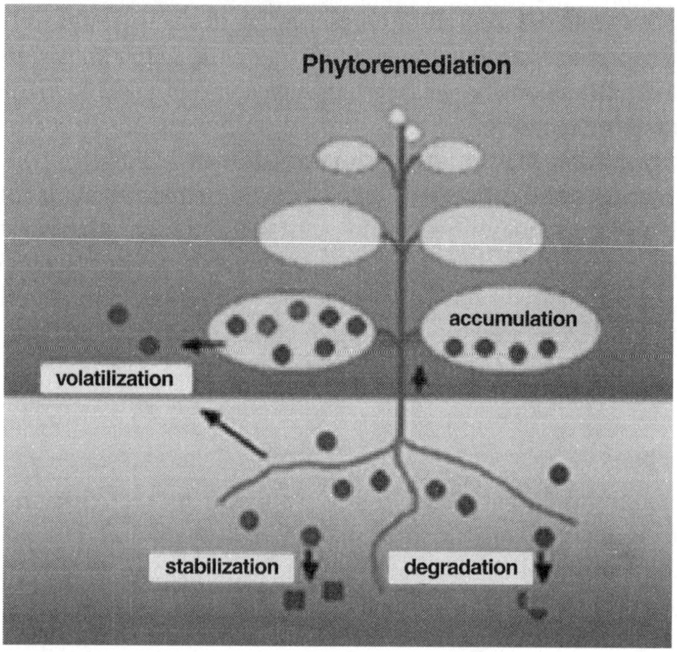

FIGURE 9.16 Bioremediation of Soil Pollutants by Plants

Metal Hyperaccumulators

It has been suggested that plants that have the ability to selectively accumulate heavy metals could be used to remove metal pollutants from soils. Such plants can accumulate 10–500 times higher levels than crops. Plants that accumulate over 1% nickel are termed hypernickleophores. In order to accumulate metals plants must be able to:

1. Tolerate high concentrations of generally toxic metal, either by vascular compartmentalization or chelation.
2. Translocate an element from the roots to the shoots. Normally, root concentrations of heavy metals exceed the shoot concentration, but in hyperaccumulators the opposite is true.
3. Rapidly accumulate the element from solution.

Hyperaccumulation has evolutionary advantage in making the plant unattractive to insects as well as reducing any chance of fungal and bacterial infection. Seaweeds are primitive non-vascular plants that adsorb nutrients and trace elements by direct adsorption from the surrounding seawater. Seaweeds have been evaluated as metal accumulators, both to clean up polluted marine environments and potentially to extract precious metals. For example:

- *Laminaria* spp. accumulates large amounts of iodine and bromine
- *Saragassum* spp. and other brownweeds accumulate arsenic
- *Fucus* accumulate uranium and strontium
- *Ulva* (Green Sea Lettuce) accumulates Vanadium

Other common metals such as Lead, Zinc, Copper and Cadmium can also be accumulated by various green and brown seaweeds.

Microbes and Air Pollution

Due to growing public demand, the government has placed huge burdens on industries to operate in a more environmentally responsible manner. To meet this challenge, the industry has had to turn to a variety of fields to meet environmental standards. To help eliminate environmental hazards, industries employ physical, chemical, thermal and biological processes. Over the last 15 years, biological treatment technologies have gained increased popularity due to their low costs, operational simplicity and because they are cleaner than alternative technologies. It is also popular with the industry because there is minimal requirement of energy and raw materials. There is also minimal production of waste materials to be disposed of. These biotechniques include biofiltration, bioscrubbers and biotrickling filters. This chapter will limit the discussion to biofiltration because it is the most common method employed and the simplest to explain.

A biofilter has a solid phase consisting of a porous medium, on which microorganisms are allowed to grow. A stream of gas containing unwanted molecules is then passed through the solid phase. The resulting microbial action leads to the destruction of contaminants. The solid phase may consist of soil, organic matter, activated carbon, clay particles or porous glass. When the volatile compound is first passed through the filter, it is sorbed by the solid phase. Then, the sorbed molecule is metabolized by the film of microorganisms that adhere to the solid phase. The microorganism uses the unwanted pollutant as a carbon and energy source and this results in changing a pollutant into an inert substance, such as carbon dioxide and water Figure 9.17 depicts a two-stage biofilter.

Volatile sulphur compounds have caused serious environmental problems due to their high odour index and toxicity. It has also been shown that sulphur-containing compounds can be oxidized in the troposphere causing acid rain and contributes to cloud formation. Researchers at the University of Nijmegen in the Netherlands have isolated and characterized a strain of Hyphomicrobium bacteria that is able to use dimethylsulphide, a major contributor to the volatile sulfur compounds released in air pollution emissions, as a carbon and energy source.

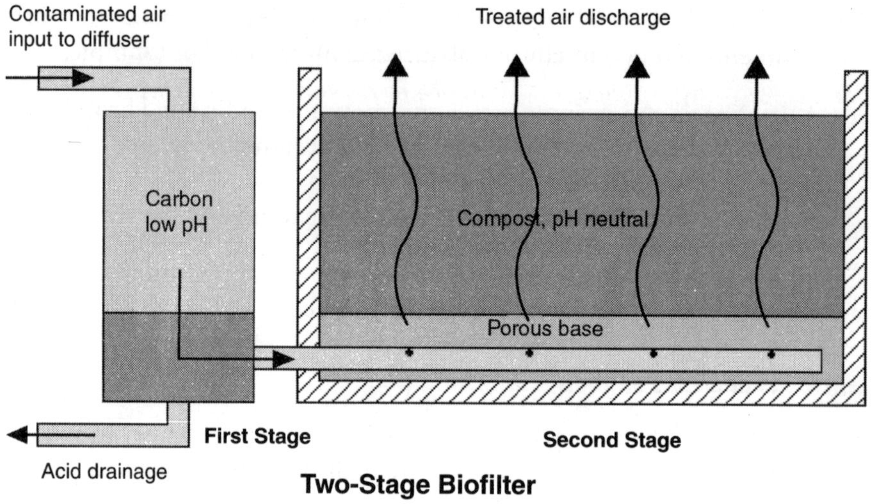

FIGURE 9.17 Two-Stage Biofilter

Bioleaching

Bioleaching (microbial leaching or biomining) is a method for extraction of specific metals from their ores using microorganisms, especially bacteria. Biomining is used by the mining industry to extract minerals such as gold, copper, zinc, uranium, nickel and cobalt from the ores with the help of biological processes. Traditional methods of metallurgy that include roasting and smelting are time-consuming and expensive. These methods also require huge concentrations of elements in ores. However, in bioleaching, minute concentrations of minerals do not pose a problem. This is because the bacteria simply ignore the waste materials surrounding the metals. This results in attaining extraction yields of over 90% in some cases. Bioleaching is therefore useful in recovery of valuable metals from low-grade ores. During bioleaching, the metals of interest are released into the solution phase during oxidation. These metals are later recovered from the solution and the solid residue is discarded.

Large quantities of low-grade ores are produced when high-grade ores are discarded into waste heaps after metal extraction and re-enter the environment. Low-grade ores contain significant amounts of metals that could be processed by microbial leaching. The microorganisms involved in bioleaching are benefited as they gain energy by breaking down minerals into their constituent elements. Bioleaching of minerals from sulphide ores is now an established industrial technology for the recovery of metals such as gold (from arsenical pyrite ores and copper).

Microorganisms Commonly Used in Bioleaching

The most important and commonly used microorganisms for bioleaching are *Thiobacillus thiooxidans* and *T. ferrooxidans*. The former oxidizes sulphur ions, while the latter oxidizes both reduced sulfur and iron compounds. The other microorganisms that can also be used in bioleaching are *Bacillus licheniformis, B. Luteus, B. megaterium, B. polymyxa, Leptospirillum ferrooxidans, Pseudomonas fluorescens, Sulfolobus acidocaldarius, Thermothrix thioparus, Thiobacillus thermophilica*. Fungi are also capable of solubilizing metals from mineral ores. The process of metal solubilization by Fungi relies on the production of organic acids and non specific chelating substances. Until now, use of species of only two fungi, *Aspergillus* and *Penicillium,* has been reported.

Bioleaching Processes

The bioleaching process is of two types:

- **Direct Bacterial Leaching:** In this process, the bacteria physically attach themselves to metal sulphide ores within the rock. The bacteria change metal sulphide ores into soluble sulfates with enzymatically catalyzed steps. For example, the iron pyrite (FeS_2) is oxidized to ferric sulphate ($Fe_2(SO_4)_3$) as shown:

$$4FeS_2 + 15O_2 + 2H_2O \xrightarrow{T.\ ferrooxidans} 2Fe_2(SO_4)_3 + 2H_2SO_4$$

- **Indirect Bacterial Leaching:** In this process, a direct contact does not exist between the bacteria and the mineral surface. The bacteria are only catalysts that accelerate oxidation of metal ores indirectly by producing leaching agents. For example, the bacteria reoxidize ferrous iron back to the ferric form, as well as oxidize elemental sulphur, in some cases. The ferric iron then chemically oxidizes the sulphide minerals producing ferrous iron. The process takes place very slowly in the absence of bacteria, as shown:

$$FeS_2 + Fe_2(SO_4) \longrightarrow 3FeSO_4 + 2S$$

$$2S + 3O_2 + 2H_2O \longrightarrow 2H_2SO_4$$

Oxidation of ferrous (Fe^{2+}) to ferric (Fe^{3+}) by *T. ferrooxidans* at low pH is given below:

$$4FeSO_4 + 2H_2SO_4 + O_2 \xrightarrow{T.\ ferrooxidans} 2Fe_2(SO_4)_3 + 2H_2O$$

Methods of Bioleaching

There are four commercial methods generally used in bioleaching:

- **Slope Leaching:** In this process, large quantities (about 10,000 tons) of ores are first ground to get fine pieces. It is then dumped in large piles down a mountainside. Water containing *Thiobacillus* inoculum is continuously sprinkled over these piles. This water is collected at the bottom and is used to extract metals.
- **Heap Leaching:** In this process, the ore is dumped in a large heap called leach dump. Further steps for treatment of ore and metal extraction is as described for slope leaching.
- ***In situ* Leaching:** In this process, ores are not displaced from their original location. The rocks containing these ores are blasted to increase permeability of water. Thereafter, water containing desired microbes is pumped through drilled passage to the ores. Acidic water seeps through the rock and collects at the bottom. From bottom the used water is pumped out, mineral is extracted from it and the water is reused.
- **Agitated Leaching:** In this process, the soil particles or rocks containing mineral ores are deposited in a tank and leached out using mechanical agitation.

Bioleaching Examples

Copper Leaching

Chalcocite (Cu_2S), chalcopyrite ($CuFeS_2$) or covellite (CuS) are some of the ores commonly used for extraction of copper. Chalcocite is oxidized to soluble form of copper (Cu^{2+}) and covellite by *T. ferroxidans.*

$$Cu_2S + O_2 \longrightarrow CuS + Cu^{2+} + H_2O$$

Covellite is then oxidized to copper sulphate either chemically or by bacteria

$$2CuFeS_2 + 81/2O_2 + H_2SO_4 \longrightarrow 2CuSO_4 + Fe_2(SO_4)_3 + H_2O$$

The subsequent reactions are strictly chemical in nature

$$CuS + 8Fe^{3+} + 4H_2O \longrightarrow Cu^{2+} + 8Fe^{2+} + SO_4^{\ 2-} + 8H^+$$

Copper is removed as below

$$Fe^0 + Cu^{2+} \longrightarrow Cu^0 + Fe^{2+}$$

Fe^{2+} is transferred in the oxidation pond

$$Fe^{2+} + 1/4O_2 + H^+ \xrightarrow{T.\ ferrooxidans} Fe^{3+} + 1/2H_2O$$

The Fe^{3+} ions produced are pumped back to the pile. Sulphuric acid is added to maintain pH. Microbial leaching of copper is widely used in the USA, Australia, Canada, Mexico, South Africa and Japan.

Nickel Leaching

Nickel is generally leached from laterites ores (0.9–3.1% nickel) using species of fungus *Penicillium* and *Aspergillus.*

Gold Leaching

Gold is extracted through bioleaching of arsenopyrite/pyrite ore and its cyanidation process. Pyrite and arsenopyrite get partially removed through biooxidation carried out by *T. ferrooxidans.* This frees the gold enough to allow efficient subsequent extraction by application of solutions of cyanide or thiourea.

$$4FeAs(III)S + 13O_2 + 6H_2O = 4H_3As(V)O_4 + 4FeSO_4$$

Advantages of Bioleaching

Some of the major advantages of bioleaching are:

- Bioleaching is effective in extraction of metals from low-grade ores where conventional methods fail.
- Bioleaching is simpler and much cheaper to operate than traditional processes, since it is not specialized like complex chemical plants. In a developing country like India, biomining becomes even more significant.
- Biomining has relatively less environmental problem than traditional extraction methods and is more environment friendly. It becomes even more important in industries since it helps in cutting down the cost, increasing the profit. For example, limiting the emission of sulphur dioxide produced during smelting is necessary but is also very expensive. This cost can be reduced by using biomiming.
- Bioleaching causes less damage to landscape because the bacteria involved grow naturally. As a result, the mine and the vicinity can be left relatively untouched. The bacteria breed in the conditions of the mine and are easily cultivated and recycled.

Disadvantages of Bioleaching

Some disadvantages associated with bioleaching are:

- The process is very slow compared to smelting and other conventional processes. This can affect the profit made. However, the cheaper cost of bacterial leaching outweighs the time it takes to extract metal.
- Sometimes, toxic chemicals are produced during the process (for instance, Sulphuric acid, arsenic, heavy metal ions such as iron, zinc). This can leak into the ground and surface water, turning it acidic and causing environmental damage. When the pH of this solution rises these ions precipitate, forming "Yellow Boy" pollution.
- So far, bioleaching has been applied only to ores of high-value metals such as gold, copper and uranium ores. Bioleaching is not yet developed to a level where it is profitable to apply it to the recovery of zinc, nickel, cobalt and other metals.

Biofertilizers

Biofertilizers are environment friendly fertilizers now used in most countries. They are ready-to-use live formulates of such beneficial microorganisms, which, on application to seed, root or soil mobilize the availability of nutrients by their biological activity in particular, and help build up the micro-flora and in turn the soil health. Environmental pollution is one of the major concerns in today's world and the use of chemical fertilizers and pesticides has caused tremendous harm to the environment. Biofertilizers can solve this problem by enriching the soil with nutrients in a natural way. The main sources of biofertilizers are bacteria, fungi and cynobacteria (blue-green algae). It can help in solving problems, such as increased salinity of soil and chemical run-offs from the agricultural fields, making them even more important to ensure a healthy future. There are many benefits of using biofertilizers:

- Some of them are eco-friendly
- Increase crop yield by 20–30%
- Replace chemical nitrogen and phosphorus by 25%
- Stimulate plant growth
- Activate soil biologically
- Restore natural soil fertility
- Provide protection against drought and some soil-borne diseases

The process of using biofertilizers is also cost-effective and it supplements chemical fertilizers. Table 9.1 displays the types of biofertilizers available:

TABLE 9.1 Types of Biofertilizers

For Nitrogen	Rhizobium for legume crops Azotobacter/Azospirillum for non-legume crops Acetobacter for sugarcane only Blue Green Algae (BGA) and Azolla for low land paddy
For Phosphorous	Phosphatika for all crops to be applied with Rhizobium, Azotobacter, Azospirillum and Acetobacter
For Enriched Compost	Cellulolytic fungal culture Phosphatika and Azotobacter culture

Some of the biofertilizers recommended for crops are displayed in Table 9.2:

TABLE 9.2 Biofertilizers

Biofertilizer	Measure	Recommended for
Rhizobium + Phosphotika	200 gm each per 10 kg of seed	pigeonpea, green gram, black gram, cowpea, etc, groundnut and soybean
Azotobacter + Phosphotika	200 gm each per 10 kg of seed	wheat, sorghum, maize, cotton, mustard
Azospirillum + Phosphotika	5 kg each per ha	transplanted rice

There are various methods for the application of biofertilizers to crops. Some of them are:

- **Seed treatment:** 200 g of nitrogenous biofertilizer and 200 g of Phosphotika are suspended in 300–400 ml of water and mixed thoroughly. Ten kilograms of seeds are treated with this paste and dried in shade. The treated seeds have to be sown as soon as possible.
- **Seedling root dip:** For rice crop, a bed is made in the field and filled with water. Recommended biofertilizers are mixed in this water and the roots of seedlings are dipped for 8–10 hrs.
- **Soil treatment:** 4 kg each of the recommended biofertilizers are mixed in 200 kg of compost and kept overnight. This mixture is incorporated in the soil at the time of sowing or planting.

Some precautions are needed to be taken while using biofertilizers. Some of them are listed:

- Store the biofertilizer packets in a cool and dry place away from direct sunlight and heat.
- Use the right combinations of biofertilizers.
- Use the biofertilizer for the specified crop alone.
- Do not mix other chemicals with the biofertilizers.
- Ensure, while purchasing, that each packet is provided with necessary information, such as the name of the product, name of the crop for which intended, name and address of the manufacturer, date of manufacture, date of expiry, batch number and instructions for use.
- Use the packet before its expiry.
- To get the best results, use both nitrogenous and phosphatic biofertilizers.
- Use biofertilizers along with chemical fertilizers and organic manures.
- Biofertilizers are not replacement of fertilizers but can supplement plant nutrient requirements.

Mycorrhiza

Mycorrhizae are a group of fungi that include a number of types based on different structures formed inside or outside the root. These are specific fungi that match with a number of favourable parameters of the host plant on which it grows. This includes soil type, presence of particular chemicals in the soil, and other conditions. These fungi grow on the roots of these plants. In fact, seedlings that have mycorrhizal fungi growing on their roots survive better after transplantation and grow faster. The fungal symbiont gets shelter and food from the plant that, in turn, acquires an array of benefits, such as better uptake of phosphorus, salinity and drought tolerance, maintenance of water balance, and an overall increase in plant growth and development. While selecting fungi, the right fungi have to be matched with the plant. There are specific fungi for vegetables, fodder crops, flowers and trees. Mycorrhizal fungi can increase the yield of a plot of land by 30–40%. They can absorb phosphorus from the soil and pass it on to the plant. Mycorrhizal plants show higher tolerance to high soil temperatures, various soil- and root-borne pathogens and heavy metal toxicity.

Legume-Rhizobium Relationship

Leguminous plants require high quantities of nitrogen compared to other plants. Nitrogen is an inert gas and its uptake is possible only in fixed form. This is facilitated by the rhizobiumbacteria present in the nodules of the root system. The bacterium lives in the soil to form root nodules (outgrowth on roots) in plants such as beans, gram, groundnut and soyabean.

Blue-Green Algae

Blue-Green Algae (BGA) are considered the simplest, living autotrophic plants, i.e., organisms capable of building up food materials from inorganic matter. They are microscopic. BGA are widely distributed in the aquatic environment. Some of them are responsible for water blooms in stagnant water. They adapt to extreme weather conditions and are found in snow and in hot springs, where the water is 85°C. Certain BGA live intimately with other organisms in a symbiotic relationship. Some are associated with the fungi in the form of lichens. The ability of BGA to photosynthesize food and fix atmospheric nitrogen accounts for their symbiotic associations and also for their presence in paddy fields.

BGA are of immense economic value as they add organic matter to the soil and increase soil fertility. Barren alkaline lands in India have been reclaimed and made productive by inducing the proper growth of certain BGA.

Biopesticides

Biopesticides, also known as biological pesticides, is a category of pesticides derived from natural materials such as animals, plants, bacteria and certain minerals, for example, garlic, mint and baking soda. By the end of 1998, there were approximately 175 registered biopesticide active ingredients and 700 products. Biopesticides, in general, have the following properties:

- Have narrow target range
- Have a very specific mode of action
- Are slow acting
- Have relatively critical application times
- Suppress, rather than eliminate, a pest population
- Have limited field persistence and a short shelf life
- Are safer to humans and the environment than conventional pesticides
- Present no residue problems.

Biopesticides fall into three major categories:

- **Microbial Pesticides:** include microorganism as the active ingredient (bacterium, fungus, virus, protozoan or alga) and help to control insects and weeds. The most widely known microbial pesticides are varieties of the bacterium *Bacillus thuringiensis*, or Bt, which are well known to control certain insects found in cabbage, potatoes and other crops. Bt produces a crystalline protein that is harmful to specific insect pests. Certain other microbial pesticides act by out-competing pest organisms. Microbial pesticides need continuous monitoring to ensure that they do not harm non-target organisms, including humans.

- **Plant Pesticides:** are produced by plants through the genetic material added to the plant. For example, scientists have inserted the gene for the Bt pesticidal protein into the plant's own genetic material. Thus, the plant manufactures the crystalline protein, which destroys the pest.
- **Biochemical Pesticides:** are naturally occurring substances that control pests through non-toxic mechanisms. Biochemical pesticides include substances that interfere with insect's growth or mating such as plant growth regulators, or substances that repel or attract pests, such as pheromones. For example, man-made pheromones are used to disrupt insect mating by creating confusion when the insects search for mates, or can be used to attract male insects to traps. Pheromones can thus be used either to detect or monitor insect populations, or to control them. Because it is sometimes difficult to determine whether a natural pesticide controls the pest by a non-toxic mode of action, EPA has established a committee to determine whether a pesticide meets the criteria for a biochemical pesticide. Table 9.3 displays some biopesticides:

TABLE 9.3 Biopesticides

Types of Biopesticides	Examples
	Insect Control
Bacteria	*Bacillus thuringiensis, B. sphaericus, B. popilliae, Serratia entomophila*
Viruses	Nuclear polyhedrosis viruses, granulosis viruses, non-occluded baculoviruses
Fungi	*Beauveria, Metarhizium, Entomophaga, Zoopthora, Paecilomyces, Normuraea*
Protozoa	*Nosema, Thelohania, Vairimorpha*
Entomopathogenic nematodes	*Steinernema, Heterorhabditis, Romanomermis*
Others	Pheromones, parasitoids, predators, microbial by-products
	Weed Control
Fungi	*Alternaria, Colletotrichum*
Bacteria	*Xanthomonas*
	Plant Disease Control
Nematode trapping fungi	*Arthrobotrys*

There are many advantages of using biopesticides. They are:

- inherently less harmful than conventional pesticides.
- designed to affect only one specific pest or a few target organisms, in contrast to conventional pesticides that may affect broad spectrum of organisms such as birds, insects and mammals.
- effective in very small quantities and decompose quickly, thereby resulting in lower exposures and largely avoiding the environmental pollution problems caused by conventional pesticides.
- greatly decrease the use of conventional pesticides, when used as a component of Integrated Pest Management (IPM) programmes, yet crop yields remain high.

TERMS TO KNOW

Reference: Steve Bentjen, Copyright 1998

Terms	Explanation
Acclimatization	The physiological and behavioral adjustments of an organism to changes in its environment.
Acid Deposition	A complex chemical and atmospheric phenomenon that occurs when emissions of sulfur and nitrogen compounds and other substances are transformed by chemical processes in the atmosphere, often far from the original sources, and then deposited on earth in either wet or dry form. The wet forms, popularly called "acid rain", can fall to earth as rain, snow, or fog. The dry forms are acidic gases or particulates.
Aerobic Treatment	Process by which microbes decompose complex organic compounds in the presence of oxygen and use the liberated energy for reproduction and growth (Such processes include extended aeration, trickling filtration and rotating biological contactors).
Air Pollutant	Any substance in air that could, in high enough concentration, harm man, other animals, vegetation, or material. Pollutants may include almost any natural or artificial composition of airborne matter capable of being airborne. They may be in the form of solid particles, liquid droplets, gases, or in combination thereof. Generally, they fall into two main groups: (1) those emitted directly from identifiable sources and (2) those produced in the air by interaction between two or more primary pollutants, or by reaction with normal atmospheric constituents, with or without photoactivation. Exclusive of pollen, fog and dust, which are of natural origin, about 100 contaminants have been identified. Air pollutants are often grouped in categories for ease in classification; some of he categories are: solids, sulfur compounds, volatile organic chemicals, particulate matter, nitrogen compounds, oxygen compounds, halogen compounds, radioactive compound and odours.
Bioaccumulants	Substances that increase in concentration in living organisms as they take in contaminated air, water, or food because the substances are very slowly metabolized or excreted.
Bioavailability	Degree of ability to be absorbed and ready to interact in organism metabolism.

Biochemical Oxygen Demand (BOD)	A measure of the amount of oxygen consumed in the biological processes that break down organic matter in water. The greater the BOD, the greater the degree of pollution.
Bioconcentration	The accumulation of a chemical in tissues of a fish or other organism to levels greater than in the surrounding medium.
Biodegradable	Capable of decomposing under natural conditions.
Biodiversity	Refers to the variety and variability among living organisms and the ecological complexes in which they occur. Diversity can be defined as the number of different items and their relative frequencies. For biological diversity, these items are organized at many levels, ranging from complete ecosystems to the biochemical structures that are the molecular basis of heredity. Thus, the term encompasses different ecosystems, species and genes.
Effluent	Wastewater—treated or untreated—that flows out of a treatment plant, sewer, or industrial outfall. Generally refers to wastes discharged into surface waters.
Emission	Pollution discharged into the atmosphere from smokestacks, other vents and surface areas of commercial or industrial facilities; from residential chimneys; and from motor vehicle, locomotive, or aircraft exhausts.
Eutrophication	The slow aging process during which a lake, estuary, or bay evolves into a bog or marsh and eventually disappears. During the later stages of eutrophication the water body is choked by abundant plant life due to higher levels of nutritive compounds such as nitrogen and phosphorus. Human activities can accelerate the process.
Global Warming	An increase in the near surface temperature of the Earth. Global warming has occurred in the distant past as the result of natural influences, but the term is most often used to refer to the warming predicted to occur as a result of increased emissions of greenhouse gases. Scientists generally agree that the Earth's surface has warmed by about 1 degree Fahrenheit in the past 140 years. The Intergovernmental Panel on Climate Change (IPCC) recently concluded that increased concentrations of greenhouse gases are causing an increase in the Earth's surface temperature and that increased concentrations of sulfate aerosols have led to relative cooling in some regions, generally over and downwind of heavily industrialized areas.

Groundwater	The supply of fresh water found beneath the Earth's surface, usually in aquifers, which supply wells and springs. Because groundwater is a major source of drinking water, there is growing concern over contamination from leaching agricultural or industrial pollutants or leaking underground storage tanks.
Phytoremediation	Low-cost remediation option for sites with widely dispersed contamination at low concentrations.
Phytotreatment	The cultivation of specialized plants that absorb specific contaminants from the soil through their roots or foliage. This reduces the concentration of contaminants in the soil, but incorporates them into biomasses that may be released back into the environment when the plant dies or is harvested.
Plume	1. A visible or measurable discharge of a contaminant from a given point of origin. Can be visible or thermal in water, or visible in the air as, for example, a plume of smoke 2. The area of radiation leaking from a damaged reactor 3. Area downwind within which a release could be dangerous for those exposed to leaking fumes
Pollution	Generally, the presence of a substance in the environment that because of its chemical composition or quantity prevents the functioning of natural processes and produces undesirable environmental and health effects. Under the Clean Water Act, for example, the term has been defined as the man-made or man-induced alteration of the physical, biological, chemical and radiological integrity of water and other media.
Pyrolysis	Decomposition of a chemical by extreme heat.
Toxic Cloud	Airborne plume of gases, vapors, fumes, or aerosols containing toxic materials.
Toxic Concentration	The concentration at which a substance produces a toxic effect.
Toxic Dose	The dose level at which a substance produces a toxic effect.
Toxic Pollutants	Materials that cause death, disease, or birth defects in organisms that ingest or absorb them. The quantities and exposures necessary to cause these effects can vary widely.
Zero Air	Atmospheric air purified to contain less than 0.1 ppm total hydrocarbons.

10 Bioethics and Biosafety

- Introduction
- Concerns in Biotechnology
- Moral and Ethical Considerations
- Health Issues
- Biosafety

Introduction

The concept of safety has been around since time immemorial. The notion that there are certain elements, which are hazardous to limb and life, prompted the society to define certain norms to check incidences of exposure to the same. Though safety is practised at various levels, it is most ubiquitous in case of technological applications. This is justified considering the fact that most of the technological applications involve the use of large amounts of power and generate tremendous energy. The concept of safety has gradually evolved from the days of industrial revolution and the steam engine. From being a policy followed by a few industries, safety has evolved into a legislation and obligation. In case of biotechnology, concerns about regulations of safety are recent but not out of place. In the current chapter, an effort would be made to address a few such issues.

Concerns in Biotechnology

There are a number of contentious issues surrounding biotechnology as far as safety is concerned. One of the issues that have been in the forefront concerns the Genetically Modified Organisms (GMO). As is evident, genetic modification is a pervasive process. Moreover, the effect of these modifications might not be immediately visible. Another aspect to consider is that most modifications that are made are done to introduce specific traits and it is these traits that are scanned in modified organisms. The genome-wide changes in these are, however, overlooked. Safety assessment of these transgenic experiments is of utmost importance because these have a significant impact on the environment.

In many cases, GMOs have competed with their wild counterparts and obliterated the same. Modified crops have ruined the natural diversity in many countries by outgrowing the natural flora in due course of time. The instances of GMOs are restricted to agricultural biotechnology at least for the time being. However, with efforts to legalize cloning in animals and humans, one might have to deal with these issues in the near future.

Moral and Ethical Considerations

No debate on safety can be complete without addressing the issue of ethics. Safety is a choice, which is made by a society. A collective sense of responsibility and awareness is essential to maintain a high level of safety, which is dictated by ethical considerations. Yardsticks for measuring ethical behaviour are never universal and the level of safety that one deems to be fit is also not constant. This is crucial since safety in technology is nothing but a trade-off between the utility that it serves and the hazards that one has to accept in the process. This is what makes the entire debate on safety so dicey. Therefore, it was essential to raise the debate at a global level and conventions like the Cartagena protocol have done just that. However, these too suffer from maladies that will be discussed subsequently.

Ethical questions are, most often than not, centred around the effects of our actions or interactions on the world around us. In most cases, the decision is guided by principles of harmfulness of a particular action, so if it is harmful one needs to abstain. This is theoretically possible. However, the same becomes extremely dicey if the consequences of that particular action are not very clear, or when it can have both harmful as well as beneficial results. This is the case with technology, more so, with biotechnology since it is still in the nascent stage and making proper evaluation is difficult.

Some of the major concerns surrounding biotechnology range from its effects on the environment and human health to impacts on socio-economic conditions and religious and moral values. Some of these issues are innate to the nature of technology, while others, such as the exploitation of poor nations' resources by the rich ones, are still in a quandary.

Generating Opinions

Generating public perceptions is crucial to developing a consensus on any technology. The average person on the street learns of biotechnology mainly through the media. In a country like ours, where science journalism is relegated to a non-existent status, half-baked stories abound the media. Even in the

West, the bulk of information flows from the industry and university sources, with the industry being the dominant force. Majority of the users of biotechnology, be it farmers or scientists outside research institutes, are rarely cited. In a nutshell, most of the reports were nothing but reports of new discoveries presented by the discoverers themselves. A cursory look at some of the reports reveals that they mainly focus on the economic and health benefits rather than discuss the regulatory issues. Even worse, none of the articles in the media are entirely negative and arguments on public awareness or ethics are rarely reported. A major issue regarding this is the one-sided portrayal of the scenario by academicians and universities, who are accused of having vested interests in creating favourable public opinion.

Glamorizing and creating hype suit scientists and institutions alike. In the process, the average man gets nothing but a shoddy representation of facts behind the technology. Academicians, on their part, blame the media and journalists, who with little scientific training, are intimidated by technical information and end up misrepresenting facts. Whatever the fact, the common man remains the loser. One of the reasons for ambiguity surrounding this technology is the lack of honesty in spreading it among masses. It is the prohibition of truth that breeds paranoia—a fact that has to be addressed soon if the potential of the technology were to be harnessed.

Whose Gene is it, Anyway?

In today's competitive environment, a successful inventor is not the one who builds a superior mousetrap but it is the one who builds a superior mouse. The issue of patenting life forms has come a long way since 1989, when Du Pont commercialized and sold mice that were carrying cancer-causing genes in their cells. These mice were used for screening anticancer drugs.

As will be elucidated in the next chapter, the regulation of granting patents has evolved over decades, however, with the advent of biotechnology, it was forced to accommodate life forms and allied products. The history goes back to the United States, where, in 1930, the Plant Patent Act gave breeders the right to grow novel strains. However, the possibility of altering single genes and switching them between species was not envisaged at the time. A new era dawned in 1980 when Anand Chakraborty got the first patent on a microorganism that was altered genetically to digest oil slicks. The most shocking aspect of this story was to come in 1995 when the U.S. patent office granted a patent to NIH (National Institute of Health) for the genetic material of a foreign national, a Haghai man from the highlands of Papua New Guinea. NIH now possesses exclusive rights to the genetic material of tribes that number around 260 and regulates any study pertaining to the use of their DNA. We have indeed come a far way from the days of colonialism where researchers used to hunt down indigenous people's resources. Now it is the people themselves who are facing a threat from Western marauders.

Under the garb of the Human Genome Project, institutes such as the NIH seek patents on human genes in various countries, with no concrete provisions to compensate the original owners. This can be seen as the beginning of a dangerous trend in which indigenous people are being viewed as raw materials for companies in developed nations. Foundations like the Rural Advancement Foundation International (RAFI) are involved in monitoring such controversial patents around the world.

The criterion for patenting is another grey area that biotechnologists have to contend with. To qualify for a patent, the invention must be novel, useful and non-obvious. The problem, which scientists encounter in biotechnology, is the determination of novelty and obviousness. In very rare cases, the value of a discovery is clear-cut and obvious. Furthermore, the application of these discoveries is even more difficult to envisage. In order to survive competition, scientists have to defend and safeguard their intellectual property and innovations. The burden of repaying heavy funds that are invested in the research also drives scientists to seek broad patents on discoveries that are still vague. It takes several years and millions of dollars to bring biotechnology applications into the market, thereby justifying the approach adopted by companies to safeguard their interests. However, if the patent is too broad in nature, such as a concept or a technique, it may restrict other researchers. This can eventually slow down the progress of science on the whole. In the new areas opened by biotechnology, adjudicating the legitimacy patent claims is as much a matter of interpretation as precedent.

So, what can be patented? The following are some:

- Genetically altered microorganisms (bacteria, fungi, algae, single-celled organisms, viruses)
- Newly discovered microbes
- Techniques for genetically manipulating microbes, plants and animals
- Proteins prepared by genetic engineering processes
- Plant, animal and human genes
- Cell lines

Taking from the Poor

The quest for unknown organisms with useful properties has set off a race among Western countries, unleashing an army of "bioprospectors" into the world's tropical forests. This has prompted a host of others to study agricultural and therapeutic uses of indigenous cultures. While Western countries reap the rewards of such endeavours, the countries where these are discovered get nothing in return.

Colonization is not a new phenomenon and neither is the transfer of valuable resources from poor countries. However, biotechnology has only added insult to injury. The global distribution of modified crops and livestock reduces the diversity of food grown around the world, increases costs to farmers and increases the dependence of populace on a few large corporations. Patenting of animals and plants means that farmers are forced to pay heavy royalties to the patent holder every time they use it. The most controversial technology in this regard is the terminator technology that renders harvested seeds unviable, forcing the farmer to buy fresh seeds before every season. Much of the situation applies to the pharmaceutical industry where the healing properties of plants used by indigenous people ends up providing profits to corporations.

Health Issues

There are a number of issues that plague the use and introduction of genetically modified products. The biggest concern has been regarding transgenic food. People with food allergies are concerned, since a chemical to which they may react badly, can be transferred to them from genetically engineered food (in which the chemical was previously absent). Accurate labeling is the need of the hour and the European Union has already imposed strict regulations regarding commercialization of such products. Another fear is that genetically altered food carrying marker genes for antibiotic resistance might transfer the same to consumers, in the process rendering antibiotics ineffective against many diseases.

The ability to diagnose genetically caused diseases has given rise to another dilemma. On one hand, such knowledge can create depression among the patients facing from incurable diseases, while on the other hand, it may enable them to prepare for the same in advance. This is also advantageous in the case of some diseases like phenylketonuria (PKU). The effects of the disease can be prevented by means of a special diet and thus prospective patients can limit its onset and the damage the disease causes.

One area that has received maximum attention is that of stem cells and the ethics behind using human embryonic cells. Ethical questions like what is self and non-self and at what stage does the embryo becomes "alive" are relentlessly pursued. In addition to this, the cost of such treatments put it out of the reach of common people, sparking debates on the class divide and issues arising therein.

Biosafety

Society has, for centuries, safely used products and processes of biotechnology. The difference between then and now is that there was no alteration of nature nor were the products and processes toxic to the environment or humans.

There are major concerns that surround the processes used in biotechnology currently. Some of these issues have been listed in Table 10.1:

TABLE 10.1 Biosafety Issues

	Concerns of Using Biotechnology Products
Health	Toxicity & Food quality/safety
	Allergies
	Pathogen drug resistance (antibiotic resistance)
Environmental	Persistency of gene or transgene or of transgene products
	Susceptibility of non-target organisms
	Increased use of chemicals in agriculture
	Unpredictable gene expression or transgene instability
Agricultural	Resistance/Tolerance of target organisms
	Weeds or Superweeds
	Alteration of nutritional value
General	Loss of familiarity
	Higher cost of agriculture
	Field trials not planned for risk assessment

Most of the organisms that are used in biotechnology for commercial purposes are relatively harmless and a few virulent organisms are used under strict containment processes. Biosafety has been an issue where stringent rules are imposed on the industries dealing with recombinant organisms. In spite of this, fear regarding the safety of the technology still exists. Vaccines and antibiotics are the two main products whose indiscriminate dispersal has to be prevented. Contaminants in otherwise safe processes may produce certain toxic molecules that are incorporated into the final products. This can lead to food poisoning.

Risk Assessment and Management

In recent years, with stringent regulations in place, the concept of risk evaluation and management has come into vogue. One can define the amount of risk associated with an event:

$$\text{Amount of risk} = \text{Hazard} \times \text{Exposure/Safeguards.}$$

It deals with the following aspects:

- Elucidation of capacity of the microorganism to have adverse effect on humans or environment.
- Establishment of probability that microorganisms might escape, either accidentally or inadvertently, from production process system.
- Evaluation of the safety of the desired products and the methods for handling the by-products.

Risk assessment vis-à-vis GMOs deals with a number of guidelines that are enumerated below:

A. Determination of the level of safety concern for the unmodified organism. This includes the following:

    1. the pest/pathogen status of the organism;
    2. its ability to become established in the accessible environment;
    3. its ecological interrelationship, function and importance in the community;
    4. its ability to transfer genetic information;
    5. the potential for monitoring and control.

B. Consideration of how genetics affect safety. This includes the following:

    1. the process of genetic modification;
    2. gene construction and expression;
    3. the degree to which knowledge of molecular biology and other information is available to predict the safety level of modified organism relative to that of the unmodified organism.

C. To combine evaluation in the steps above.

The risk assessment concerning GMOs is made by integration of the information on the unmodified organism and on the genetic modification.

It is important to have well defined foreign DNA inserts in the transformed organisms. By use of available transformation methods it is still difficult to obtain a transformed individual species containing single or low-copy number of foreign genes. Determination of possibilities of genes movement from transgenic plants to surroundings is important. Gene escape from transgenic plants can be precipitated by one of several events (Love, 1994):

- transfer of modified genes into microorganisms or pathogen population;
- horizontal movement of genes into other plant species, due to latent infection with a bacterial vector;

- escape and proliferation of transgenic plants in to the wild;
- hybridization with and transgene introduction into native wild species (Evenhuis et al., 1991; Kapteijns, 1993; Regal,1994).

However, it was soon realized that these guidelines were not all encompassing and biosafety was an issue that had to be addressed globally and at various levels. The Cartegena Protocol that was held in 2000 adopted a special resolution on the issue of biosafety. It sought to protect biological diversity from potential risks posed by living modified organisms resulting from modern biotechnology and issued a number of regulations pertaining to risk assessment and management. The objective of risk assessment, under this Protocol, was to identify and evaluate the adverse effects of modified organisms. It also studied the conservation and sustainable use of biological diversity, taking into account the threats posed to human health. Some of the guidelines for risk assessment are:

- Risk assessment should be carried out in a scientifically sound and transparent manner, and can take into account expert advice of, and guidelines developed by, relevant international organizations.
- Lack of scientific knowledge or scientific consensus should not necessarily be interpreted as indicating a particular level of risk, an absence of risk, or an acceptable risk.
- Risks associated with living modified organisms or products thereof, namely, processed materials that are of living modified organism origin, containing detectable novel combinations of replicable genetic material obtained through the use of modern biotechnology, should be considered in the context of the risks posed by the non-modified recipients or parental organisms in the likely potential receiving environment.
- Risk assessment should be carried out on a case-by-case basis. The required information may vary in nature and level of detail from case to case, depending on the living modified organism concerned, its intended use and the likely potential receiving environment.

The various steps that are enumerated for the risk assessment are:

1. An identification of any novel genotypic and phenotypic characteristics associated with the living modified organism that may have adverse effects on biological diversity in the likely potential receiving environment, taking also into account risks to human health;
2. An evaluation of the likelihood of these adverse effects being realized, taking into account the level and kind of exposure of the likely potential receiving environment to the living modified organism;

3. An evaluation of the consequences should these adverse effects be realized;
4. An estimation of the overall risk posed by the living modified organism based on the evaluation of the likelihood and consequences of the identified adverse effects being realized;
5. A recommendation as to whether or not the risks are acceptable or manageable, including, where necessary, identification of strategies to manage these risks; and
6. Where there is uncertainty regarding the level of risk, it may be addressed by requesting further information on the specific issues of concern or by implementing appropriate risk management strategies and/or monitoring the living modified organism in the receiving environment.

The issue of biosafety has more to do with enforcement than legislation. It is an ironical state that legislators of these guidelines (read Western countries) are the most frequent violators. In addition to this, the guidelines serve as a handle to stifle the growth of developing countries in areas such as biotechnology where the West enjoys a distinct advantage.

Promethean Pursuits

The Greek god Prometheus was punished for his presumption of stealing fire, which is akin to what many people think about biotechnology. Humans have since long both benefited and suffered as a result of their curiosity. Risk assessment is partly a matter of data, partly a matter of interpretation and temperament. It is also affected by familiarity. We are more willing to live with familiar risks than new ones. Our wait for any tangible signs of harm caused by biotechnology should not let us abandon the same. If risk is the sole criterion, vehicles should be banned considering the average number of lives that are lost due to road accidents. The fact that this does not happen is enough to state that the path to social change is not dictated by logic alone. When benefits far outweigh the risks, one chooses to remain relatively uncritical of the technology regardless of the weight of evidence against it. The problem with biotechnology is that since we have little or no personal experience of the risks and benefits, risks tend to occupy our psyche.

The standard argument against biotechnology is "tampering against nature", which falls flat on its face if one were to consider the harm we have done to nature using seemingly mundane technologies. Whether it is building dams or polluting the environment, humans have already outdone them in the pursuit of destroying the nature. Ethical issues can never be perceived by nature. These are human issues and depend on human nature. The choices have been ours and not created by biotechnology, whether we use fire to warm ourselves and cook or burn down someone's house, the pronouncement has been with us.

11 Intellectual Property Rights

- HISTORY
- INDIAN SCENARIO
- THE INDIAN PATENTS ACT 2005
- INDIAN BIODIVERSITY ACT
- PROTECTION OF PLANT VARIETIES AND FARMERS' RIGHTS

HISTORY

The basic tenets of ownership and rights on the property of concern are protected by certain laws and legislations. These laws are relatively easier to establish and regulate in case of tangible assets. However, it becomes ambiguous and difficult to define for intangible property. Prominent among these are the intellectual properties of an individual. Biotechnology is an area that has been driven by innovation and significant individual contributions. The need to recognize their contributions and offer incentives for innovations triggered the catalysis of certain legal proceedings. This was the genesis of patent regulations. In addition to patents, IPR (Intellectual Property Rights) can be protected by means of copyrights, trade secrets and trademarks. Creators can be given the right to prevent others from using their inventions, designs or other creations and in order to use them, creators can negotiate payment in return. These are "Intellectual Property Rights". Intellectual property can be divided into two:

- Industrial Property
- Copyright

Industrial Property

It can be divided into two main areas:

- One area can be characterized as the protection of distinctive signs, in particular trademarks (which distinguish the goods or services of one undertaking from those of other undertakings) and geographical indications (which identify a good as originating in a place where a given characteristic of the good is essentially attributable to its geographical origin).

The protection of such distinctive signs aims to stimulate and ensure fair competition and to protect consumers, by enabling them to make informed choices between various goods and services. The protection may last indefinitely, provided the sign in question continues to be distinctive.

- Other types of industrial property are protected primarily to stimulate innovation, design and the creation of technology. In this category fall inventions (protected by patents), industrial designs and trade secrets.

  The social purpose is to provide protection for the results of investment in the development of new technology, thus giving the incentive and means to finance research and development activities.

Copyright

The rights of authors of literary and artistic works (such as books and other writings, musical compositions, paintings, sculpture, computer programs and films) are protected by copyright, for a minimum period of 50 years after the death of the author. The rights associated with copyright and related rights include those of performing artists in their performances, producers of phonograms in their recordings, and those of broadcasters in their radio and television programs. The main social purpose of protection of copyright and related rights is to encourage and reward creative work.

With time and the changing face of the world economy, a need arose to have a common platform with uniform regulatory processes to facilitate trade without impeding the process of innovation and development. This was also important keeping in mind that disputes were beginning to raise among nations vis-à-vis intellectual property laws owing to lack of any consensus. It was in 1948 that the General Agreement on Trade and Tariff (GATT) was framed by the developed countries to settle these issues. GATT was far from being the perfect solution since developed countries largely dictated the framework. In addition, the draconian legislations passed by the U.S. in 1988 (read Omnibus Trade and Competitiveness Act) gave the developed nations a handle to dictate terms to the lesser developed nations. GATT was beginning to serve the interests of these few nations and not the world as it was envisaged.

GATT Chronology

1944	Creation of the World Bank (WB), the International Monetary Fund (IMF) and the International Trade Organisation (ITO).
1946–47	Negotiation between 20 countries regarding the reduction of trade tariffs and subsequent ratification of the first agreements of the GATT (General Agreement on Tariffs and Trade).
1948	The GATT comes into effect.
1949	Second cycle of negotiations between the contracting parties of the GATT held in Annecy (15 countries participate).
1951	Third cycle of the GATT negotiations (38 countries participate).
1956	Fourth cycle of the GATT negotiations (26 countries participate).
1960–61	Fifth cycle of the GATT negotiations (26 countries participate).
1964–67	Sixth cycle of the GATT negotiations–the Kennedy Round (62 countries participate).
1973–79	Seventh cycle of the GATT negotiations–the Tokyo Round (102 countries participate).
1986	The Uruguay Round is inaugurated in Punta del Este, eighth cycle of the GATT negotiations (123 countries participate), the issue of intellectual property rights discussed along with a host of other topics.
1994	The Marrakech Agreements are signed.
1995	The Marrakech Agreements come into effect. The World Trade Organization (WTO) is created.
1996	First WTO ministerial conference.
1998	Second WTO ministerial conference.
1999	Third WTO ministerial conference is convened.
2001	Fourth WTO ministerial conference is convened at Doha.
2003	Fifth WTO ministerial conference is convened at Cancun.

The WTO is an organization for liberalizing trade and a forum for governments to negotiate trade agreements. Whereas, GATT had mainly dealt with trade in goods, the WTO and its agreements now cover trade in services, and in traded inventions, creations and designs (intellectual property). Some of the benefits of the WTO (as listed on the official website) clearly try to highlight the mandate on which it works.

1. The system helps promote peace;
2. Disputes are handled constructively;

3. Rules make life easier for all;
4. Free trade cuts the costs of living;
5. It provides more choice of products and qualities;
6. Trade raises incomes;
7. Trade stimulates economic growth;
8. The basic principles make life more efficient;
9. Governments are shielded from lobbying;
10. The system encourages good government.

TRIPS: A Snapshot

The TRIPS agreement, which came into effect on 1 January 1995, is to date the most comprehensive multilateral agreement on intellectual property. The broad issues covered under the agreement are:

- Application of basic principles of the trading system and other international intellectual property.
- Adequate protection to intellectual property rights.
- Enforcement of these rights in their own territories.
- Settlement of disputes on intellectual property between members of the WTO.
- Transitional arrangements during the period when the new system is being introduced.

What is a Patent?

A patent is a property right granted by the government to a person that allows the holder to prohibit others from making, using, offering for sale, or selling his invention. A patent can be transferred to another party by a written agreement. Patent protection is for inventions. There are three types of patents that can be granted:

- **Utility:** Utility patents are granted to any person who invents or discovers any new and useful process, machine, manufacture, or compositions of matter, or any new and useful improvement thereof. "Process" as used in the patent statute primarily refers to technical processes. The term "manufacture" refers to all manufactured articles. "Composition of matter" refers to chemical compositions.
- **Design:** The emphasis of this type of patent is on the design of the invention, not on its functionality. What is important with this type of patent are the invention's unique ornamental and aesthetic properties.
- **Plant:** This type of patent is used for asexually reproduced, non-tuberous plants (Adapted from Kenneth Sibley 1994). Asexually propagated

plants are those that are reproduced by means other than from seeds, such as by the rooting of cuttings, budding, grafting, etc. Design and plant patents are granted for shorter terms than utility patents.

A very important consideration for the grant of patents is that the invention should not be already in use or documented. This is true even if the said invention is used in a foreign country other than from where the patent application is filed. Another criterion to be kept in mind is that the invention should not be *obvious*. Hence, the important criteria for consideration of a patent grant are novelty and non-obviousness of the invention.

Indian Scenario

The Indian patent scenario has a long history. The British who had a long history of awarding patents for novel inventions introduced it with an aim to protect their interests within their colonies. The first enactment of the patents act was in the year 1856, which provided patents for inventions in India. In 1859, a new enactment was introduced in India that granted the inventor exclusive privileges of making, selling and using the invention or authorizing others to do so for a period of 14 years. The subsequent years saw a spate of enactments concerned with protecting the inventor's right in India, which ultimately culminated into the Indian designs and Patents Act in 1911. After Independence, an independent evaluating committee reviewed the colonial law on the subject, which resulted in the Indian Patent Law of 1970. The original patents Act of 1970 was a balanced act, which aided the growth of the industrial sector and also covered public interests. Some of the important features of the Act were:

- Absence of product patent for pharmaceutical and food products, only process patents were granted.
- The term of patent was 7 years from the date of application or 5 years from the date of sealing.
- There were no export constraints.
- Agriculture was excluded from the scope of patentability.

The WTO agreements came as a package deal and after a long deliberations India decided to adopt the Final Act of the Uruguay Negotiations in 1994 and became one of the original signatories. In order to comply with the TRIPs agreement, the Indian Patent Act of 1970 was amended twice in 1999 and 2002. A third amendment was introduced by means of an ordinance on

26 December 2004, before subsequent modifications, and was passed by the parliament in the year 2005.

THE INDIAN PATENTS ACT 2005

With the assent of the President the Patent (amendment) Act 2005 is now an integral part of the national patent legislation. The process of amendment of the national patent legislation has been a controversial issue, even necessitating the division of the entire process of amendment into three installments. The third and final was the most controversial and was cleared after a prolonged deliberation. Fortunately, some of the controversial clauses that were introduced in the previous amendments and the Third Amendment ordinance were reversed when the draft patents bill was discussed in the parliament. These include some of the contentious issues such as the criteria for patentability, compulsory licensing and pre-grant opposition. The amendments also incorporate a new definition for "new invention" which is "any invention or technology which has not been anticipated by publication in any document or used in the country or elsewhere in the world before the date of filing of patent application with complete specification, i.e., the subject matter has not fallen in public domain or that it does not form part of the state of art." The amendments also give a definition for "pharmaceutical substance" as being "a new entity involving one or more inventive steps". The new amendments have now also restored pre-grant opposition to patents. These modifications were critical to minimize the damage to Indian interests.

Mailbox Items

During the grace period (1995–2005) for the implementation of WTO obligations, the Indian government had to provide for a 'mailbox' to receive applications for product patents as an interim measure, so that the applicants can enjoy the priority date when the product patent regime eventually comes into force. However, some Indian companies are producing these drugs based on the disclosures from the patents granted to those applicants in other countries, without inviting infringement. The fate of these Indian companies, after the product patents came into force in India, was not clear. The latest amendments have now clarified that Indian companies that are already producing drugs for which applications for product patents are pending in the 'mailbox' can continue to produce them after payment of a royalty even if the drug is placed under a patent. Specifically, it states: " . . . the patent holder shall only be entitled to receive reasonable royalty from such enterprises which have made significant investment and were producing and marketing the concerned product prior to 1.1.2005 and which continue to manufacture the product covered by the patent on the date of grant of the patent, and no infringement proceedings shall be instituted against such enterprises". This came as a big relief not only to those companies producing these drugs at affordable prices, but also to patients who need them. However, whether the drugs will continue to be affordable after product patent comes into force will depend upon how the term "reasonable royalty" is defined.

Compulsory Licensing

Compulsory licensing has been one of the most contentious issues in the WTO agreement. This refers to the right of the government of a country, say India, to allow the production of a patented drug by another company, by forcing the patent holder to license it to that company for a royalty fixed by the government. This may become essential to meet epidemics or other national contingencies at affordable costs. However, this provision was allowed by WTO only under exceptional situations, national calamities and emergencies, and the earlier amendments to the Indian act did not clarify whether the licensees could export them. The latest amendments have now provided that when patented drugs are produced under compulsory license in India by Indian companies: "the license is granted with a predominant purpose of supply in the domestic market and that the licensee may also export the patented product, if need be in accordance with Section 84(7) (a) (iii)". The act has allowed the exports of patented drugs produced through compulsory license in the country, to developing countries with no manufacturing capacity. This comes as a big relief to Indian companies such as those supplying patented AIDS drugs under compulsory license to South Africa.

Further Improvements

It is possible to make further improvements to take advantage of the flexibilities available under TRIPS. For example, in case of mailbox applications, the patent regime should be applicable from the prospective date of filing to minimize the impact of payment of royalties. Similarly, rather than using the vague term 'reasonable royalty' to the Patent holders in cases of compulsory licensing, which can lead to steep hikes in the prices of medicines, royalty percentage should have been fixed. For instance, Canada has fixed royalty at 2 per cent, leaving no scope for any ambiguity or wishful interpretations. The Indian Patents (Amendment) Act 2005 does not allow Indian residents to file patents overseas without the approval of the Indian Controller of Patents, which is a major setback to Indian researchers. The amendments also lack a clear definition of new entities, and defining it can solve the problem of evergreening of patents by companies. The reversal of burden of proof has also not been adequately addressed. As opposed to the normal principles of justice where an accused is deemed innocent unless proven guilty and the burden of proof is on the accuser (plaintiff), the current provisions of WTO have shifted the burden of proof to the accused. This could mean that large companies with huge patent portfolios can intimidate smaller competitors with accusations of infringement, and the accused have to prove their innocence. Interestingly, in all the other areas of dispute settlement in WTO, all available jurisprudence clearly shows that the burden of proof is on the plaintiff and not on the accused. Therefore, there is no objective reason for India to accept this reversal of burden of proof in the case of patents, however subtle it may be.

Indian Biodiversity Act

Human communities have acquired traditional knowledge over millennia through their interaction with their surrounding environments, including the use of available biological resources for various benefits. The indigenous knowledge relating to categories such as agricultural knowledge, medicinal knowledge and biodiversity-related knowledge forms a monumental resource of a country. Any commercial appropriation of this traditional knowledge is largely attributable to the knowledge, innovations and practices of local communities and hence, a share of benefits must accrue to the creators and holders of traditional knowledge. Similarly, ever since the international Convention of Biological Diversity (CBD) was signed on 29 December 1993 at Rio de Janeiro, biodiversity is also a national resource of a country, just like other resources and hence a commercial value arising out of it must benefit the country or locality. CBD has three major objectives: the conservation of biological diversity, the sustainable use of its components and fair and equitable sharing of benefits arising out of the utilization of genetic resources. Article 8(j) of the CBD also provides for respecting, protecting and rewarding the Knowledge, Innovations and Practices (KIP) of local communities.

Biodiversity and traditional knowledge have a special relevance for India. India is one of the major centres for world biodiversity and has been recognized as one of the 12 megadiversity countries of the world with 47,000 known species of plants, 500 varieties of mammals, 2,000 species of birds, 30,000 types of insects and a wide variety of fish, amphibians and reptiles, innumerable microbial species (tens of thousands). Some of the areas like the Himalayan region and the Andaman and Nicobar islands may be far richer in diversity with hitherto unknown biological diversity. Many of the cultivated plants have also originated from India. India also has an affluent traditional and indigenous knowledge coded as well as uncoded and being sustained by informal means. Considering the declining rate at which patentable new drugs are discovered and the huge time lags and financial uncertainties in the modern drug discovery process, traditional medicinal systems and other forms of traditional knowledge have suddenly acquired immense prominence among multinational corporations as a goldmine for discovering new drugs. Similarly, prior to CBD, foreign companies or researchers could freely use the native Indian biological diversity for various patentable inventions and commercial exploitation. Improper or inadequate documentation of traditional knowledge in India has also made it difficult to check patents based on traditional knowledge in public domain, due to lack of availability of information with patent examiners. Even when there has been available documentation, the time, effort and money involved in getting individual patents examined and revoked in foreign patent offices is prohibitive.

These issues were addressed through the Biodiversity Bill, 2000, followed by the Indian Biodiversity Act, 2002. Section 36(iv) of this act provides for protection of knowledge of local people relating to biodiversity through measures such as registration of such knowledge and development of a *sui generis* system. In order to ensure equitable sharing of benefits arising from the use of biological resources and associated knowledge, Sections 19 and 21 stipulate prior approval of the National Biodiversity Authority (NBA) before their access. While granting approval, NBA will impose terms and conditions, which secure equitable sharing of benefits. Section 6 provides that anybody seeking any kind of intellectual property rights on a research based upon biological resource or knowledge obtained from India, need to obtain prior approval of the NBA. Section 18(iv) stipulates that one of the functions of NBA is to take measures to oppose the grant of IPRs in any country outside India on any biological resource obtained from India or knowledge associated with such biological resource. The grounds for rejection as well as revocation of a patent application were extended to non disclosure or wrongful disclosure of the source of origin of biological resource or knowledge in the patent application as well as incorporating provisions to include anticipation of invention by available local knowledge, including oral knowledge, as one of the grounds for opposition as also for revocation of patents, if granted. Indian efforts have also been concerted to biodiversity identification and documentation, establishment of gene banks, biodiversity conservation with 13 bio-reserves already in place and 13 more proposed, and the documentation of traditional knowledge and prior art.

Protection of Plant Varieties and Farmers' Rights

India used the WTO provision of *sui generis* system of IPR protection in the case of plants and adopted the Plant Variety Protection and Farmers' Rights Act (PVPFR). This provides for registration of new plant varieties instead of allowing plant patents, and also secures the farmers' rights to save and reuse or exchange the seeds of the protected varieties, as well as the breeders' rights to use them for the development of new varieties. This was extremely critical to save the Indian agriculture from the vagaries of patenting. However, the controversial Seed Bill pending before the parliament is likely to undo some of the farmer-friendly provisions of PVPFR. Without adequate provisions for enforcement, this act introduces compulsory registration of seed producers, processors and traders with state authorities and has a provision for self-certification of seed varieties. It does not comply with farmers' rights and plant breeders' rights on traditional varieties provided under the PVPFR and also deviates from it on other key issues like disclosure of the origin of the variety. For example, The PVPFR allows Plant Breeders Right (PBR) for a variety on voluntary registration, but the compulsory registration requirement of the Seed Bill can lead to a situation where one can procure PBR, that is, exclusive commercialization right for a variety unregistered under PVPFR by registering it under the Seed Bill. Moreover, the Seed Bill allows registration for multiple parties, allowing free-for-all seed trade for any variety, including those that enjoy exclusive commercialization rights under PVPFR. In case of hybrid seeds whose parentage is not disclosed, registration under the Seed Bill can result in monopoly commercial rights. Moreover, the bill does not contain any mechanism to ensure seed supply and price control, like the compulsory licensing provision under PVPFR. The Seed Bill allows patent-like monopoly for hybrid and transgenic seeds and therefore can be used to deny the right of free access to all plant varieties for breeding new varieties provided under PVPFR.

The Seed Bill also creates loopholes in some of the provisions of the Biodiversity Act and the Environment Protection Act to benefit the seed industry. For example, since the Seed Bill does not seek clarification on the origin of variety, it can be used as a platform by seed companies to bypass the disclosure provisions made in the Indian Biodiversity act meant to prevent biopiracy. This provision in the Seed Bill can be used to export seeds without enforcing plant breeder rights, thus legitimizing piracy. The disclosure of origin is also critical to ensure the benefit sharing arising from seed sale of traditional varieties of tribals and farmers. The Seed Bill also allows the provisional registration of transgenic plant varieties for two years even before official approval under the Environment Protection Act. This will result in compromise in environment and human safety if the seed is subsequently found to be unsafe, as the new varieties released into the environment on the basis of provisional registration can never be fully retracted.

The self-certification provision under the Seed Bill requires multi-location test data, for which no competent institutional framework has been established. This places the financial burden of these tests on the seed variety developer. Moreover, private companies can use this provision for exaggerated performance claims. Though the Seed Bill allows compensation to the farmer if agronomic performance claimed by the seed provider is not met, this can be done only through District Consumer Courts. This is very difficult for small, illiterate farmers residing in remote villages. Therefore, adequate mechanism to easily claim compensation should be incorporated in the Seed Bill. The PVPFR has provisions for pre-grant opposition and therefore the Seed Bill should also incorporate such provisions.

Terms to Know

Apparatus Claim	This refers to a patent claim, which describes structurally a piece of equipment and is embraced by the expression "machine" in the definition of patentable subject matter in the U.S. Patent Statute.
Basic Patent	The first member of a Derwent patent family. This may or may not be the first published patent (which is usually the first to be documented by services such as Derwent World Patents Index).
Citations	Citations may be made by the examiner or author. They comprise a list of references that are believed to be relevant prior art and which may have contributed to the "narrowing" of the original application. The examiner can also cite references from technical journals, textbooks, handbooks and sources.
Claim(s)	A claim is one of the numbered paragraphs that appear at the end of a patent and defines the scope of protection given to the owner of the patent (i.e., the right to prevent others from making, using, selling, offering for sale, or importing the claimed invention). Each claim is treated separately for purposes of determining validity and infringement. For example, claims may be directed toward apparatus, methods, products and compositions of matter and new and useful improvements thereof.

Design Patent	This type of patent covers the original and ornamental aspects of a product. The prime focus is on the overall ornamental appearance.
Double Patenting	An inventive entity cannot obtain claims in two patents directed either to the same invention or to an obvious variation of the same invention. Where the inventions are identical, the second patent is invalid. With respect to obvious variations, this rejection can be overcome by filing a terminal disclaimer that, in effect, states that the second or any additional patents to issue will expire on the same date as the first patent, thereby eliminating any improper extension of the period of exclusivity resulting from the second or additional patents. A challenge based upon double patenting cannot be made where a second patent has issued as a result of a U.S. Patent and Trademark Office restriction requirement and the filing of a divisional application. In such a situation, the only reason more than one patent was issued is that the U.S. Patent and Trademark Office compelled limitation of a prior case to one embodiment.
European Patent Convention (EPC)	Nineteen European countries are parties to the European Patent Convention. A patent application filed under this convention will, when granted, usually automatically be effective in each of the countries designated by the applicant.
File Wrapper	This is the collection of documents located in the U.S. Patent and Trademark Office including a patent application and communications by the applicant and the U.S. Patent and Trademark Office with respect to the application.
First to File	The applicant who is the first to file an application for an invention will be awarded the patent over all others. This law is becoming increasingly the standard for countries adhering to Trade-Related aspects of Intellectual Property (TRIPs) guidelines.
Jepson Claim	This is a claim form wherein the preamble acknowledges the prior art teachings. The words "the improvement comprising" or "the improvement consisting of" or "the improvement consisting essentially of" are generally employed to form a border between the preamble and the body of the claim. It is followed in the body of the claim by what is believed to be unique.

Novelty	The concept that the claims must be totally new. The invention must never have been made public in any way, anywhere, before the date on which the application for a patent is filed.
Obviousness	The concept that the claims defining an invention in a patent application must involve an inventive step if, when compared with what is already known (i.e., prior art), it would not be obvious to someone skilled in the art.
Paris Convention	The Paris Convention was established 20 March 1883, effective 7 July 1884, and amended 2 June 1934 and 14 July 1967. Signatories to the Paris Convention are allowed one year from first filing their patent application (usually in their own country) in which to make further applications in member countries and claim the original priority date.
Patent	A grant by the federal government to an inventor of the right to exclude others from making, using, selling, offering for sale, or importing a patented invention. Patents that cover structural or functional aspects of products, composition, and processes are utility patents. Other types of patents include design patents covering ornamental designs of useful objects and plant patents covering new varieties of living plants.
Patentee	The inventor to whom the patent is granted and all persons to whom the patent is subsequently assigned. The term is slightly broader than the owner of the patent as the inventor would still be considered a patentee even if he or she has assigned his or her invention to someone else.
Prior Art	The existing body of technological information against which an invention is judged to determine if it is patentable as being novel and unobvious. It must be early enough in time to be cited against the application.
Research Disclosure	Defensive-type publications which are published, often anonymously, to give companies and inventors "freedom of use" rather than legal protection. Once research disclosures are published the invention described cannot be patented.
Royalty (Reasonable Royalty)	A hypothetical royalty fee used as a measure of damages to compensate for the infringement of a patent. It is the statutory minimum amount of recovery for compensatory damages for patent infringement.

Specification	The part of the patent application that precedes the claims and in which the inventor specifies, describes, illustrates and discloses the invention in detail. The term does not imply that the invention is necessarily new or was ever protected.
Term of Patent	The maximum number of years that the monopoly rights conferred by the grant of a patent may last.
Utility	Fitness for some desirable practical or commercial purpose.
Willful Infringement	Conduct that constitutes an illegal use of another's patented invention where the infringer has no reasonable basis for believing that its actions are legal (i.e., infringement when the infringer knew of the patent and had no reasonable basis for believing that it either did not infringe the patent or that the patent was invalid or unenforceable). A finding of willful infringement is a basis for increasing damages up to treble the actual amount and/or the award of attorneys' fees to the prevailing party. Good-faith reliance on the advice of counsel can be relevant to the issue of whether an infringement was willful.

12 Real-time PCR: Opportunities and Pitfalls

- Methods of Quantifying Transcription
- RT-PCR: A Rapid and Reliable Technique
- Quantification RT-PCR
- Instrumentation
- Applications
- Conclusions

Many cellular decisions concerning survival, growth and differentiation are reflected in altered patterns of gene expression and the ability to quantitate transcription levels of specific genes has always been central to any research into gene function (Zamorano et al. 1996). More recently, the emergence of molecular medicine has resulted in the increased use of techniques able to quantitate levels of RNA in clinical diagnostics. Such applications are wide-ranging and include procedures designed to quantify the regulation and expression of drug resistance markers in tumour cells (Ramachandran & Melnick 1999), monitor responses to chemotherapy (Desjardin et al. 1999), measure the biodistribution and of transcription of gene-encoded therapeutics (Fairman et al. 1999), provide a molecular assessment of tumour stage (Bustin & Dorudi 1998), detect circulating tumour cells in cancer patients (Ghossein & Rosai 1996), and detect bacterial (Hill 1996) and viral (Holodniy 1994) pathogens.

Methods of Quantifying Transcription

Four methods are in common use for the quantification of transcription: northern blotting and in situ hybridization (Parker & Barnes 1999), RNAse protection assays (Hod 1992, Saccomanno et al. 1992) and the reverse transcription polymerase chain reaction (RT-PCR) (Weis et al. 1992). A fifth method, cDNA arrays, is still limited in its use by cost considerations (Bucher 1999). Northern analysis is the only method providing information about mRNA size, alternative splicing and the integrity of RNA samples. The RNAse protection assay is most useful for mapping transcript initiation and termination sites and intron/exon boundaries, and for discriminating between related mRNAs of similar size, which would migrate at similar positions on a northern blot. In situ hybridization is the most complex method of all, but is the only one that allows localization of transcripts to specific cells within a tissue. The main limitation of these three techniques is their comparatively low sensitivity (Melton et al. 1984).

RT-PCR: A Rapid and Reliable Technique

RT-PCR is an in vitro method for enzymatically amplifying defined sequences of RNA (Rappolee et al. 1988) and permits the analysis of different samples from as little as one cell in the same experiment. It is the most sensitive and the most flexible of the quantification methods (Wang & Brown 1999) and can be used to compare the levels of mRNAs in different sample populations, to characterize patterns of mRNA expression, to discriminate between closely related mRNAs and to analyse RNA structure. RT-PCR can also circumvent time-consuming and technically demanding cloning steps and generate reagents, such as full-length complementary DNA (cDNA) inserts for cloning (Borson et al. 1992), or arbitrarily primed enhanced sequence tag cDNA libraries (Neto et al. 1997). The focus on maximizing sensitivity has led to the development of ever-more complex procedures. Semi-nested (Wasserman et al. 1999), nested (Israeli et al. 1994, Soeth et al. 1996), and even three-step nested (Funaki et al. 1997, 1998) RT-PCR techniques increase the sensitivity, but compromise the specificity of the reaction (Henke et al. 1997). They increase the likelihood of contamination and of detection of false positives (Kwok & Higuchi 1989) and are unable to distinguish genuine, low-level transcription from illegitimate transcription (Chelly et al. 1989).

Basic Considerations

The RT-PCR is a complex assay and all physical and chemical components of the reaction are interdependent. They must be considered carefully when optimizing the specificity, sensitivity, reproducibility or fidelity of the reaction. Successful performance of the RT-PCR depends on a clear understanding of the primary aim of the assay. Specificity, sensitivity and reproducibility will be the most important qualitative characteristics for clinical diagnostic uses.

Conversely, if the purpose of the experiment is mutation analysis, cloning or expression, fidelity will be the prime consideration. As RNA cannot serve as a template for PCR, the first step in an RT-PCR assay is the reverse transcription of the RNA template into cDNA, followed by its exponential amplification in a PCR reaction. Usually, this involves the use of dedicated RNA- and DNA-dependent DNA polymerases, either in separate or in single reactions, as the use of dedicated enzymes with different properties allows for the customization of individual RT-PCR assays. Separation of the RT and PCR steps has the advantage of generating a stable cDNA pool that can be stored virtually indefinitely. Alternatively, a single polymerase able to function both as an RNA and DNA-dependent DNA polymerase can be used in a single reaction and minimizes hands-on time in addition to the risk of contamination.

Factors Affecting the RT-PCR

Reverse transcription RNA transcripts can exhibit significant secondary structure that affects the ability of the RNA dependent DNA polymerase (reverse transcriptase, RT) to generate transcripts (Buell et al. 1978). This can affect RT-PCR quantification and should be minimized when comparing or quantifying diverse mRNA populations (Shimomaye & Salvato 1989). The two commonly used RTs are avian myeloblastosis virus reverse transcriptase (AMV-RT) and Moloney murine leukaemia virus reverse transcriptase (MMLV-RT). AMV-RT is more robust than MMLV-RT (Brooks et al. 1995), retains significant polymerization activity up to 55°C (Freeman et al. 1996) and can help eliminate problems associated with RNA secondary structure. In contrast, MMLV-RT and engineered derivatives (Kotewicz et al. 1988) have significantly less RNAse H activity than AMV-RT (Gerard et al. 1997). As even reduced RNAse H activity can interfere with the synthesis of long amplicons (DeStefano et al. 1991), MMLV-RT may be a better choice if the aim of the experiment is the amplification of full-length cDNA molecules. The RT step can be primed using specific primers, random hexamers or oligo-dT primers, and the choice of primers requires careful consideration. The use of mRNA-specific primers decreases background priming, whereas the use of random and oligo-dT primers maximizes the number of mRNA molecules that can be analysed from a small sample of RNA. Primers can cause marked variation in calculated mRNA copy numbers and, compared with specific downstream primers, random hexamers can overestimate mRNA copy numbers by up to 19-fold (Zhang & Byrne 1999).

Polymerase Chain Reaction

In an analogous manner, the choice of thermostable DNA-dependent DNA polymerases is determined by the aims of the experiment. There is a wide choice of enzymes that differ in their processivity, fidelity, thermal stability and ability to read modified triphosphates such as deoxyuridine and deoxyinosine in the template strand (Adams & Kelly 1994, Perler et al. 1996). The most commonly used enzyme, Taq DNA polymerase, has a 5′ nuclease activity but lacks a 3′ proofreading exonuclease activity. This makes its use less advisable when fidelity is the main consideration; instead, proofreading exonucleases such as Vent and Deep Vent (New England Biolabs, Beverly, MA, USA) or Pfu (Stratagene, La Jolla, CA, USA) should be used (Cline et al. 1996), although their Michaelis constants (Km(DNA)) can differ by more than two orders of magnitude from that of Taq polymerase, which has a direct effect on the effective primer annealing temperature.

As the specificity of the PCR step is crucially dependent on the primers (Zhang & Byrne 1999), reaction conditions must be re-optimized every time a different enzyme is used and results obtained may not be directly comparable. Furthermore, the proofreading activity can lead to primer degradation from the 3′ end. The rate of degradation is dependent on the length of the primer: a 60 mer is degraded much faster than a 30 mer. This can be counteracted by synthesizing primers with phosphorothioate linkages on the last two 3′ bases, as this linkage renders the primers extendable but not degradable (de Noronha & Mullins 1992). In addition, Mg2+ and dNTP concentrations require strict control, because Mg2+ affects enzyme activity, and imbalanced dNTP mixtures will reduce polymerase fidelity (Eckert & Kunkel 1991).

In addition, Mg2+ increases the melting temperature (Tm) of double-stranded DNA, and forms soluble complexes with dNTPs to produce the actual substrate that the polymerase recognizes. Therefore, high concentrations of dNTPs interfere with polymerase activity and affect primer annealing by reducing free Mg2+. In addition, the Km(dNTP) values of different polymerases vary significantly and must be taken into consideration when planning an RT-PCR assay.

One-enzyme RT-PCR

In situations in which convenience is the main priority or the target RNA contains extensive secondary structure, it is advisable to use a single tube system utilizing Thermus thermophilus (Tth) polymerase, a DNA polymerase with intrinsic RT, but no RNAse H, activity (Myers & Gelfand 1991). This assay uses bicine buffers containing Mn2+ ions that are compatible with both RT and subsequent PCR (Chiocchia & Smith 1997). This permits the RT reaction to be carried out at increased temperatures using primers with significantly higher Tm, while reducing both hands-on time and the likelihood of introducing contaminants into reaction mixtures. There is also evidence that RT-PCR reactions using Tth polymerase may be more robust and resistant to inhibitors present in biological specimen (Poddar et al. 1998). The main drawbacks of this system are that:

(i) because the sensitivity of Tth polymerase is two orders of magnitude lower than that of Taq polymerase (Grabko et al. 1996), this system is not as sensitive as two-enzyme procedures (Cusi et al. 1994, Leutenegger et al. 1999);
(ii) the presence of Mn2+ ions reduces the fidelity of nucleotide incorporation (Beckman et al. 1985, Fromant et al. 1995);
(iii) Tth polymerase lacks a 3' exonuclease proofreading activity; and
(iv) first-strand synthesis must be initiated from specific primers, rather than using random primers.

It is found that reducing the denaturation temperature to 92°C can improve the sensitivity of the assay, probably because Tth polymerase is somewhat less thermostable than Taq polymerase. However, for some templates, especially if they are long or G/C rich, this reduction may result in lower amplification efficiencies, as incompletely melted DNA will mask primer and probe binding sites.

Quantification RT-PCR

Quantification of mRNA can be by semiquantitative or quantitative methods (Ferre 1992). Semi-quantitative RT-PCR results are typically generated by frequent sampling of the RT-PCR reaction mixture followed by dot-blot analysis (Montgomery & Dallman 1997). This additional complexity does not lend itself to easy automation and, as the name implies, this method is at best semi-quantitative and its use should be avoided. Quantitative procedures use non-competitive or, more commonly, competitive RT-PCR techniques (Wang et al. 1989, Vanden Heuvel et al. 1993). They depend on the spiking into the RNA samples before RT of known amounts of RT-PCR-amplifiable competitors that, ideally, should be synthetic RNA molecules. Noncompetitive RT-PCR refers to the co-amplification of the target with a second RNA molecule under reagent concentrations and conditions so that there is no competition between target and standard, and with which it shares neither the primer recognition sites nor any internal sequence (Reischl & Kochanowski 1999).

However, this means that quantification can be unreliable, as the efficiencies of both the RT and PCR steps are likely to vary even when monitoring relative changes of transcription within a sample (Ferre 1992). In addition, as standard RNA is added in a linear fashion within one order of magnitude of the target, noncompetitive RT-PCR requires some approximate knowledge of target amounts (Freeman et al. 1999). In competitive RT-PCR, the internal standard shares the same primer recognition and internal sequences with the primary target, leading to competition for reagents. Because they are virtually identical, both should be amplified with the same efficiency and their amplicons can be distinguished by the addition of a restriction enzyme site to the standard (Gilliland et al. 1990, Becker-Andre 1993) or by varying its size (Gilliland et al. 1992, Scheuermann & Bauer 1993).

A series of PCR tubes containing the target are spiked with serial dilutions of known copy numbers of the internal standard. The greater the concentration of the internal standard, the more likely it is that the primers will bind and amplify it, rather than the target. After gel electrophoresis, a comparison of the intensities of ethidium-bromide-stained standard and target amplicons allow target quantification (Raeymaekers 1999). However, this may still not result in absolute quantification because, for example, stable differences in amplification efficiency between target and competitor will remain undetected (Raeymaekers 1993). There have been numerous attempts to enhance the accuracy of quantification by improving RNA isolation (Fernandez et al. 1997), reaction conditions (Corey et al. 1997), and detection methods (Borson et al. 1998, Hayward et al. 1998, Marchand et al. 1999). In spite of, or perhaps because of, this, competitive quantification procedures are, without exception, labour- and reagent-intensive. The minimum error in competitive RT-PCR reactions is approximately 10%, even when competitor and target are present in equal amounts (Souaze et al. 1996), and it is usually significantly greater.

There are several reasons for this.

(i) The amplification efficiencies of exogenous competitor, which must be synthesized for each gene being quantified, are not always equivalent.
(ii) Competitor RNA must be transcribed in vitro, which adds an additional enzymatic step; furthermore, once transcribed it may degrade during long-term storage.
(iii) Heteroduplex formation between the nearly identical standard and target can result in variable sensitivity and accuracy (Henley et al. 1996); use of DNA competitors with primer binding sites flanking unrelated internal sequences (mimics) (Siebert & Larrick 1993) eliminates that problem, but causes new ones.
(iv) The quantification of competitor fragments generated by restriction digestion requires the dilution of PCR products before digestion, introducing one source of error; a second source of error is the restriction digest itself, as it may not completely digest the competitor.
(v) Finally, the technique is not amenable to high throughput because, typically, for each unknown target a large series of reactions with different

competitor: target ratios has to be run (Raeymaekers 1993); therefore, conventional RTPCR-based procedures are not suitable for routine use, especially when several targets from many samples are being quantified.

INSTRUMENTATION

The application of fluorescence techniques to the RT-PCR, together with suitable instrumentation capable of combining amplification, detection and quantification, has led to the development of kinetic RT-PCR methodologies that are revolutionizing the possibilities for quantitating nucleic acids (Orlando et al. 1998). There are three competing instruments on the market; all three are run as closed-tube systems and quantification requires no post-amplification manipulation. This avoids problems of contamination, results in short turn-around times for data acquisition and analysis and minimizes hands-on time. The entire process, starting at the RT and ending with full quantification, is automated, which makes these instruments ideally suited for high-throughput screening applications (Martell et al. 1999, Wattjes et al. 2000). The ABI Prism 7700 (Perkin-Elmer Applied Biosystems, Foster City, CA, USA) contains a built-in thermal cycler with 96-well positions, and is able to detect fluorescence between 500 nm and 660 nm.

Fluorescence is induced during the RTPCR by distributing laser light to all 96 samples contained in thin-walled reaction tubes via a multiplexed array of optical fibres. The resulting fluorescent emission returns via the fibres and is directed to a spectrograph with a charge-coupled device (CCD) camera. Because each well is irradiated sequentially, the dimensions of the CCD array can be used for spectral resolution of the fluorescent light. This instrument can be used for assays based on DNA-binding dyes, molecular beacons and hydrolysis probes. RT-PCR reactions typically take 2 h to complete. Its description by the manufacturer as a real-time technique is not strictly correct as, unlike competing systems, the progress of the RT-PCR reaction can currently be analysed only after termination of the amplification run.

Lightcycler

Lightcycler (Roche Molecular Biochemicals, Mannheim, Germany) performs the RT-PCR in small-volume glass capillary tubes, contained within a rotor-like carousel that are heated and cooled in an airstream (Wittwer et al. 1989). The carousel is rotated past a blue light-emitting diode, and fluorescence is read by three photodetection diodes with different wavelength filters that allow the use of spectrally distinct fluorescent probes (Wittwer et al. 1997b). The performance of the Lightcycler in terms of sensitivity and specificity is equal to that of Taqman (Nitsche et al. 1999). In addition to being able to use assays based on

DNA-binding dyes, hydrolysis probes and molecular beacons, it can be used for dual hybridization probes. Up to 32 reactions are typically carried out (Wittwer et al. 1990) and RT-PCR reactions are completed in less than 20 min (Wittwer & Garling 1991). The Lightcycler is a truly versatile technique, as the fluorescence readings taken at every cycle of the PCR reaction are displayed immediately after each measurement, allowing amplification runs to be terminated or extended, as appropriate, during individual runs. Biorad Instruments have recently launched an optical module that fits their standard thermal cycler and transforms it into a real-time RT-PCR system. This instrument is capable of generating and detecting a wider range of excitation frequencies than either the ABI 7700 or the Lightcycler. At present, it can monitor up to four different fluorescent reporters at any one time and can be used for any one of the alternative fluorescent RT-PCR strategies. Furthermore, unlike the ABI 7700 that scans its 96 samples sequentially, this instrument can scan up to 96 samples simultaneously, with a sampling frequency that can be defined by the user.

Chemistries

There are currently four competing techniques available that detect amplified product with about the same sensitivity (Wittwer et al. 1997a). They use fluorescent dyes and combine the processes of amplification and detection of an RNA target to permit the monitoring of PCR reactions in real-time during the PCR; their high sensitivity eliminates the need for a second-round amplification, and decreases opportunities for generating false-positive results (Morris et al. 1996). The simplest method uses fluorescent dyes that bind specifically to double-stranded-DNA. The other three rely on the hybridization of fluorescence-labelled probes to the correct amplicon. The methods differ in their specificity, although at later amplification cycles all can show artefacts that do not correlate to specific product accumulation. As amplicon detection in the molecular beacon, hydrolysis and hybridization probe assays depend on successful hybridization of the probe, these RT-PCR procedures obviate the need for post-PCR Southern analysis or sequencing to confirm the identity of the amplicon.

Molecular Beacons

Molecular beacons (Stratagene) are DNA hybridization probes that form a stem-and-loop structure; the loop portion of the molecule is complementary to the target nucleic acid molecule and the stem is formed by the annealing of complementary arm sequences on the ends of the probe sequence (Tyagi & Kramer 1996) (Figure 3). A fluorescent marker is attached to the end of one arm and a quencher is attached to the end of the other arm.

The quencher is a non-fluorescent chromophore that dissipates the energy that it receives from the fluorophore as heat. In solution, free molecular beacons adopt a hairpin structure and the stem keeps the arms in close proximity, resulting in efficient quenching of the fluorophore. When molecular beacons encounter a complementary target at the annealing temperature, they undergo a conformational transition that forces the stem apart and results in the formation of a probe/target hybrid that is longer and more stable than the stem (Bonnet et al. 1999). This separates the fluorophore and the quencher, leading to the restoration of fluorescence, which can be detected, whereas any free molecular beacons remain closed and non-fluorescent. If the target DNA sequence does not exactly match the molecular beacon sequence, hybridization and fluorescence will not occur.

This is because the thermodynamic properties of the molecular beacons favour the formation of a hairpin form rather than continued hybridization to a less than perfectly matched target sequence. Molecular beacons require the use of a spectrofluorimetric thermal cycler for real-time detection, but can also be used in an endpoint assay using a fluorimeter. The specificity of this system is such that it can readily detect single nucleotide differences (Giesendorf et al. 1998, Marras et al. 1999), something that can be difficult to achieve with the Taqman system. Molecular beacons can also be used in non-PCR amplification assays that utilize isothermal nucleic acid amplification schemes (Guatelli et al. 1990, Tyagi et al. 1996) such as strand-displacement amplification (Walker et al. 1992), nucleic acid sequence-based amplification (Leone et al. 1998) and rolling-circle amplification (Lizardi et al. 1998).

The main drawback with molecular beacons is associated with the design of the hybridization probe. Optimal design of the molecular beacon stem is crucial, as the molecular beacon may fold into alternate conformations that do not place the fluorophore in the immediate vicinity of the quencher, resulting in a sub-population that is not quenched well and in large background signals. Alternatively, if the stem of a molecular beacon is too strong, it can interfere with the hybridization and the fluorescence of the beacon annealed to its target may be incompletely restored. Therefore, accurate thermal denaturation profiles to determine their melting characteristics must be established for each molecular beacon and different target sequences will have to be matched with different stem sequences.

DNA-binding Dyes

This method involves detection of the binding of a fluorescent dye (SYBR Green) to DNA (Morrison et al. 1998). The unbound dye exhibits little fluorescence in solution, but during elongation increasing amounts of dye bind to the nascent double-stranded DNA. When monitored in real-time, this results in an increase in the fluorescence signal that can be observed during the polymerization step, and that falls off when the DNA is denatured. Consequently, fluorescence measurements at the end of the elongation step of every PCR cycle are performed to monitor the increasing amount of amplified DNA.

This method obviates the need for target-specific fluorescent probes, but its specificity is determined entirely by its primers. As the presence of any double-stranded DNA generates fluorescence, this assay is no more specific than conventional RT-PCR. However, additional specificity and RT-PCR product verification can be achieved by plotting fluorescence as a function of temperature to generate a melting curve of the amplicon (Ririe et al. 1997). This is done by slowly increasing the temperature above the Tm of the amplicon and measuring the fluorescence. As the Tm of the amplicon depends markedly on its nucleotide composition, it is possible to identify the signal obtained from the correct product.

A characteristic melting peak at the melting temperature (Tm) of the amplicon will distinguish it from amplification artefacts that melt at lower temperatures in broader peaks. A second drawback of this procedure is that multiple dye molecules bind to a single amplified molecule and, consequently, the amount of signal generated after irradiation is dependent on the mass of double-stranded DNA produced in the reaction. Assuming the same amplification efficiencies, amplification of a longer product will generate more signal than a shorter one. If amplification efficiencies are different, quantification will be even more inaccurate.

This is in contrast to the use of fluorogenic probes, in which the signal from a single fluorophore is observed for each amplified molecule synthesized, regardless of its length.

Hybridization Probes

This method uses two hybridization probes to maximize specificity (Wittwer et al. 1997a). One of the probes carries at its 3′ end a fluorescein donor, which emits green fluorescent light when excited by the Lightcyclers light source. Its emission spectrum overlaps the excitation spectrum of an acceptor fluorophore that is attached to the 5′ end of the second probe.

This probe must be blocked at its 3′ end to prevent its extension during the annealing step. Excitation of the donor results in fluorescence resonance energy transfer to the acceptor and the emission of red fluorescent light. In solution, the two dyes are apart, and because the energy transfer depends on the spacing between the two dye molecules, only background fluorescence is emitted by the donor. After the denaturation step, both probes hybridize to their target sequence in a head-to-tail arrangement during the annealing step. This brings the two dyes in close proximity to one another and the fluorescein can transfer its energy at high efficiency. The intensity of the light of longer wavelength emitted by the second dye is measured, with increasing amounts of measured fluorescence proportional to the amount of DNA synthesized during the PCR reaction.

A fluorescent signal is detected only as a result of two independent probes hybridizing to their correct target sequence. This increases specificity and generates additional flexibility for probe design. Furthermore, as the probes are not hydrolyzed, fluorescence is reversible and allows the generation of melting curves.

Hydrolysis Probes

The Taqman assay (Perkin-ElmerApplied Biosystems) utilizes the 5′ nuclease activity of the DNA polymerase to hydrolyze a hybridization probe bound to its target amplicon. It usually utilizes either Taq or Tth polymerase, but any enzyme with an equivalent 5′ nuclease activity properties (e.g. Tfl), can be used (Gut et al. 1999). After the reverse transcription step (Fig. 5A), successful quantification requires the annealing of three oligonucleotides to the DNA. Two template-specific primers define the endpoints of the amplicon and provide the first level of specificity.

Problems associated with DNA contamination are minimized by choosing primers that span at least one intron of the genomic sequence. The additional specificity of this assay is provided by the use of a third oligonucleotide probe that hybridizes to the amplicon during the annealing/extension phase of the PCR. The probe contains a fluorescent reporter dye at its 5′ end, the emission spectrum of which is quenched by a second fluorescent dye at its 3′ end. If no amplicon complementary to the probe is amplified during the PCR, the probe remains unbound. As the 5′-exonuclease activity of Taq and Tth polymerase is double-strand-specific (Heid et al. 1996), unbound probe remains intact and no reporter fluorescence is detected. Conversely, if the correct amplicon has been amplified, the probe can hybridize to that amplicon after the denaturation step.

It remains hybridized while the polymerase extends the primers until it reaches the hybridized probe, when it displaces its 5′ end to hold it in a forked structure. The enzyme continues to move from the now free end to the bifurcation of the duplex, where cleavage takes place (Lyamichev et al. 1993). This separates the reporter and quencher dyes and releases quenching of reporter fluorescence emission. The largest fluorescence signal is obtained when the two labels are at the extreme 5′ and 3′ ends of the probe, probably because of more efficient cleavage by the polymerase (Livak et al. 1995). As the polymerase will cleave the probe only while it remains hybridized to its complementary strand, the temperature conditions of the polymerization phase of the PCR must be adjusted to ensure probe binding. Most probes have a Tm of around 70°C; therefore, the Taqman system uses a combined annealing and polymerization step at 60–62°C. This ensures that the probe remains bound to its target during the primer extension step. It also ensures maximum 5′-3′ exonuclease activity of the Taq and Tth DNA polymerases (Tombline et al. 1996). However, it also reduces the processivity of the polymerases, which are required to function at sub-optimal temperatures. For longer amplicons, this may require a longer annealing/polymerization step and/or increased Mn2+ or Mg2+ concentration to stabilize probe binding. It also makes this system potentially less efficient and flexible than other real-time systems, especially when mutation-specific RT-PCR is attempted.

Amplicon Design

The optimal amplicon length for real-time RTPCR is less than 100 bp, and we aim to generate amplicons no longer than 80 bp, but have used the Taqman system to detect amplicons as long as 400 bp (Bustin et al. 1999). Shorter amplicons amplify more efficiently than longer ones and are more tolerant of reaction conditions. This is because they are more likely to be denatured during the 92–95°C step of the PCR, allowing the probes and primers to compete more effectively for binding to their complementary targets. As the extension rate of Taq polymerase is between 30 and 70 bases/second (Jeffreys et al. 1988), it also means that polymerization times as short as 15 s are sufficient to replicate the amplicon, making amplification of genomic DNA contaminants less likely and reducing the time it takes to complete the assay. The length of primers and probe required to generate a specific product defines the minimum practical length of the amplicon, which in hands is 63 bp (Bustin et al. 1999).

Primer Design

Primers should bind to separate exons to avoid false positive results arising from amplification of contaminating genomic DNA. If the intron/exon boundaries are unknown, or when targeting an intron-less gene, it is necessary to treat the RNA sample with RNAse-free DNAse. Use of Mn2+ rather than Mg2+ minimizes any problems caused by amplification of re-annealed DNA fragments (Bauer et al. 1997). Primer (and probe) selection is based on estimated Tm, the desire for small amplicon size and the location of the probe.

They must be designed to very exacting specifications and the Taqman system provides its own primer/probe design program, Primer Express. The current version usually generates appropriate primer/probe sets, but it contains several bugs that can make primer design difficult, and requires manual fine-tuning. There are several alternative software tools available, with the most useful ones calculating Tm thermodynamically using the nearest-neighbour algorithm (Rychlik et al. 1990 OLIGO (Molecular Biology Insights, Cascade, CO, USA, http://oligo.net) or the online oligonucleotide Tm calculator (http://alces.med.umn.edu/rawtm.html). Recently, a program has been developed for Tm calculations of matched and mismatched probes designed to run on Microsoft Excel (Schutz & von Ahsen 1999) and is available for academic users as freeware from http://server1.medikc.med. unigoettingen.de/ meltcalc.htm.

The optimal length for single-stranded primers is about 15–20 bases, and the G/C content should be between 20% and 70%. Their Tm should not differ by more than 1–2°C, with minimum and maximum Tm of 58° and 60°C respectively, for Taqman primers. Molecular beacons- and hybridization probe associated primers can have a wider range of Tm, but the Tm of any one pair should be similar. Primers are generally used in the 50–200 nM range. Too great a primer concentration may promote mispriming and accumulation of non-specific product. Too low a primer concentration is probably less of a problem for real-time assays, as target copy numbers will have been calculated well before the primer supply is exhausted. Non-specific priming can be minimized by selecting primers that have only one or two G/Cs within the last five nucleotides at the 3′ end. This relative instability at their 3′ ends makes them less likely to hybridize transiently, and available for non-specific extension by the DNA polymerase.

Probe Design

The optimum concentration of fluorigenic probes depends on background fluorescence in addition to primer concentration. The fluorescence of nonhybridized molecular beacons and hybridization probes is quenched more efficiently than is the fluorescence of uncleaved hydrolysis probes, resulting in significantly reduced background fluorescence and enhanced ability to distinguish the fluorescence of hybridized molecular beacon and hybridization probes. Probe concentration is usually in the 100 nM range and they must compete for binding to their target with the amplification primers, which are present in a greater concentration.

As amplification primers are extended as soon as they bind to their targets, the hybridization target sequence is rapidly masked with newly synthesized DNA. Therefore, the Tm of the probes must be significantly greater (approximately 10°C) than that of the primers, to ensure that they hybridize before the primers. The fluorescence of hybridization probes is directly correlated to probe hybridization, hence they must remain hybridized during the annealing step. Conversely, the fluorescence of exonuclease probes is correlated with probe hydrolysis, and they must be available for cleavage during the polymerization step, to allow fluorescence measurement.

An additional requirement of the two hybridization probe system is that the target sequence should be located towards the 3′ end of the amplicon, close to the reverse primer-binding site on the opposite strand. As fluorescence is measured during the annealing phase, this arrangement allows time for the measurement to be made before the probes are displaced by the polymerase. Paired probes used for quantification should have equal Tm, whereas for mutational analysis the Tm of the sensor probe should be less than that of the anchor probe. This ensures that the sensor probe controls the generation of the fluorescent signal. Most of the probes are about 30 bases in length, and aim for a G/C content of around 50%. This ensures good specificity and a Tm high enough to be useful under PCR conditions. It also minimizes the potential for inter- or intramolecular complementarity, which can interfere with the PCR reaction. The probes should not overlap with, or have sequence complementarity with either of the primers, and Taqman probes should not contain a G at their 5′ ends, because such an arrangement quenches reporter fluorescence, even after cleavage.

Threshold Cycle

The concept of the threshold cycle (Ct) is at the heart of accurate and reproducible quantification using fluorescence-based RT-PCR (Higuchi et al. 1993). Fluorescence values are recorded during every cycle and represent the amount of product amplified to that point in the amplification reaction. The more template present at the beginning of the reaction, the fewer number of cycles it takes to reach a point in which the fluorescent signal is first recorded as statistically significant above background (Gibson et al. 1996). This point is defined as the Ct, and will always occur during the exponential phase of amplification.

Therefore, quantification is not affected by any reaction components becoming limited in the plateau phase, which results in a systematic bias against the more abundant templates and makes any quantification based on measurements of overall product yield intrinsically unreliable. The reporter signal is normalized to the fluorescence of an internal reference dye (Taqman) or between the three dyes (Lightcycler), to allow for corrections in fluorescent fluctuations caused by changes in concentration or volume, and a Ct value is reported for each sample. This value can be translated into a quantitative result by constructing a standard curve.

Standard Curves

Standard curves using fluorescence are easily generated, thanks to the linear response over a large dynamic range. Quantification of mRNA transcription can be either relative or absolute. Relative quantification determines the changes in steadystate transcription of a gene and is often adequate. A relative standard consists of a sample, the calibrator, which is used to create a dilution series with arbitrary units. The calibrator can be any nucleic acid, as long as its concentration and length of amplicon are known.

During the RT-PCR assay, the target Ct is compared directly with the calibrator Ct and is recorded as containing either more or less mRNA. In contrast, absolute quantification of transcription allows the precise determination of copy number per cell, total RNA concentration, or unit mass of tissue. It requires the construction of an absolute standard curve for each individual amplicon to ensure accurate reverse transcription and PCR amplification profiles.

There are two ways to prepare an absolute standard curve. The first method involves subcloning the amplicon behind a T7 or SP6 RNA polymerase promoter in a plasmid vector. An in vitro-transcribed sense RNA transcript is generated, the sample is digested with RNAse-free DNAse and the RNA is quantitated accurately. It is important to ensure that the standard RNA template is a single, pure species free from DNA contamination. We take triplicate measurements of its concentration on a spectrophotometer such as the Genequant II (Amersham Pharmacia Biotech, Little Chalfont, UK) and convert the absorbance to a\target copy number per $_{r}$g RNA.

Performing three independent serial dilutions of the RNA standard generates the standard curve and assaying each dilution in duplicate, together with positive and negative control reactions. To maximise accuracy, the dilutions should be made over the range of copy numbers that include the amount of target mRNA expected in the experimental RNA samples. The high and low Ct values are discarded to correct for pipetting error and the remaining four values are averaged to give the final Ct value for that dilution. The Ct value is inversely proportional to the log of the initial copy number (Higuchi et al. 1993). Therefore, a standard curve is generated by plotting the Ct values, with 95% confidence intervals, against the logarithm of the initial copy numbers. The copy numbers of experimental RNAs can be calculated after real-time amplification from the linear regression of that standard curve. The second method using single-stranded sense-strand oligodeoxynucleotides is an acceptable alternative to the T7RNA polymerase-generated standard amplicon. Upon comparing the sensitivity (y intercept) and amplification efficiency (slope) of a standard amplicon prepared from T7-transcribed RNA and a sense oligodeoxynucleotide they are found to be virtually identical. In contrast, the sensitivity of a doublestranded DNA amplicon standard is increased, although the amplification efficiency is unchanged (Overbergh et al. 1999). The advantage of using a single-stranded sense oligodeoxynucleotide is that it significantly simplifies the process of obtaining a standard curve for amplicons up to 100 bp, which encompasses most real-time RT-PCR amplicons.

Reproducibility

The RT-PCR reaction is fraught with hazards (Freeman et al. 1999) and is characterized by significant variation and non-reproducibility, even with identical samples between different laboratories (Keilholz et al. 1998). Its reproducibility is inevitably compromised by the variable efficiency of the RT itself and the need for two sequential enzymatic steps (Bishop et al. 1997). The exponential nature of PCR amplification, together with the small quantities of target molecules, means that trivial variations in reaction components and thermal cycling conditions and mispriming events during the early stages of the PCR can greatly influence the final yield of the amplified product (Wu et al. 1991).

This has created considerable uncertainty about the usefulness and reliability of the RT-PCR assay (Sokoloff et al. 1996), especially when used as a routine clinical diagnostic assay (Gala et al. 1998, de la Taille et al. 1999). It is a particular problem when cancer patients are screened for the presence of micrometastatic disease (Dingemans et al. 1997, Zippelius et al. 1997, Henke & Loening 1998) and severely limits its usefulness for prognosis and patient surveillance (Lambrechts et al. 1998). Clearly, there is an urgent requirement for stringent quality control (Keilholz 1998), standardization of the sample acquisition (Jung et al. 1997) and processing processes (Schittek et al. 1999), the RT-PCR procedure (Raj et al. 1998) and the detection of the RT-PCR products (Hughes & Goldman 1990). Optimization and consistency are as critical for obtaining reproducible results using real-time RTPCR as they are for conventional methods. However, there can be no doubt that real-time RT-PCR assays are significantly less variable than any conventional RT-PCR procedure. Where measured, the coefficient of variation for Ct data has been shown to be very low, at less than 2% for the Taqman (Heid et al. 1996, Gerard et al. 1998) and as little as 0\X{2022}4% for the Lightcycler (Wittwer et al. 1996a), which is significantly better than the 14% reported for conventional RT-PCR (Zhang et al. 1997). Using Taqman, the variability between duplicates and triplicates within the same run, and between different runs, is usually between 0 and 5%. Reproducibility is influenced by parameters such as distribution statistics (de Vries et al. 1999), and Ct data are less reproducible when working with very low copy numbers, because of the stochastic effects in the quantification of small numbers of target molecules (Peccoud & Jacob 1996).

Particle distribution statistics predict that it will require a much greater number of replicates to differentiate five from 10 copies of RNA than for the differentiation of 5,000 from 10,000 copies. Of course, this emphasizes the importance of repetitive testing in clinical samples. Some evidence exists that Ct values obtained from short amplicons show greater reproducibility than those obtained from long ones. It is also likely that amplification efficiency influences reproducibility. Regardless, one of the strengths of these assays is the ease with which it is possible to determine multiple Ct values for every sample, which encourages replicate determinations of the same sample and permits the application of statistical analyses to the quantification procedure.

Normalization

RT-PCR-specific errors in the quantification of mRNA transcripts are easily compounded by any variation in the amount of starting material between samples. This is especially relevant when the samples have been obtained from different individuals, and will result in the misinterpretation of the expression profiles of the target genes. Consequently, the question of what constitutes an appropriate standard arises (Thellin et al. 1999), and constitutes an important aspect of experimental design. The accepted method for minimizing these errors and correcting for sample-to-sample variation is to amplify, simultaneously with the target, a cellular RNA that serves as an internal reference against which other RNA values can be normalized (Karge et al. 1998).

The ideal internal standard should be expressed at a constant level among different tissues of an organism, at all stages of development, and should be unaffected by the experimental treatment. In addition, an endogenous control should also be expressed at roughly the same level as the RNA under study. In the absence of any one single RNA with a constant level of expression in all these situations (Haberhausen et al. 1998, Thellin et al. 1999), three RNAs are most commonly used to normalize patterns of gene expression. The mRNAs specifying the housekeeping genes glyceraldehyde-3-phosphate-dehydrogenase (GAPDH) and β-actin and are ribosomal RNAs (rRNA), although other mRNAs have been used occasionally, for example, histone H3 (Kelley et al. 1993) and cyclophilin (Haendler et al. 1987).

β-Actin

β-Actin mRNA is expressed at moderately abundant levels in most cell types and encodes a ubiquitous cytoskeleton protein. It was one of the first RNAs to be used as an internal standard, and it is still advocated as a quantitative reference for RT-PCR assays (Kreuzer et al. 1999). This is despite widespread evidence that its levels of transcription can vary widely in response to experimental manipulation in human breast epithelial cells (Spanakis 1993), and blastomeres (Krussel et al. 1998), as well as in various porcine tissues (Foss et al. 1998) and canine myocardium (Carlyle et al. 1996). In addition, the presence of pseudogenes interferes with the interpretation of results (Dirnhofer et al. 1995, Raff et al. 1997, Mutimer et al. 1998) and primers commonly used for detecting β-actin mRNA amplify DNA also (Dakhama et al. 1996).

GAPDH

The RNA encoding GAPDH is a ubiquitously expressed, moderately abundant message. It is frequently used as an endogenous control for quantitative RT-PCR analysis because, in some experimental systems, its expression is constant at different times and after experimental manipulation (Edwards & Denhardt 1985, Winer et al. 1999). However, there is overwhelming evidence suggesting that its use as an internal standard is inappropriate (Oliveira et al. 1999, Thellin et al. 1999). GAPDH concentrations vary significantly between different individuals (Bustin et al. 1999), during pregnancy (Cale et al. 1997), with developmental stage (Puissant et al. 1994, Calvo et al. 1997), during the cell cycle (Mansur et al. 1993) and after the addition of the tumour promoter 12-Otetradecanoyl-phorbol-13-acetate (Spanakis 1993), dexamethasone (Oikarinen et al. 1991) and carbon tetrachloride (Goldsworthy et al. 1993).

Insulin stimulates GAPDH transcription (Rolland et al. 1995, Barroso et al. 1999) through multiple insulinresponsive elements in its promoter (Alexander et al. 1988, Nasrin et al. 1990), at least one of which is recognized by C/EBP (Alexander-Bridges et al. 1992). Similarly, the calcium ionophore A23187 induces GAPDH transcription through specific motifs in its promoter (Chao et al. 1990). Growth hormone (Freyschuss et al. 1994), vitamin D (Desprez et al. 1992), oxidative stress (Ito et al. 1996), hypoxia (Graven et al. 1994, Zhong & Simons 1999), manganese (Hazell et al. 1999) and the tumour suppressor TP53 (Chen et al. 1999), have all been shown to activate its transcription, which is also inducible in endothelial cells (Rimarachin et al. 1992) and during apoptosis (Ishitani et al. 1997). Furthermore, it has been shown to be upregulated in cancer, for example, in rat hepatomas (Chang et al. 1998), malignant murine cell lines (Bhatia et al. 1994) and human prostate carcinoma (Ripple & Wilding 1995). Conversely, food deprivation (Yamada et al. 1997) and retinoic acid (Barroso et al. 1999) down regulate GAPDH transcription in the gut and in adipocytes, respectively. Finally, numerous transcription regulatory domains have been identified in the yeast GAPDH promoter, again suggesting that this gene is subject to complex transcriptional regulation (Yagi et al. 1994). Previous use of real-time RT-PCR has demonstrated that there is wide variation in the levels of transcription of GAPDH mRNA in nucleated blood cells, between individuals and between samples taken from the same individual at different time points (Bustin et al. 1999). This confirms that the use of GAPDH to normalize RNA levels in experiments that involve using mRNA from different individuals is inaccurate and should be discouraged. In light of all this evidence, it is a mystery why GAPDH continues to find favour as an internal standard.

rRNA

rRNAs, which constitute 85–90% of total cellular RNA, are useful internal controls, as the various rRNA transcripts are generated by a distinct polymerase (Paule & White 2000) and their levels are less likely to vary under conditions that affect the expression of mRNAs (Barbu & Dautry 1989). Indeed, they have been shown to be more reliable than either of the housekeeping genes in rat livers (de Leeuw et al. 1989), human skin fibroblasts (Mansur et al. 1993) and human (Zhong & Simons 1999) and mouse (Bhatia et al. 1994) malignant cell lines. However, rRNA transcription can be affected by biological factors and drugs (Spanakis 1993) and the variation in levels of transcription between samples taken from different individuals has not been quantified. In addition, there are two other drawbacks to its use: rRNA cannot be used for normalization when quantitating targets that have been enriched for mRNA because it is lost during mRNA purification, and rRNA is expressed at much greater levels than target mRNAs. Clearly, this raises questions about how to compare copy numbers between samples obtained from different individuals or from different tissues. For RNA extracted from tissue culture cells or from blood for example when haematological malignancies are under investigation target gene copy numbers can be readily expressed as copy numbers/cell. This is not possible when the RNA has been extracted from solid tissue or cancers. Therefore, it is necessary to find another means of normalizing their copy numbers. Until one identifies a more appropriate measure, one can relate all copy numbers to total RNA concentration, which can be measured accurately using a Genequant II spectrophotometer (Amersham Pharmacia Biotech). Copy numbers of target genes are then recorded as copy number per $_{r}$g total RNA.

There are obvious limitations to this approach. Total RNA levels may be increased in highly proliferating cells and this will affect the accuracy of any comparison of absolute copy numbers between normal and tumour cells. As rRNA makes up the bulk of total RNA, the reservations expressed earlier also apply. In addition, the accuracy of the quantification will be critically dependent on the accuracy of RNA measurement. The major disadvantages of the A260 method are the large relative contribution that proteins and free nucleotides make to the signal, the inability to distinguish between DNA and RNA, the interference caused by contaminants commonly found in nucleic acid preparations, and the relative insensitivity of the assay.

However, the use of fluorescent nucleic acid stains for example RiboGreen RNA quantitation (Molecular Probes, Eugene, OR, USA) \X{2013} alleviates many of these problems and can detect as little as 5 ng/ml RNA (Jones et al. 1998). There seems to be a good agreement between levels of transcription of various mRNAs calculated on the basis of total RNA concentration and published data using other techniques, such as immunocytochemical detection of their translation products. Nevertheless, it must be clearly stated that none of the methods used for standardization between samples is completely satisfactory, and individual experimental conditions will dictate what the most appropriate normaliser should be. Clearly, the ability to achieve absolute quantification of target nucleic acids makes the search for a relevant and universal means of comparing and normalising between samples very urgent.

APPLICATIONS

Quantification

Straightforward quantification is easily accomplished by any of the methods described above, and there is probably little to choose between them in terms of their sensitivity, although hybridization-based assays are more specific than those dependent on DNA-binding dyes and there is some evidence that molecular beacons are a little more specific than linear probes (Tyagi et al. 1998). The Taqman assay has been used to quantitate transcription from individual cells (Fink et al. 1998) and paraffin-embedded tissue (Sheils & Sweeney 1999), to investigate tissue-distribution proteins associated with signal transduction (Bustin & McKay 1999), for cytokine mRNA profiling from fresh (Overbergh et al. 1999) and stored frozen cells (Kruse et al. 1997) and for quantitating the RNA component of human telomerase (Yajima et al. 1998).

Its potential applications in medical diagnostics are wide-ranging and it has been used to screen serum for the diagnosis of genetic diseases (Kaplan et al. 1992), the presence of viral (Morris et al. 1996, Berger et al. 1998, Mercier et al. 1999) and bacterial (Norton & Batt 1999) pathogens, detection of colorectal (Bustin et al. 1999) and melanoma (de Vries et al. 1999) cells in patients blood, and quantification of c-myc (Bieche et al. 1999a) or ERBB2 (Bieche et al. 1999b) transcription in breast cancer. Molecular beacons have been used to detect RNAs from a single cell (Steuerwald et al. 1999) and from within living cells (Matsuo 1998, Sokol et al. 1998) and for other standard research (Lewin et al. 1999) and diagnostic (Arnold et al. 1999) RT-PCR applications. The range of applications for the Lightcycler are similarly wide, and include detection of minimal residual disease (Eckert et al. 2000), cancer cells in the peritoneal cavity of patients with gastrointestinal and gynaecological malignancies (Nakanishi et al. 1999) and pathogens (Mathews et al. 1999), and monitoring delivery, expression and maintenance of transgenes (Thorpe & Porteous 1999).

Mutation/Allele Detection

The flexibility of the Lightcycler design currently gives it an advantage over the Taqman system for the detection of mutations or allelic variants and it has been used extensively in molecular diagnostics (Aslanidis et al. 1999, Nakao et al. 2000). Two alternative procedures depending on melting-curve analysis have been developed. The first uses a sense primer internally labelled with an acceptor dye together with a donor-labelled probe that is completely complementary to the wild-type sense strand (Lay & Wittwer 1997, Bernard et al. 1998a). After asymmetric amplification, the probe is annealed to the extension product of the labelled primer. This brings the fluorophores into contact close enough for resonance energy transfer to occur. As the temperature increases, the probe melts from its target, decreasing the resonance energy transfer. If there is no mismatch, melting occurs at a characteristic temperature that is greater than if a single base mutation is present (Bernard et al. 1998b, Cane et al. 1999, Aoshima et al. 2000). The second procedure is similar, but uses two hybridization probes (Bernard et al. 1999, Bollhalder et al. 1999). One, the sensor, covers the site of the mutation, and the other one, the anchor, is designed to be longer, with a greater Tm and generates the fluorescent signal.

The probe hybridization status is monitored continuously as the temperature is slowly increased and characteristic, sequence-specific probe melting curves are generated. A single base mismatch under the probe decreases the Tm by as little as 3°C for GT and as much as 10°C for AC mismatches. The recent introduction of peptide nucleic acids (PNA) (Corey 1997) offers additional scope for this technology. These DNA mimics can bind DNA independently of salt concentration and demonstrate significantly greater specificity in binding to complementary DNA, with the effect that a RNA-DNA mismatch is more destabilizing than a DNA-DNA mismatch. The Taqman system has also been used for allelic discrimination by using probes labelled with different fluorescence reporter dyes (Lee et al. 1993, Livak 1999).

The assay is performed under competitive conditions with both probes present in the same tube. Under the correct conditions, there is preferential binding of the exact-match probe, which blocks hybridization of the mismatched probe. In addition, the mismatch itself has a disruptive effect on hybridization and speeds up the complete dissociation from the template. Both effects result in inefficient cleavage of the mismatch probe. Consequently, an increase in either fluorescent signal indicates homozygosity, whereas an increase in both signals indicates heterozygosity. A recent improvement involves the use of probes containing 5-propyne-2_-deoxyuridine in place of thymidine, which increases the Tm relative to standard Taqman probes and is useful when amplifying sequences with high A/T content. Nevertheless, the need to carry out the PCR reaction between 60°C and 62°C makes it more difficult to achieve perfect discrimination between mismatch and exact-match probes. This is of particular relevance when attempting mutant allele specific-type amplifications from samples in which only a minority of the templates will contain the targeted mutation, for example, detecting mutant K-ras in the blood of patients with colorectal cancer.

Multiplex RT-PCR

Detailed diagnostic and scientific analyses are often restricted by the limited availability of biological material. Multiplex RT-PCR is a time- and reagent-saving amplification technique in which multiple primer sets are used to amplify multiple specific targets simultaneously (Edwards & Gibbs 1994) from the same sample. In principle, all real-time detection systems are capable of simultaneous detection of multiple fluorescent dyes. However, successful multiplexing is never trivial, and requires careful consideration about the suitability of both chemistry and instrumentation.

One limitation arises from the limited availability of multiple fluorescent reporters with good spectral resolution. Both the hydrolysis and hybridization probe methodologies require the interaction of two different fluorescent dyes per target, and each probe or probe set requires a different pair of fluorophores. Consequently, fewer probes can be distinguished from each other in a multiplex assay. Molecular beacons, in contrast, use a universal, non-fluorescent quencher and require only a single fluorophore per probe. Therefore, use of molecular beacons has the potential to be used to identify more targets per reaction.

A second limitation is that the use of fluorophores is dependent on the excitation and detection characteristics of the individual spectrofluorimetric thermal cyclers. The ABI 7700 uses a laser with a fixed excitation wavelength, which limits the fluorophores to those that can be efficiently excited by its blue laser, preventing the use of fluorophores that fluoresce in the far-red wavelengths. In addition, it can detect fluorescence only between 500 nm and 660 nm.

The three-channel fluorimeter design of the Lightcycler restricts it to detecting a maximum of three fluorescent dyes simultaneously, whereas the Biorad instrument has the potential to excite and detect emission from the widest range of fluorophores; however, the latter has been launched only recently. There are relatively few reports of multiplexing using any of these systems. The Taqman system has been used successfully to quantitate up to three (Sharma et al. 1999, Shin et al. 1999) targets, with no apparent reduction in sensitivity compared with the individual reactions (Moody et al. 2000).

The non-fluorescent quencher used by molecular beacons allows the quantification of four pathogenic retroviruses simultaneously in blood (Vet et al. 1999) and the Lightcycler has been used to detect two independent targets simultaneously (Bernard et al. 1999, Lee et al. 1999a, von Ahsen et al. 2000). The potential for combining several dyes with instrumentation that allows synchronous scanning of excitation and emission wavelengths has been demonstrated with the detection of six PCR products in one reaction (Lee et al. 1999b).

In addition, the recent introduction of the Biorad instrument with a variable-wavelength light source and a high-resolution detector should allow the use of a wider range of fluorophores and the accommodation of more probes in the same assay. The final difficulty is associated with limitations caused by mutual interference of multiple sets of PCR primers, which can reduce the dynamic range of the sensitivity and make quantification unreliable. Furthermore, the efficiency of multiplex detection is significantly affected by the extension time and the concentrations of dNTPs, primers and MgCl2 (Seoh et al. 1998, James 1999). Vastly different levels of target mRNAs will cause quantification problems even for real-time detection, as the exponential phase of amplification of the less abundant mRNA will not overlap with that of the highly abundant target. Therefore, in practice, if accurate quantification is the main aim, it is probably best to limit multiplexing to the detection of two or three transcripts.

Detection of mRNA Splice Variants

Alternative splicing of RNAs is an important component of cell-specific gene regulation that contributes to the fine-tuning of gene expression and the functional diversification of proteins (Lopez 1998). It permits an additional tissue-specific level of regulation, as the mRNA can be transcribed from the same promoter, but in each tissue the RNA transcript can be spliced in a particular way to produce a different mRNA and hence protein. Alternative RNA splicing is widespread, with the Alternative Splicing Database (ASD) currently listing 1922 protein and 2486 DNA sequences (Dralyuk et al. 2000).

The combined amplification/hybridization approach of real-time RT-PCR assays is ideally suited for the detection and quantification of known splice variants. For example, hormonal regulation of gene expression can occur through the direct regulation of mRNA splicing (Chew 1997). The role of these alternatively spliced products in normal physiology has been difficult to determine, partly because detection of alternative transcripts using conventional procedures is time-consuming (Winkler et al. 1999) and their quantification, although reasonably accurate, requires extensive additional processing (Robinson et al. 1997), as discussed previously.

The specificity of the real-time RT-PCR assay allows the detection of multiple mRNA splice variants, as they can be distinguished reliably by a combination of exon-specific primers and an appropriate probe (Kafert et al. 1999). Probes designed to hybridize to specific exon splice junctions will detect and allow quantification of individual splice variants. Alternatively, probes designed to hybridize wholly within an exon will result in the more general identification of all mRNAs defined by the exon-specific upstream and downstream primer sequences. Therefore, judicious arrangement of primer/probe combinations allows maximum information about the presence, and the levels of, any number of mRNA splice variants to be gathered with a minimum of complexity and trouble. Importantly, because the results obtained are ratios of one splice variant to another, the problems of normalization become irrelevant.

Cost Considerations

In technical terms, there is very little difference between the hybridization-dependent real-time RT-PCR systems and they easily outperform conventional RT-PCR assays. Their main problem is the requirement for dedicated, sophisticated spectrofluorimetric thermal cyclers, which are expensive to purchase and maintain. The ABI 7700 Prism is significantly more expensive to purchase than either the Lightcycler or Biorad \X{2019}s instrument, and all three are more likely to be part of a core facility than on the bench in every laboratory. However, as the opportunities for using quantitative RT-PCR in a wide variety of settings continue to expand, it is likely that more cost-effective solutions will be developed that will make this assay more easily accessible. DNA dye-binding assays are the cheapest option for performing kinetic RT-PCR assays, as they do not require an amplicon-specific probe.

Hybridisation analyses require a separate primer set, together with fluorescent-labelled probe for each amplicon under investigation, which can make gene expression profiling experiments costly. In addition, assays carried out on the ABI 7700 have an absolute requirement for a reference dye in the reaction mixture. That reaction buffer can be obtained only as part of a complete kit, resulting in additional expense. In combination, all this adds significantly to the costs of performing quantitative real-time RT-PCR assays.

However, kinetic RT-PCR primers and probes are designed according to very strict criteria, with narrowly defined divalent cation and Tm ranges. Together with the reliability and ease of these assays, this reduces the need to waste reagents for assay optimisation. Furthermore, their specificity obviates the need to sequence or use Southern blotting to confirm the identity of the amplicons, and produces significant saving on researchers time. The ability to obtain immediate results also compensates for the increased consumables cost incurred, as does the ease with which numerous templates can be quantitated in a very short period.

Conclusions

Kinetic, or real-time, fluorescence RT-PCR significantly simplifies and accelerates the process of producing reproducible quantification of mRNAs. With the use of appropriate standard curves, absolute copy numbers of mRNA can easily be calculated. Several chemistries are available and their combination of amplification with simultaneous hybridization means that the high sensitivity of the assay is not compromised by concerns about its specificity. In addition, kinetic RT-PCR assays are readily standardized, making the comparison of results from different laboratories easier and more reliable. Therefore, real-time RT-PCR must be the method of choice for any experiments requiring sensitive, specific and reproducible quantification of mRNA.

References

1. Adams MW & Kelly RM 1994 Thermostability and thermoactivity of enzymes from hyperthermophilic Archaea. *Bioorganic and Medicinal Chemistry* **2:** 659–667.
2. von Ahsen N, Oellerich M & Schutz E 2000 Use of two reporter dyes without interference in a single-tube rapidcycle PCR: alpha(1)-antitrypsin genotyping by multiplex real-time fluorescence PCR with the LightCycler. *Clinical Chemistry* **46:** 156–161.
3. Alexander MC, Lomanto M, Nasrin N & Ramaika C 1988 Insulin stimulates glyceraldehyde-3-phosphate dehydrogenase gene expression through cis-acting DNA sequences. *PNAS* **85:** 5092–5096.
4. Alexander-Bridges M, Dugast I, Ercolani L, Kong XF, Giere L & Nasrin N 1992 Multiple insulin-responsive elements regulate transcription of the GAPDH gene. *Advances in Enzyme Regulation* **32:** 149–159.
5. Aoshima T, Sekido Y, Miyazaki T, Kajita M, Mimura S, Watanabe K, Shimokata K & Niwa T 2000 Rapid detection of deletion mutations in inherited metabolic diseases by melting curve analysis with LightCycler. *Clinical Chemistry* **46:** 119–122.
6. Arnold SF, Tims E & Mcgrath BE 1999 Identification of bone morphogenetic proteins and their receptors in human breast cancer cell lines: importance of bmp2. *Cytokine* **11:** 1031–1037.

7. Aslanidis C, Nauck M & Schmitz G 1999 High-speed prothrombin G A 20210 and methylenetetrahydrofolate reductase C T 677 mutation detection using real-time fluorescence PCR and melting curves. *Biotechniques* **27:** 234–238.
8. Barbu V & Dautry F 1989 Northern blot normalization with a 28S rRNA oligonucleotide probe. *Nucleic Acids Research* **17:** 7115.
9. Barroso I, Benito B, Garci-Jimenez C, Hernandez A, Obregon MJ & Santisteban P 1999 Norepinephrine, tri-iodothyronine and insulin upregulate glyceraldehyde-3-phosphate dehydrogenase mRNA during Brown adipocyte differentiation. *European Journal of Endocrinology* **141:** 169–179.
10. Bauer P, Rolfs A, Regitz-Zagrosek V, Hildebrandt A & Fleck E 1997 Use of manganese in RT-PCR eliminates PCR artifacts resulting from DNase I digestion. *Biotechniques* **22:** 1128–1132.
11. Becker-Andre M 1993 Absolute levels of mRNA by polymerase chain reaction-aided transcript titration assay. *Methods in Enzymology* **218:** 420–445.
12. Beckman RA, Mildvan AS & Loeb LA 1985 On the fidelity of DNA replication: manganese mutagenesis *in vitro*. *Biochemistry* **24:** 5810–5817.
13. Berger A, Braner J, Doerr HW & Weber B 1998 Quantification of viral load: clinical relevance for human immunodeficiency virus, hepatitis B virus and hepatitis C virus infection. *Intervirology* **41:** 24–34.
14. Bernard PS, Ajioka RS, Kushner JP & Wittwer CT 1998*a* Homogeneous multiplex genotyping of hemochromatosis mutations with fluorescent hybridization probes. *American Journal of Pathology* **153:** 1055–1061.
15. Bernard PS, Lay MJ & Wittwer CT 1998*b* Integrated amplification and detection of the C677T point mutation in the methylenetetrahydrofolate reductase gene by fluorescence resonance energy transfer and probe melting curves. *Analytical Biochemistry* **255:** 101–107.
16. Bernard PS, Pritham GH & Wittwer CT 1999 Color multiplexing hybridization probes using the apolipoprotein E locus as a model system for genotyping. *Analytical Biochemistry* **273:** 221–228.
17. Bhatia P, Taylor WR, Greenberg AH & Wright JA 1994 Comparison of glyceraldehyde-3-phosphate dehydrogenase and 28S-ribosomal RNA gene expression as RNA loading controls for northern blot analysis of cell lines of varying malignant potential. *Analytical Biochemistry* **216:** 223–226.
18. Bieche I, Laurendeau I, Tozlu S, Olivi M, Vidaud D, Lidereau R & Vidaud M 1999*a* Quantitation of MYC gene expression in sporadic breast tumors with a real-time reverse transcription-PCR assay. *Cancer Research* **59:** 2759–2765.
19. Bieche I, Onody P, Laurendeau I, Olivi M, Vidaud D, Lidereau R & Vidaud M 1999*b* Real-time reverse transcription-PCR assay for future management of ERBB2-based clinical applications. *Clinical Chemistry* **45:** 1148–1156.
20. Bishop GA, Rokahr KL, Lowes M, McGuinness PH, Napoli J, DeCruz DJ, Wong WY & McCaughan GW 1997 Quantitative reverse transcriptase-PCR amplification

of cytokine mRNA in liver biopsy specimens using a non-competitive method. *Immunology and Cell Biology* **75:** 142–147.

21. Bollhalder M, Mura C, Landt O & Maly FE 1999 LightCycler PCR assay for simultaneous detection of the H63D and S65C mutations in the HFE hemochromatosis gene based on opposite melting temperature shifts. *Clinical Chemistry* **45:** 2275–2278.
22. Bonnet G, Tyagi S, Libchaber A & Kramer FR 1999 Thermodynamic basis of the enhanced specificity of structured DNA probes. *PNAS* **96:** 6171–6176.
23. Borson ND, Salo WL & Drewes LR 1992 A lock-docking oligo(dT) primer for 5_ and 3_ RACE PCR. *PCR Methods and Applications* **2:** 144–148.
24. Borson ND, Strausbauch MA, Wettstein PJ, Oda RP, Johnston SL & Landers JP 1998 Direct quantitation of RNA transcripts by competitive single-tube RT-PCR and capillary electrophoresis. *Biotechniques* **25:** 130–137.
25. Brooks EM, Sheflin LG & Spaulding SW 1995 Secondary structure in the 3_ UTR of EGF and the choice of reverse transcriptases affect the detection of message diversity by RT-PCR. *Biotechniques* **19:** 806–815.
26. Bucher P 1999 Regulatory elements and expression profiles. *Current Opinion in Structural Biology* **9:** 400–407.
27. Buell GN, Wickens MP, Payvar F & Schimke RT 1978 Synthesis of full length cDNAs from four partially purified oviduct mRNAs. *Journal of Biological Chemistry* **253:** 2471–2482.
28. Bustin SA & Dorudi S 1998 Molecular assessment of tumour stage and disease recurrence using PCR-based assays. *Molecular Medicine Today* **4:** 389–396.
29. Bustin SA & McKay IA 1999 The product of the primary response gene BRF1 inhibits the interaction between 14–3–3 proteins and cRaf-1 in the yeast trihybrid system. *DNA and Cell Biology* **18:** 653–661.
30. Bustin SA, Gyselman VG, Williams NS & Dorudi S 1999 Detection of cytokeratins 19/20 and guanylyl cyclase C in peripheral blood of colorectal cancer patients. *British Journal of Cancer* **79:** 1813–1820.
31. Cale JM, Millican DS, Itoh H, Magness RR & Bird IM 1997 Pregnancy induces an increase in the expression of glyceraldehyde-3-phosphate dehydrogenase in uterine artery endothelial cells. *Journal of the Society for Gynecologic Investigation* **4:** 284–292.
32. Calvo EL, Boucher C, Coulombe Z & Morisset J 1997 Pancreatic GAPDH gene expression during ontogeny and acute pancreatitis induced by caerulein. *Biochemical and Biophysical Research Communications* **235:** 636–640.
33. Cane PA, Cook P, Ratcliffe D, Mutimer D & Pillay D 1999 Use of real-time PCR and fluorimetry to detect lamivudine resistance-associated mutations in hepatitis B virus. *Antimicrobial Agents and Chemotherapy* **43:** 1600–1608.
34. Carlyle WC, Toher CA, Vandervelde JR, McDonald KM, Homans DC & Cohn JN 1996 Changes in beta-actin mRNA expression in remodeling canine myocardium. *Journal of Molecular and Cellular Cardiology* **28:** 53–63.

35. Chang TJ, Juan CC, Yin PH, Chi CW & Tsay HJ 1998 Up-regulation of beta-actin, cyclophilin and GAPDH in N1S1 rat hepatoma. *Oncology Reports* **5:** 469–471.
36. Chao CC, Yam WC & Lin-Chao S 1990 Coordinated induction of two unrelated glucose-regulated protein genes by a calcium ionophore: human BiP/GRP78 and GAPDH. *Biochemical and Biophysical Research Communications* **171:** 431–438.
37. Chelly J, Concordet JP, Kaplan JC & Kahn A 1989 Illegitimate transcription: transcription of any gene in any cell type. *PNAS* **86:** 2617–2621.
38. Chen RW, Saunders PA, Wei H, Li Z, Seth P & Chuang DM 1999 Involvement of glyceraldehyde-3-phosphate dehydrogenase (GAPDH) and p53 in neuronal apoptosis: evidence that GAPDH is upregulated by p53. *Journal of Neuroscience* **19:** 9654–9662.
39. Chew SL 1997 Alternative splicing of mRNA as a mode of endocrine regulation. *Trends in Endocrinology and Metabolism* **8:** 405–413.
40. Chiocchia G & Smith KA 1997 Highly sensitive method to detect mRNAs in individual cells by direct RT-PCR using Tth DNA polymerase. *Biotechniques* **22:** 312–318.
41. Cline J, Braman JC & Hogrefe HH 1996 PCR fidelity of pfu DNA polymerase and other thermostable DNA polymerases. *Nucleic Acids Research* **24:** 3546–3551.
42. Corey DR 1997 Peptide nucleic acids: expanding the scope of nucleic acid recognition. *Trends in Biotechnology* **15:** 224–229.
43. Corey E, Arfman EW, Liu AY & Vessella RL 1997 Improved reverse transcriptase-polymerase chain reaction protocol with exogenous internal competitive control for prostate-specific antigen mRNA in blood and bone marrow. *Clinical Chemistry* **43:** 443–452.
44. Cusi MG, Valassina M & Valensin PE 1994 Comparison of M-MLV reverse transcriptase and Tth polymerase activity in RT-PCR of samples with low virus burden. *Biotechniques* **17:** 1034–1036.
45. Dakhama A, Macek V, Hogg JC & Hegele RG 1996 Amplification of human beta-actin gene by the reverse transcriptase-polymerase chain reaction: implications for assessment of RNA from formalin-fixed, paraffin-embedded material. *Journal of Histochemistry and Cytochemistry* **44:** 1205–1207.
46. Desjardin LE, Perkins MD, Wolski K, Haun S, Teixeira L, Chen Y, Johnson JL, Ellner JJ, Dietze R, Bates J, Cave MD & Eisenach KD 1999 Measurement of sputum Mycobacterium tuberculosis messenger RNA as a surrogate for response to chemotherapy. *American Journal of Respiratory and Critical Care Medicine* **160:** 203–210.
47. Desprez PY, Poujol D & Saez S 1992 Glyceraldehyde-3-phosphate dehydrogenase (GAPDH, E.C. 1·2·1·12.) gene expression in two malignant human mammary epithelial cell lines: BT-20 and MCF-7. Regulation of gene expression by 1,25-dihydroxyvitamin D3 (1,25-(OH)2D3). *Cancer Letters* **64:** 219–224.
48. DeStefano JJ, Buiser RG, Mallaber LM, Myers TW, Bambara RA & Fay PJ 1991 Polymerization and RNase H activities of the reverse transcriptases from avian

myeloblastosis, human immunodeficiency, and Moloney murine leukemia viruses are functionally uncoupled. *Journal of Biological Chemistry* **266:** 7423–7431.

49. Dingemans AM, Brakenhoff RH, Postmus PE & Giaccone G 1997 Detection of cytokeratin-19 transcripts by reverse transcriptase-polymerase chain reaction in lung cancer cell lines and blood of lung cancer patients. *Laboratory Investigation* **77:** 213–220.
50. Dirnhofer S, Berger C, Untergasser G, Geley S & Berger P 1995 Human beta-actin retropseudogenes interfere with RT-PCR. *Trends in Genetics* **11:** 380–381.
51. Dralyuk I, Brudno M, Gelfand MS, Zorn M & Dubchak I 2000 ASDB: database of alternatively spliced genes. *Nucleic Acids Research* **28:** 296–297.
52. Eckert KA & Kunkel TA 1991 DNA polymerase fidelity and the polymerase chain reaction. *PCR Methods and Applications* **1:** 17–24.
53. Eckert C, Landt O, Taube T, Seeger K, Beyermann B, Proba J & Henze G 2000 Potential of LightCycler technology for quantification of minimal residual disease in childhood acute lymphoblastic leukemia. *Leukemia* **14:** 316–323.
54. Edwards DR & Denhardt DT 1985 A study of mitochondrial and nuclear transcription with cloned cDNA probes. Changes in the relative abundance of mitochondrial transcripts after stimulation of quiescent mouse fibroblasts. *Experimental Cell Research* **157:** 127–143.
55. Edwards MC & Gibbs RA 1994 Multiplex PCR: advantages, development, and applications. *PCR Methods and Applications* **3:** S65–S75.
56. Fairman J, Roche L, Pieslak I, Lay M, Corson S, Fox E, Luong C, Koe G, Lemos B, Grove R, Fradkin L & Vernachio J 1999 Quantitative RT-PCR to evaluate *in vivo* expression of multiple transgenes using a common intron. *Biotechniques* **27:** 566–574.
57. Fernandez PM, Pluta LJ, Fransson-Steen R, Goldsworthy TL & Fox TR 1997 Reverse transcription-polymerase chain reaction-based methodology to quantify differential gene expression directly from microdissected regions of frozen tissue sections. *Molecular Carcinogenesis* **20:** 317–326.
58. Ferre F 1992 Quantitative or semi-quantitative PCR: reality versus myth. *PCR Methods and Applications* **2:** 1–9.
59. Fink L, Seeger W, Ermert L, Hanze J, Stahl U, Grimminger F, Kummer W & Bohle RM 1998 Real-time quantitative RT-PCR after laser-assisted cell picking. *Nature Medicine* **4:** 1329–1333.
60. Foss DL, Baarsch MJ & Murtaugh MP 1998 Regulation of hypoxanthine phosphoribosyltransferase, glyceraldehyde-3-phosphate dehydrogenase and beta-actin mRNA expression in porcine immune cells and tissues. *Animal Biotechnology* **9:** 67–78.
61. Freeman WM, Vrana SL & Vrana KE 1996 Use of elevated reverse transcription reaction temperatures in RT-PCR. *Biotechniques* **20:** 782–783.
62. Freeman WM, Walker SJ & Vrana KE 1999 Quantitative RT-PCR: pitfalls and potential. *Biotechniques* **26:** 112–115.

63. Freyschuss B, Sahlin L, Masironi B & Eriksson H 1994 The hormonal regulation of the oestrogen receptor in rat liver: an interplay involving growth hormone, thyroid hormones and glucocorticoids. *Journal of Endocrinology* **142:** 285–298.
64. Fromant M, Blanquet S & Plateau P 1995 Direct random mutagenesis of gene-sized DNA fragments using polymerase chain reaction. *Analytical Biochemistry* **224:** 347–353.
65. Funaki NO, Tanaka J, Itami A, Kasamatsu T, Ohshio G, Onodera H, Monden K, Okino T & Imamura M 1997 Detection of colorectal carcinoma cells in circulating peripheral blood by reverse transcription-polymerase chain reaction targeting cytokeratin-20 mRNA. *Life Sciences* **60:** 643–652.
66. Funaki NO, Tanaka J, Ohshio G, Onodera H, Maetani H & Imamura M 1998 Cytokeratin 20 mRNA in peripheral venous blood of colorectal carcinoma patients. *British Journal of Cancer* **77:** 1327–1332.
67. Gala JL, Heusterspreute M, Loric S, Hanon F, Tombal B, Van Cangh P, De Nayer P & Philippe M 1998 Expression of prostate-specific antigen and prostate-specific membrane antigen transcripts in blood cells: implications for the detection of hematogenous prostate cells and standardization. *Clinical Chemistry* **44:** 472–481.
68. Gerard GF, Fox DK, Nathan M & D'Alessio JM 1997 Reverse transcriptase. The use of cloned Moloney murine leukemia virus reverse transcriptase to synthesize DNA from RNA. *Molecular Biotechnology* **8:** 61–77.
69. Gerard CJ, Olsson K, Ramanathan R, Reading C & Hanania EG 1998 Improved quantitation of minimal residual disease in multiple myeloma using real-time polymerase chain reaction and plasmid–DNA complementarity determining region III standards. *Cancer Research* **58:** 3957–3964.
70. Ghossein RA & Rosai J 1996 Polymerase chain reaction in the detection of micrometastases and circulating tumor cells. *Cancer* **78:** 10–16.
71. Gibson UE, Heid CA & Williams PM 1996 A novel method for real time quantitative RT-PCR. *Genome Research* **6:** 995–1001.
72. Giesendorf BA, Vet JA, Tyagi S, Mensink EJ, Trijbels FJ & Blom HJ 1998 Molecular beacons: a new approach for semiautomated mutation analysis. *Clinical Chemistry* **44:** 482–486.
73. Gilliland G, Perrin S, Blanchard K, Bunn HF 1990 Analysis of cytokine mRNA and DNA: detection and quantitation by competitive polymerase chain reaction. *PNAS* **87:** 2725–2729.
74. Gilliland G, Perrin S & Bunn HF 1992 Competitive PCR for quantitation of mRNA. In *PCR Protocols: a Guide to Methods and Applications*, pp 60–69. Eds MA Innis, DH Gelfand, JJ Sninsky & TJ White. San Diego, CA: Academic Press.
75. Goldsworthy SM, Goldsworthy TL, Sprankle CS & Butterworth BE 1993 Variation in expression of genes used for normalization of Northern blots after induction of cell proliferation. *Cell Proliferation* **26:** 511–518.

76. Grabko VI, Chistyakova LG, Lyapustin VN, Korobko VG & Miroshnikov AI 1996 Reverse transcription, amplification and sequencing of poliovirus RNA by Taq DNA polymerase. *FEBS Letters* **387:** 189–192.
77. Graven KK, Troxler RF, Kornfeld H, Panchenko MV & Farber HW 1994 Regulation of endothelial cell glyceraldehyde-3-phosphate dehydrogenase expression by hypoxia. *Journal of Biological Chemistry* **269:** 24446–24453.
78. Guatelli JC, Whitfield KM, Kwoh DY, Barringer KJ, Richman DD & Gingeras TR 1990 Isothermal, *in vitro* amplification of nucleic acids by a multienzyme reaction modeled after retroviral replication. *PNAS* **87:** 1874–1878.
79. Gut M, Leutenegger CM, Huder JB, Pedersen NC & Lutz H 1999 One-tube fluorogenic reverse transcription-polymerase chain reaction for the quantitation of feline coronaviruses. *Journal of Virological Methods* **77:** 37–46.
80. Haberhausen G, Pinsl J, Kuhn CC & Markert-Hahn C 1998 Comparative study of different standardization concepts in quantitative competitive reverse transcription-PCR assays. *Journal of Clinical Microbiology* **36:** 628–633.
81. Haendler B, Hofer-Warbinek R & Hofer E 1987 Complementary DNA for human T-cell cyclophilin. *EMBO Journal* **6:** 947–950.
82. Hayward AL, Oefner PJ, Sabatini S, Kainer DB, Hinojos CA & Doris PA 1998 Modeling and analysis of competitive RT-PCR. *Nucleic Acids Research* **26:** 2511–2518.
83. Hazell AS, Desjardins P & Butterworth RF 1999 Increased expression of glyceraldehyde-3-phosphate dehydrogenase in cultured astrocytes following exposure to manganese. *Neurochemistry International* **35:** 11–17.
84. Heid CA, Stevens J, Livak KJ & Williams PM 1996 Real time quantitative PCR. *Genome Research* **6:** 986–994.
85. Henke W & Loening SA 1998 Detection of illegitimate transcripts of prostate-specific antigen mRNA in blood by reverse transcription-polymerase chain reaction. *International Journal of Cancer* **77:** 164–165.
86. Henke W, Jung M, Jung K, Lein M, Schlechte H, Berndt C, Rudolph B, Schnorr D & Loening SA 1997 Increased analytical sensitivity of RT-PCR of PSA mRNA decreases diagnostic specificity of detection of prostatic cells in blood. *International Journal of Cancer* **70:** 52–56.
87. Henley WN, Schuebel KE & Nielsen DA 1996 Limitations imposed by heteroduplex formation on quantitative RT-PCR. *Biochemical and Biophysical Research Communications* **226:** 113–117.
88. Higuchi R, Fockler C, Dollinger G & Watson R 1993 Kinetic PCR analysis: real-time monitoring of DNA amplification reactions. *Biotechnology* **11:** 1026–1030.
89. Hill WE 1996 The polymerase chain reaction: applications for the detection of foodborne pathogens. *Critical Reviews in Food Science and Nutrition* **36:** 23–173.
90. Hod Y 1992 A simplified ribonuclease protection assay. *Biotechniques* **13:** 852–854.
91. Holodniy M 1994 Clinical application of reverse transcription-polymerase chain reaction for HIV infection. *Clinical and Laboratory Medicine* **14:** 335–349.

92. Hughes T & Goldman JM 1990 Improved results with PCR for chronic myeloid leukaemia. *Lancet* **336:** 812.
93. Ishitani R, Sunaga K, Tanaka M, Aishita H & Chuang DM 1997 Overexpression of glyceraldehyde-3-phosphate dehydrogenase is involved in low K+-induced apoptosis but not necrosis of cultured cerebellar granule cells. *Molecular Pharmacology* **51:** 542–550.
94. Israeli RS, Miller WH Jr, Su SL, Powell CT, Fair WR, Samadi DS, Huryk RF, DeBlasio A, Edwards ET, Wise GJ, *et al.* 1994 Sensitive nested reverse transcription polymerase chain reaction detection of circulating prostatic tumor cells: comparison of prostate-specific membrane antigen and prostate-specific antigen-based assays. *Cancer Research* **54:** 6306–6310.
95. Ito Y, Pagano PJ, Tornheim K, Brecher P & Cohen RA 1996 Oxidative stress increases glyceraldehyde-3-phosphate dehydrogenase mRNA levels in isolated rabbit aorta. *American Journal of Physiology* **270:** H81–H87.
96. James D 1999 A simple and reliable protocol for the detection of apple stem grooving virus by RT-PCR and in a multiplex PCR assay. *Journal of Virological Methods* **83:** 1–9.
97. Jeffreys AJ, Wilson V, Neumann R & Keyte J 1988 Amplification of human minisatellites by the polymerase chain reaction: towards DNA fingerprinting of single cells. *Nucleic Acids Research* **16:** 10953–10971.
98. Jones LJ, Yue ST, Cheung CY & Singer VL 1998 RNA quantitation by fluorescence-based solution assay: RiboGreen reagent characterization. *Analytical Biochemistry* **265:** 368–374.
99. Jung R, Ahmad-Nejad P, Wimmer M, Gerhard M, Wagener C & Neumaier M 1997 Quality management and influential factors for the detection of single metastatic cancer cells by reverse transcriptase polymerase chain reaction. *European Journal of Clinical Chemistry and Clinical Biochemistry* **35:** 3–10.
100. Kafert S, Krauter J, Ganser A & Eder M 1999 Differential quantitation of alternatively spliced messenger RNAs using isoform-specific real-time RT-PCR. *Analytical Biochemistry* **269:** 210–213.
101. Kaplan JC, Kahn A & Chelly J 1992 Illegitimate transcription: its use in the study of inherited disease. *Human Mutation* **1:** 357–360.
102. Karge WH, Schaefer EJ & Ordovas JM 1998 Quantification of mRNA by polymerase chain reaction (PCR) using an internal standard and a nonradioactive detection method. *Methods in Molecular Biology* **110:** 43–61.
103. Keilholz U 1998 New prognostic factors in melanoma: mRNA tumour markers. *European Journal of Cancer* **34:** (Suppl 3) S37–S41.
104. Keilholz U, Willhauck M, Rimoldi D, Brasseur F, Dummer W, Rass K, de Vries T, Blaheta J, Voit C, Lethe B & Burchill S 1998 Reliability of reverse transcriptionpolymerase chain reaction (RT-PCR)-based assays for the detection of circulating tumour cells: a quality-assurance initiative of the EORTC Melanoma Cooperative Group. *European Journal of Cancer* **34:** 750–753.

105. Kelley MR, Jurgens JK, Tentler J, Emanuele NV, Blutt SE & Emanuele MA 1993 Coupled reverse transcriptionpolymerase chain reaction (RT-PCR) technique is comparative, quantitative, and rapid: uses in alcohol research involving low abundance mRNA species such as hypothalamic LHRH and GRF. *Alcohol* **10:** 185–189.
106. Kotewicz ML, Sampson CM, D'Alessio JM & Gerard GF 1988 Isolation of cloned Moloney murine leukemia virus reverse transcriptase lacking ribonuclease H activity. *Nucleic Acids Research* **16:** 265–277.
107. Kreuzer KA, Lass U, Landt O, Nitsche A, Laser J, Ellerbrok H, Pauli G, Huhn D & Schmidt CA 1999 Highly sensitive and specific fluorescence reverse transcription-PCR assay for the pseudogene-free detection of beta-actin transcripts as quantitative reference. *Clinical Chemistry* **45:** 297–300.
108. Kruse N, Pette M, Toyka K & Rieckmann P 1997 Quantification of cytokine mRNA expression by RT PCR in samples of previously frozen blood. *Journal of Immunological Methods* **210:** 195–203.
109. Krussel JS, Huang HY, Simon C, Behr B, Pape AR, Wen Y, Bielfeld P, Polan ML 1998 Single blastomeres within human preimplantation embryos express different amounts of messenger ribonucleic acid for beta-actin and interleukin-1 receptor type I. *Journal of Clinical Endocrinology and Metabolism* **83:** 953–959.
110. Kwok S & Higuchi R 1989 Avoiding false positives with PCR. *Nature* **339:** 237–238.
111. Lambrechts AC, 't Veer LJ & Rodenhuis S 1998 The detection of minimal numbers of contaminating epithelial tumor cells in blood or bone marrow: use, limitations and future of RNA-based methods. *Annals of Oncology* **9:** 1269–1276.
112. Lay MJ & Wittwer CT 1997 Real-time fluorescence genotyping of factor V Leiden during rapid-cycle PCR. *Clinical Chemistry* **43:** 2262–2267.
113. Lee LG, Connell CR & Bloch W 1993 Allelic discrimination by nick-translation PCR with fluorogenic probes. *Nucleic Acids Research* **21:** 3761–3766.
114. Lee MA, Brightwell G, Leslie D, Bird H & Hamilton A 1999*a* Fluorescent detection techniques for real-time multiplex strand specific detection of Bacillus anthracis using rapid PCR. *Journal of Applied Microbiology* **87:** 218–223.
115. Lee LG, Livak KJ, Mullah B, Graham RJ, Vinayak RS & Woudenberg TM 1999*b* Seven-color, homogeneous detection of six PCR products. *Biotechniques* **27:** 342–349.
116. de Leeuw WJ, Slagboom PE & Vijg J 1989 Quantitative comparison of mRNA levels in mammalian tissues: 28S ribosomal RNA level as an accurate internal control. *Nucleic Acids Research* **17:** 10137–10138.
117. Leone G, van Schijndel H, van Gemen B, Kramer FR & Schoen CD 1998 Molecular beacon probes combined with amplification by NASBA enable homogeneous, real-time detection of RNA. *Nucleic Acids Research* **26:** 2150–2155.
118. Leutenegger CM, Mislin CN, Sigrist B, Ehrengruber MU, Hofmann-Lehmann R & Lutz H 1999 Quantitative real-time PCR for the measurement of feline cytokine mRNA. *Veterinary Immunology and Immunopathology* **71:** 291–305.

119. Lewin SR, Vesanen M, Kostrikis L, Hurley A, Duran M, Zhang L, Ho DD & Markowitz M 1999 Use of real-time PCR and molecular beacons to detect virus replication in human immunodeficiency virus type 1-infected individuals on prolonged effective antiretroviral therapy. *Journal of Virology* **73:** 6099–6103.
120. Livak KJ 1999 Allelic discrimination using fluorogenic probes and the 5_ nuclease assay. *Genetic Analysis* **14:** 143–149.
121. Livak KJ, Flood SJ, Marmaro J, Giusti W & Deetz K 1995 Oligonucleotides with fluorescent dyes at opposite ends provide a quenched probe system useful for detecting PCR product and nucleic acid hybridization. *PCR Methods and Applications* **4:** 357–362.
122. Lizardi PM, Huang X, Zhu Z, Bray-Ward P, Thomas DC & Ward DC 1998 Mutation detection and single-molecule counting using isothermal rolling-circle amplification. *Nature Genetics* **19:** 225–232.
123. Lopez AJ 1998 Alternative splicing of pre-mRNA: developmental consequences and mechanisms of regulation. *Annual Review of Genetics* **32:** 279–305.
124. Lyamichev V, Brow MA & Dahlberg JE 1993 Structurespecific endonucleolytic cleavage of nucleic acids by eubacterial DNA polymerases. *Science* **260:** 778–783.
125. Mansur NR, Meyer-Siegler K, Wurzer JC & Sirover MA 1993 Cell cycle regulation of the glyceraldehyde-3-phosphate dehydrogenase/uracil DNA glycosylase gene in normal human cells. *Nucleic Acids Research* **21:** 993–998.
126. Marchand J, Bord A, Penarier G, Laure F, Carayon P & Casellas P 1999 Quantitative method to determine mRNA levels by reverse transcriptase-polymerase chain reaction from leukocyte subsets purified by fluorescence-activated cell sorting: application to peripheral cannabinoid receptors. *Cytometry* **35:** 227–234.
127. Marras SA, Kramer FR & Tyagi S 1999 Multiplex detection of single-nucleotide variations using molecular beacons. *Genetic Analysis* **14:** 151–156.
128. Martell M, Gomez J, Esteban JI, Sauleda S, Quer J, Cabot B, Esteban R & Guardia J 1999 High-throughput real-time reverse transcription-PCR quantitation of hepatitis C virus RNA. *Journal of Clinical Microbiology* **37:** 327–332.
129. Mathews SA, Volp KM & Timms P 1999 Development of a quantitative gene expression assay for Chlamydia trachomatis identified temporal expression of sigma factors. *FEBS Letters* **458:** 354–358.
130. Matsuo T 1998 *In situ* visualization of messenger RNA for basic fibroblast growth factor in living cells. *Biochimica et Biophysica Acta* **1379:** 178–184.
131. Melton DA, Krieg PA, Rebagliati MR, Maniatis T, Zinn K & Green MR 1984 Efficient *in vitro* synthesis of biologically active RNA and RNA hybridization probes from plasmids containing a bacteriophage SP6 promoter. *Nucleic Acids Research* **12:** 7035–7056.
132. Mercier B, Burlot L & Ferec C 1999 Simultaneous screening for HBV DNA and HCV RNA genomes in blood donations using a novel TaqMan PCR assay. *Journal of Virological Methods* **77:** 1–9.

133. Montgomery RA & Dallman MJ 1997 Semi-quantitative polymerase chain reaction analysis of cytokine and cytokine receptor gene expression during thymic ontogeny. *Cytokine* **9:** 717–726.

134. Moody A, Sellers S & Bumstead N 2000 Measuring infectious bursal disease virus RNA in blood by multiplex real-time quantitative RT-PCR. *Journal of Virological Methods* **85:** 55–64.

135. Morris T, Robertson B & Gallagher M 1996 Rapid reverse transcription-PCR detection of hepatitis C virus RNA in serum by using the TaqMan fluorogenic detection system. *Journal of Clinical Microbiology* **34:** 2933–2936.

136. Morrison TB, Weis JJ & Wittwer CT 1998 Quantification of low-copy transcripts by continuous SYBR Green I monitoring during amplification. *Biotechniques* **24:** 954–962.

137. Mutimer H, Deacon N, Crowe S & Sonza S 1998 Pitfalls of processed pseudogenes in RT-PCR. *Biotechniques* **24:** 585–588.

138. Myers TW, Gelfand DH 1991 Reverse transcription and DNA amplification by a Thermus thermophilus DNA polymerase. *Biochemistry* **30:** 7661–7666.

139. Nakanishi H, Kodera Y, Yamamura Y, Kuzuya K, Nakanishi T, Ezaki T & Tatematsu M 1999 Molecular diagnostic detection of free cancer cells in the peritoneal cavity of patients with gastrointestinal and gynecologic malignancies. *Cancer Chemotherapy and Pharmacology* **43:** (Suppl) S32–S36.

140. Nakao M, Janssen JW, Seriu T & Bartram CR 2000 Rapid and reliable detection of N-ras mutations in acute lymphoblastic leukemia by melting curve analysis using LightCycler technology. *Leukemia* **14:** 312–315.

141. Nasrin N, Ercolani L, Denaro M, Kong XF, Kang I & Alexander M 1990 An insulin response element in the glyceraldehyde-3-phosphate dehydrogenase gene binds a nuclear protein induced by insulin in cultured cells and by nutritional manipulations *in vivo*. *PNAS* **87:** 5273–5277.

142. Neto ED, Harrop R, Correa-Oliveira R, Wilson RA, Pena SD & Simpson AJ 1997 Minilibraries constructed from cDNA generated by arbitrarily primed RT-PCR: an alternative to normalized libraries for the generation of ESTs from nanogram quantities of mRNA. *Gene* **186:** 135–142.

143. Nitsche A, Steuer N, Schmidt CA, Landt O & Siegert W 1999 Different real-time PCR formats compared for the quantitative detection of human cytomegalovirus DNA. *Clinical Chemistry* **45:** 1932–1937.

144. de Noronha CM & Mullins JI 1992 Amplimers with 3_-terminal phosphorothioate linkages resist degradation by vent polymerase and reduce Taq polymerase mispriming. *PCR Methods and Applications* **2:** 131–136.

145. Norton DM & Batt CA 1999 Detection of viable Listeria monocytogenes with a 5_ nuclease PCR assay. *Applied and Environmental Microbiology* **65:** 2122–2127.

146. Oikarinen A, Makela J, Vuorio T & Vuorio E 1991 Comparison of collagen gene expression in the developing chick embryo tendon and heart. Tissue and

development time-dependent action of dexamethasone. *Biochimica et Biophysica Acta* **1089:** 40–46.

147. Oliveira JG, Prados RZ, Guedes AC, Ferreira PC & Kroon EG 1999 The housekeeping gene glyceraldehyde-3-phosphate dehydrogenase is inappropriate as internal control in comparative studies between skin tissue and cultured skin fibroblasts using Northern blot analysis. *Archives of Dermatological Research* **291:** 659–661.
148. Orlando C, Pinzani P & Pazzagli M 1998 Developments in quantitative PCR. *Clinical Chemistry and Laboratory Medicine* **36:** 255–269.
149. Overbergh L, Valckx D, Waer M & Mathieu C 1999 Quantification of murine cytokine mRNAs using real time quantitative reverse transcriptase PCR. *Cytokine* **11:** 305–312.
150. Parker RM, Barnes NM 1999 mRNA: detection by *in situ* and northern hybridization. *Methods in Molecular Biology* **106:** 247–283.
151. Paule MR & White RJ 2000 Survey and summary: transcription by RNA polymerases I and III. *Nucleic Acids Research* **28:** 1283–1298.
152. Peccoud J & Jacob C 1996 Theoretical uncertainty of measurements using quantitative polymerase chain reaction. *Biophysics Journal* **71:** 101–108.
153. Perler FB, Kumar S & Kong H 1996 Thermostable DNA polymerases. *Advances in Protein Chemistry* **48:** 377–435.
154. Poddar SK, Sawyer MH & Connor JD 1998 Effect of inhibitors in clinical specimens on Taq and Tth DNA polymerase-based PCR amplification of influenza A virus. *Journal of Medical Microbiology* **47:** 1131–1135.
155. Puissant C, Bayat-Sarmadi M, Devinoy E, Houdebine LM 1994 Variation of transferrin mRNA concentration in the rabbit mammary gland during the pregnancy–lactation–weaning cycle and in cultured mammary cells. A comparison with the other major milk protein mRNAs. *European Journal of Endocrinology* **130:** 522–529.
156. Raeymaekers L 1993 Quantitative PCR: theoretical considerations with practical implications. *Analytical Biochemistry* **214:** 582–585.
157. Raeymaekers L 1999 General principles of quantitative PCR. In *Quantitative PCR Protocols*, pp 31–41. Eds B Kochanowski & U Reischl. Totowa, NJ: Humana Press. Raff T, van der Giet M, Endemann D, Wiederholt T & Paul M 1997 Design and testing of beta-actin primers for RT-PCR that do not co-amplify processed pseudogenes. *Biotechniques* **23:** 456–460.
158. Raj GV, Moreno JG, Gomella LG 1998 Utilization of polymerase chain reaction technology in the detection of solid tumors. *Cancer* **82:** 1419–1442.
159. Ramachandran C & Melnick SJ 1999 Multidrug resistance in human tumors—molecular diagnosis and clinical significance. *Molecular Diagnosis* **4:** 81–94.
160. Rappolee DA,Mark D & BandaMJ & Werb Z 1988 Wound macrophages express TGF-alpha and other growth factors *in vivo*: analysis by mRNA phenotyping. *Science* **241:** 708–712.

161. Reischl U & Kochanowski B 1999 Quantitative PCR. In *Quantitative PCR Protocols*, pp 3–30. Eds B Kochanowski & U Reischl. Totowa, NJ: Humana Press.
162. Rimarachin JA, Norcross J, Szabo P & Weksler BB 1992 GAPDH acts as an inducible not constitutive gene in cultured endothelial cells. *In Vitro Cellular and Developmental Biology* **28A:** 705–707.
163. Ripple MO & Wilding G 1995 Alteration of glyceraldehyde-3-phosphate dehydrogenase activity and messenger RNA content by androgen in human prostate carcinoma cells. *Cancer Research* **55:** 4234–4236.
164. Ririe KM, Rasmussen RP & Wittwer CT 1997 Product differentiation by analysis of DNA melting curves during the polymerase chain reaction. *Analytical Biochemistry* **245:** 154–160.
165. Robinson CA, Hayward-Lester A, Hewetson A, Oefner PJ, Doris PA & Chilton BS 1997 Quantification of alternatively spliced RUSH mRNA isoforms by QRT-PCR and IP-RPHPLC analysis: a new approach to measuring regulated splicing efficiency. *Gene* **198:** 1–4.
166. Rolland V, Dugail I, Le L, X & LavauM1995 Evidence of increased glyceraldehyde-3-phosphate dehydrogenase and fatty acid synthetase promoter activities in transiently transfected adipocytes from genetically obese rats. *Journal of Biological Chemistry* **270:** 1102–1106.
167. Rychlik W, Spencer WJ & Rhoads RE 1990 Optimization of the annealing temperature for DNA amplification *in vitro*. *Nucleic Acids Research* **18:** 6409–6412.
168. Saccomanno CF, Bordonaro M, Chen JS & Nordstrom JL 1992 A faster ribonuclease protection assay. *Biotechniques* **13:** 846–850.
169. Scheuermann RH & Bauer SR 1993 Polymerase chain reaction based mRNA quantification using an internal standard: analysis of oncogene expression. *Methods in Enzymology* **218:** 446–473.
170. Schittek B, Blaheta HJ, Urchinger G, Sauer B & Garbe C 1999 Increased sensitivity for the detection of malignant melanoma cells in peripheral blood using an improved protocol for reverse transcription-polymerase chain reaction. *British Journal of Dermatology* **141:** 37–43.
171. Schutz E & von Ahsen N 1999 Spreadsheet software for thermodynamic melting point prediction of oligonucleotide hybridization with and without mismatches. *Biotechniques* **27:** 1218–1224.
172. Seoh ML, Wong SM & Zhang L 1998 Simultaneous TD/RTPCR detection of cymbidium mosaic potexvirus and odontoglossum ringspot tobamovirus with a single pair of primers. *Journal of Virological Methods* **72:** 197–204.
173. Sharma VK, Dean-Nystrom EA & Casey TA 1999 Semiautomated fluorogenic PCR assays (TaqMan) for rapid detection of Escherichia coli O157:H7 and other Shiga toxigenic E. coli. *Molecular and Cellular Probes* **13:** 291–302.
174. Sheils OM & Sweeney EC 1999 TSH receptor status of thyroid neoplasms—TaqMan RT-PCR analysis of archival material. *Journal of Pathology* **188:** 87–92.

175. Shimomaye E & Salvato M 1989 Use of avian myeloblastosis virus reverse transcriptase at high temperature for sequence analysis of highly structured RNA. *Gene Analysis Techniques* **6:** 25–28.

176. Shin JH, Nolte FS, Holloway BP & Morrison CJ 1999 Rapid identification of up to three Candida species in a single reaction tube by a 5_ exonuclease assay using fluorescent DNA probes. *Journal of Clinical Microbiology* **37:** 165–170.

177. Siebert PD & Larrick JW 1993 PCR mimics: competitive DNA fragments for use as internal standards in quantitative PCR. *Biotechniques* **14:** 244–249.

178. Soeth E, Roder C, Juhl H, Kruger U, Kremer B & Kalthoff H 1996 The detection of disseminated tumor cells in bone marrow from colorectal-cancer patients by a cytokeratin-20-specific nested reverse-transcriptase-polymerase-chain reaction is related to the stage of disease. *International Journal of Cancer* **69:** 278–282.

179. Sokol DL, Zhang X, Lu P & Gewirtz AM 1998 Real time detection of DNA·RNA hybridization in living cells. *PNAS* **95:** 11538–11543.

180. Sokoloff MH, Tso CL, Kaboo R, Nelson S, Ko J, Dorey F, Figlin RA, Pang S, deKernion J & Belldegrun A 1996 Quantitative polymerase chain reaction does not improve preoperative prostate cancer staging: a clinicopathological molecular analysis of 121 patients. *Journal of Urology* **156:** 1560–1566.

181. Souaze F, Ntodou-Thome A, Tran CY, Rostene W & Forgez P 1996 Quantitative RT-PCR: limits and accuracy. *Biotechniques* **21:** 280–285.

182. Spanakis E 1993 Problems related to the interpretation of autoradiographic data on gene expression using common constitutive transcripts as controls. *Nucleic Acids Research* **21:** 3809–3819.

183. Steuerwald N, Cohen J, Herrera RJ & Brenner CA 1999 Analysis of gene expression in single oocytes and embryos by real-time rapid cycle fluorescence monitored RT-PCR. *Molecular Human Reproduction* **5:** 1034–1039.

184. de la Taille A, Olsson CA & Katz AE 1999 Molecular staging of prostate cancer: dream or reality? *Oncology* **13:** 187–194.

185. Thellin O, Zorzi W, Lakaye B, De Borman B, Coumans B, Hennen G, Grisar T, Igout A & Heinen E 1999 Housekeeping genes as internal standards: use and limits. *Journal of Biotechnology* **75:** 291–295.

186. Thorpe PH & Porteous DJ 1999 Rapid quantitation of gene therapy specific CFTR expression using the amplification refractory mutation system. *Biotechniques* **27:** 122–126.

187. Tombline G, Bellizzi D & Sgaramella V 1996 Heterogeneity of primer extension products in asymmetric PCR is due both to cleavage by a structure-specific exo/endonuclease activity of DNA polymerases and to premature stops. *PNAS* **93:** 2724–2728.

188. Tyagi S & Kramer FR 1996 Molecular beacons: probes that fluoresce upon hybridization. *Nature Biotechnology* **14:** 303–308.

189. Tyagi S, Landegren U, Tazi M, Lizardi PM & Kramer FR 1996 Extremely sensitive, background-free gene detection using binary probes and beta replicase. *PNAS* **93:** 5395–5400.

190. Tyagi S, Bratu DP & Kramer FR 1998 Multicolor molecular beacons for allele discrimination. *Nature Biotechnology* **16:** 49–53.

191. Vanden Heuvel JP, Tyson FL & Bell DA 1993 Construction of recombinant RNA templates for use as internal standards in quantitative RT-PCR. *Biotechniques* **14:** 395–398.

192. Vet JA, Majithia AR, Marras SA, Tyagi S, Dube S, Poiesz BJ & Kramer FR 1999 Multiplex detection of four pathogenic retroviruses using molecular beacons. *PNAS* **96:** 6394–6399.

193. de Vries TJ, Fourkour A, Punt CJ, van de Locht LT, Wobbes T, van den BS, de Rooij MJ, Mensink EJ, Ruiter DJ & van Muijen GN 1999 Reproducibility of detection of tyrosinase and MART-1 transcripts in the peripheral blood of melanoma patients: a quality control study using real-time quantitative RT-PCR. *British Journal of Cancer* **80:** 883–891.

194. Walker GT, Fraiser MS, Schram JL, Little MC, Nadeau JG & Malinowski DP 1992 Strand displacement amplification—an isothermal, *in vitro* DNA amplification technique. *Nucleic Acids Research* **20:** 1691–1696.

195. Wang T & BrownMJ 1999 mRNA quantification by real time TaqMan polymerase chain reaction: validation and comparison with RNase protection. *Analytical Biochemistry* **269:** 198–201.

196. Wang AM, Doyle MV & Mark DF 1989 Quantitation of mRNA by the polymerase chain reaction. *PNAS* **86:** 9717–9721.

197. Wasserman L, Dreilinger A, Easter D & Wallace A 1999 A seminested RT-PCR assay for HER2/neu: initial validation of a new method for the detection of disseminated breast cancer cells. *Molecular Diagnostics* **4:** 21–28.

198. Wattjes MP, Krauter J, Nagel S, Heidenreich O, Ganser A & Heil G 2000 Comparison of nested competitive RT-PCR and real-time RT-PCR for the detection and quantification of AML1/MTG8 fusion transcripts in t(8;21) positive acute myelogenous leukemia. *Leukemia* **14:** 329–335.

199. Weis JH, Tan SS, Martin BK & Wittwer CT 1992 Detection of rare mRNAs via quantitative RT-PCR. *Trends in Genetics* **8:** 263–264.

200. Winer J, Jung CK, Shackel I & Williams PM 1999 Development and validation of real-time quantitative reverse transcriptase-polymerase chain reaction for monitoring gene expression in cardiac myocytes *in vitro*. *Analytical Biochemistry* **270:** 41–49.

201. Winkler A, Mahal B, Zieglgansberger W & Spanagel R 1999 Accurate quantification of the mRNA of NMDAR1 splice variants measured by competitive RT-PCR. *Brain Research Brain Research Protocols* **4:** 69–81.

202. Wittwer CT & Garling DJ 1991 Rapid cycle DNA amplification: time and temperature optimization. *Biotechniques* **10:** 76–83.

203. Wittwer CT, Fillmore GC & Hillyard DR 1989 Automated polymerase chain reaction in capillary tubes with hot air. *Nucleic Acids Research* **17:** 4353–4357.
204. Wittwer CT, Fillmore GC, Garling DJ 1990 Minimizing the time required for DNA amplification by efficient heat transfer to small samples. *Analytical Biochemistry* **186:** 328–331.
205. Wittwer CT, Herrmann MG, Moss AA & Rasmussen RP 1997*a* Continuous fluorescence monitoring of rapid cycle DNA amplification. *Biotechniques* **22:** 130–138.
206. Wittwer CT, Ririe KM, Andrew RV, David DA, Gundry RA & Balis UJ 1997*b* The LightCycler: a microvolume multisample fluorimeter with rapid temperature control. *Biotechniques* **22:** 176–181.
207. Wu DY, Ugozzoli L, Pal BK, Qian J & Wallace RB 1991 The effect of temperature and oligonucleotide primer length on the specificity and efficiency of amplification by the polymerase chain reaction. *DNA and Cell Biology* **10:** 233–238.
208. Yagi S, Yagi K, Fukuoka J & Suzuki M 1994 The UAS of the yeast GAPDH promoter consists of multiple general functional elements including RAP1 and GRF2 binding sites. *Journal of Veterinary Medical Science* **56:** 235–244.
209. Yajima T, Yagihashi A, Kameshima H, Kobayashi D, Furuya D, Hirata K & Watanabe N 1998 Quantitative reverse transcription-PCR assay of the RNA component of human telomerase using the TaqMan fluorogenic detection system. *Clinical Chemistry* **44:** 2441–2445.
210. Yamada H, Chen D, Monstein HJ & Hakanson R 1997 Effects of fasting on the expression of gastrin, cholecystokinin, and somatostatin genes and of various housekeeping genes in the pancreas and upper digestive tract of rats. *Biochemical and Biophysical Research Communications* **231:** 835–838.
211. Zamorano PL, Mahesh VB & Brann DW 1996 Quantitative RT-PCR for neuroendocrine studies. A minireview. *Neuroendocrinology* **63:** 397–407.
212. Zhang J & Byrne CD 1999 Differential priming of RNA templates during cDNA synthesis markedly affects both accuracy and reproducibility of quantitative competitive reverse-transcriptase PCR. *Biochemical Journal* **337:** 231–241.
213. Zhang J, Desai M, Ozanne SE, Doherty C, Hales CN & Byrne CD 1997 Two variants of quantitative reverse transcriptase PCR used to show differential expression of alpha-, beta and gamma-fibrinogen genes in rat liver lobes. *Biochemical Journal* **321:** 769–775.
214. Zhong H & Simons JW 1999 Direct comparison of GAPDH, beta-actin, cyclophilin, and 28S rRNA as internal standards for quantifying RNA levels under hypoxia. *Biochemical and Biophysical Research Communications* **259:** 523–526.
215. Zippelius A, Kufer P, Honold G, Kollermann MW, Oberneder R, Schlimok, G, Riethmuller G & Pantel K 1997 Limitations of reverse-transcriptase polymerase chain reaction analyses for detection of micrometastatic epithelial cancer cells in bone marrow. *Journal of Clinical Oncology* **15:** 2701–2708.

13 Functional Genomics and Microarrays

- Functional Genomics
- Microarrays: An Overview
- Conclusion

Functional Genomics

Biological and biomedical research is in the midst of a significant transition that is being driven by two primary factors: the massive increase in the amount of DNA sequence information and the development of technologies to exploit its use. Consequently, we find ourselves at a time when new types of experiments are possible, and observations, analyses and discoveries are being made on an unprecedented scale. Over the past few years, more than 30 organisms have had their genomes completely sequenced, with another 100 or so in progress (see www.tigr.org or genomes@ncbi.nlm.nih.gov for a list). At least partial sequence has been obtained for tens of thousands of mouse, rat and human genes, and the sequence of two entire human chromosomes (chromosomes 21 and 22) has been determined. Within the year, a large proportion of the human genome will be deciphered, in both public and private efforts, and the complete sequence of the mouse and other animal and plant genomes will undoubtedly follow close behind. Unfortunately, the billions of bases of DNA sequence do not tell us what all the genes do, how cells work, how cells form organisms, what goes wrong in disease, how we age or how to develop a drug. This is where functional genomics comes into play. The purpose of genomics is to understand biology, not simply to identify the component parts, and the experimental and computational methods take advantage of as much sequence information as possible. In this sense, functional genomics is less a specific project or programme than it is a mindset and general approach to problems. The goal is not simply to provide a catalogue of all the genes and information about their functions, but to understand how the components work together to comprise functioning cells and organisms. To take full advantage of the large and rapidly increasing body of sequence information, new technologies are required.

Tools Employed

Among the most powerful and versatile tools for genomics are high-density arrays of oligonucleotides or complementary DNAs. Nucleic acid arrays work by hybridization of labelled RNA or DNA in solution to DNA molecules attached at specific locations on a surface. The hybridization of a sample to an array is, in effect, a highly parallel search by each molecule for a matching partner on an affinity matrix, with the eventual pairings of molecules on the surface determined by the rules of molecular recognition. Arrays of nucleic acids have been used for biological experiments for many years. Traditionally, the arrays consisted of fragments of DNA, often with unknown sequence, spotted on a porous membrane (usually nylon). The arrayed DNA fragments often came from cDNA, genomic DNA or plasmid libraries, and the hybridized material was often labelled with a radioactive group. Recently, the use of glass as a substrate and fluorescence for detection, together with the development of new technologies for synthesizing or depositing nucleic acids on glass slides at very high densities, have allowed the miniaturization of nucleic acid arrays with concomitant increases in experimental efficiency and information content. While making arrays with more than several hundred elements was until recently a significant technical achievement, arrays with more than 250,000 different oligonucleotide probes or 10,000 different cDNAs per square centimetre can now be produced in significant numbers. Although it is possible to synthesize or deposit DNA fragments of unknown sequence, the most common implementation is to design arrays based on specific sequence information, a process sometimes referred to as downloading the genome onto a chip.

There are several variations on this basic technical theme discussed later on in the chapter: the hybridization reaction may be driven (for example, by an electric field) by other detection methods besides fluorescence can be used. The surface may be made of materials other than glass such as plastic, silicon, gold, a gel or membrane, or may even be comprised of beads at the ends of fibre-optic bundles. Nonetheless, the key elements of parallel hybridization to localized, surface-bound nucleic acid probes and subsequent counting of bound molecules are ubiquitous, and high-density arrays of nucleic acids on glass (often called DNA microarrays, oligonucleotide arrays, GeneChip arrays, or simply chips) and their biological uses will be the focus of this section.

Global Gene Expression Experiments

One of the most important applications for arrays so far is the monitoring of gene expression (mRNA abundance). The collection of genes that are expressed or transcribed from genomic DNA, sometimes referred to as the expression profile or the transcriptome, is a major determinant of cellular phenotype and function. The transcription of genomic DNA to produce mRNA is the first step in the process of protein synthesis, and differences in gene expression are responsible for both morphological and phenotypic differences as well as indicative of cellular responses to environmental stimuli and perturbations. Unlike the genome, the transcriptome is highly dynamic and changes rapidly and dramatically in response to perturbations or even during normal cellular events such as DNA replication and cell division. In terms of understanding the function of genes, knowing when, where and to what extent a gene is expressed is central to understanding the activity and biological roles of its encoded protein. In addition, changes in the multi-gene patterns of expression can provide clues about regulatory mechanisms and broader cellular functions and biochemical pathways. In the context of human health and treatment, the knowledge gained from these types of measurements can help determine the causes and consequences of disease, how drugs and drug candidates work in cells and organisms, and what gene products might have therapeutic uses themselves or may be appropriate targets for therapeutic intervention. Past discussions of arrays have often centred on technical issues and specific performance characteristics. Now that nucleic acid arrays have been constructed for many different organisms and used successfully to measure transcript abundance in a host of different experiments, the focus of interest has thankfully shifted. Investigators are now more concerned with questions concerning experimental design, data analysis, the use of small amounts of mRNA from limited sources, the best ways to extract biological meaning from the results, pathway and cell-circuitry modelling, and medical uses of expression patterns.

Array-Based Gene Expression Monitoring

One way to think of measurements with arrays is that they are simply a more powerful substitute for conventional methods of evaluating mRNA abundance. For some early experiments, only a relatively small set of genes, which were thought to be important to a process, were included on the arrays. However, such experiments did not capitalize on the arrays potential: a key advantage of using arrays, especially those that contain probes for tens of thousands of different genes, is that it is not necessary to guess what the important genes or mechanisms are in advance. Instead of looking only under the proverbial lamppost, a broader, more complete and less biased view of the cellular response is obtained. The breadth of array-based observations almost guarantees that surprising findings will be made. A recent study measured the transcriptional changes that occur as cells progress through the normal cell-division cycle in humans for approximately 40,000 genes (R. J. Cho et al., unpublished results). In addition to the induction of DNA replication genes and genes involved with cell-cycle control and chromosome segregation that would be expected at specific stages in the cell cycle, a large collection of genes involved with smooth muscle function, apoptosis and intercellular adhesion and cell motility were found to be upregulated during a specific phase. The expected results act effectively as internal controls that provide a certain amount of validation (and comfort), while new information is obtained by a systematic search of a larger part of gene space. In addition, because arrays often contain probes for genes of unknown function (and often with only partial sequence information), any outcome for these could be considered, in some sense, both surprising and novel (although clearly requiring further characterization).

Other Gene Expression Methods

Not surprisingly, there are other ways to measure mRNA abundance, gene expression and changes in gene expression. For measuring gene expression at the level of mRNA, northern blots, polymerase chain reaction after reverse transcription of RNA (RT-PCR), nuclease protection, cDNA sequencing, clone hybridization, differential display, subtractive hybridization, cDNA fragment fingerprinting and serial analysis of gene expression (SAGE) have all been put to good use to measure the expression levels of specific genes, characterize global expression profiles or to screen for significant differences in mRNA abundance. But if messenger RNA is only an intermediate on the way to production of the functional protein products, why measure mRNA at all? One reason is simply that protein-based approaches are generally more difficult, less sensitive and have a lower throughput than RNA-based ones. But more importantly, mRNA levels are immensely informative about cell state and the activity of genes, and for most genes, changes in mRNA abundance are related to changes in protein abundance. Because of its importance, however, many methods have been developed for monitoring protein levels either directly or indirectly. These include western blots, two-dimensional gels, methods based on protein or peptide chromatographic separation and mass spectrometric detection, methods that use specific protein-fusion reporter constructs and colorimetric readouts, and methods based on characterization of actively translated, polysomal mRNA. The importance of the protein-based methods is that they measure the final expression product rather than an intermediate. In addition, some of them enable the detection of post-translational protein modifications (for example, phosphorylation and glycosylation) and protein complexes, and in some cases, yield information about protein localization, none of which are obtained directly by measurements of mRNA. There is no question that protein- and RNA-based measurements are complementary, and that protein-based methods are important as they measure observables that are not readily detected in other ways.

MICROARRAYS: AN OVERVIEW

The degree of miniaturization of the sample determines whether the array is a macroarray and microarray. Currently available macroarrays on nylon membranes allow analysis of several hundred genes at once; however, significant improvements in substrate materials, robotics, and signal detection have made possible miniaturization of arrays that have hundreds of thousands of oligonucleotides arrayed on a cm2 chip. Microarrays make it possible to survey gene expression in a systematic, comprehensive and efficient manner. The use of microarrays to study gene expression profiles in biologic samples began in 1995. This breakthrough in microarray technology came about through two key innovations: First was the use of solid supports, such as glass or silicon chips that facilitated miniaturization with fluorescence-based detection that allows analysis of tens of thousands of genes in a few cm. The second major breakthrough was the development of methods for high-density oligonucleotide synthesis directly on the microarray chip. The development of microarray technology is dependent on both advances in the acquisition of genetic sequence information and the ability to study small quantities of nucleic acid. Although microarrays can be used to study DNA, RNA or protein expression, the term generally refers to cDNA testing for RNA expression rather than to protein for which the term protein array is more common or for DNA for which the term genomic array is used. The term expression arrays is usually used to denote testing for RNA expression.

Microarray Fabrication and How Arrays Work

Despite the availability of expression arrays in many varieties, generally two categories are employed; customized cDNA microarrays that are composed of cDNA or oligonucleotides and commercially produced high-density arrays that contain synthesized oligonucleotides. Whether they are developed by a company or non-commercially by a scientist, arrays depend on the same basic principle: The ability of single stranded sequences of nucleotides to stick to- or hybridize to its complementary sequence and form double-stranded DNA. Spotted arrays such as shown in can compare RNAs from two samples on a single chip. To do this they require a source of genes spotted onto the chip. These are usually furnished as Expressed Sequence Tags (EST) clones or oligonucleotides. Briefly, microarray systems are usually comprised of cDNA probes formatted in a micro-scale on glass surfaces (chip) plus the instruments needed to handle samples (automated robotics) and to read the reporter molecules (done by scanners) and to analyse the data (using Bioinformatics tools). Commercially available high-density arrays provide analysis of the expression of a large number of genes (e.g., 12,000 human genes on the chip); yet they can only analyse a single sample per chip. This means considerable cost and makes them unsuitable for large-scale experiments in most laboratories.

Microarray Systems

The basic types of DNA microarrays include sequencing chips, expression chips and chips for comparative genomic hybridization. Technologies used in automated microarray production are photolithography, mechanical micro spotting and ink jets. Generally, the applications of this technology are either identification of gene mutation or determination of levels of gene expression. Monitoring of gene expression or the occurrence of polymorphisms in genomic DNA is made through the hybridization of RNA- or DNA-derived samples on chips. The hybridization of labeled nucleic acid molecules to an array of complementary sequences immobilized on a solid substrate is the key unifying principle of all microarray experiments. In greater detail, a simple cDNA array experiment has five basic steps. First, the target cDNA is spotted or printed onto a solid substrate (glass or plastic surface). Then the sample RNA is isolated. Third, the cDNA is synthesized including labeling it for later detection. Fourth, the labeled probe cDNA is hybridized to target the cDNA on the substrate. When sample DNA or RNA is applied to the array any sequences in the sample that find a match will bind to a specific spot on the array. Finally the hybridization results are imaged and analysed. Using specialized computer programs the amount of sample bound to each spot on the microarray is determined. The results are color coded, so that the most active genes are colored red, and genes that are repressed are colored green. Alternatively, the experiment can be set up so that genes that are over represented in cancer cells are labeled red and genes that are missing in the tumor cells are labeled green. In summary, DNA microarrays use the same DNA probe detection method as in single gene studies but on a much larger scale that allows thousands of specific DNA-or RNA sequences to be tested/detected simultaneously.

Chip Construction

Although several methods for building chips have been developed, two have especially prevailed. In the more commonly used method microarrays are constructed by physically attaching DNA fragments such as library clones or polymerase chain reaction (PCR) products to a solid substrate. By using a robotic arrayer and capillary printing tips at least 23,000 gene fragments can be printed on a microscope slide. In the other method arrays are constructed by synthesizing single-strand oligonucleotides in situ by using photolithographic techniques. The advantages of the former method include relatively low cost and substantial flexibility, while the advantage of the latter method is the higher density (>280,000/1.28X1.28-cm array) and elimination of the need to collect and store the cloned DNA or PCR products. Customized cDNA microarrays are fabricated by selecting the genes to be printed on the array from a (public) databases/repositories or institutional sources. A high-throughput DNA preparation, usually done by robotic systems, performs the necessary tens of thousands of PCR reactions. Purified PCR products representing specific genes are spotted onto the chip (matrix) by a robot, which deposits a nanoliter of PCR product onto the matrix in a retrievable order. Nylon filter arrays have largely been replaced by glass-based arrays, typically microscope slides, which have the advantage of two-color fluorescence labeling with low inherent background fluorescence. DNA adherence to the slide is enhanced by treatment with polylysine or other cross-linking chemical coatings. Spotted samples are cross-linked to the matrix by ultraviolet irradiation and denatured by exposure to either heat or alkali.

Molecular Biology and Microarrays

In the mid-1970s molecular biology started a new era with their major focus on developing molecular cloning and DNA sequencing. Five years later, automated DNA sequencing took the world of biotechnology by storm with the dream of sequencing the entire human genome. This dream later seemed to be possible with the development of automated sequencing techniques and the invention of the polymerase chain reaction (PCR), which produced a surge in new experiments. The year 2001 witnessed a major achievement through the declaration of the completion of the first draft of the human genome sequence. Science observers described this as the greatest scientific achievement since the discovery of the atom and the gene. It was even mentioned as the greatest accomplishment since the discovery of agriculture. Having the human genome gave a profound impetus to the strategies biological scientists can use to identify disease-associated genes. The Human Genome Project has created a massive amount of DNA sequence information and influenced many research and educational fields. Besides the human genome, other prokaryote and eukaryotic complete genome sequences have also become available. This massive growth of information created a new challenge in the form of new gene identifications, determining the function of these genes, defining disease associations, and elucidating the correlation between genotype and phenotype. To overcome these challenges a major aim for bioscience workers has been to study genetic variation in large sets of genes; not one gene at a time. There is hope that this will illuminate key single or clustered genes, expose important disease mechanisms, identify novel drug targets and also enable the prediction of the therapeutic responses to these drugs. With the massive increase in molecular information and gene sequences, the tradition of analysing one gene at a time in not able to give the comprehensive answers for many questions such as gene co-regulation and interactions. There is need for developing technology by which clusters of genes can be analysed simultaneously. New techniques and tools for conducting research have been developed to overcome these technical obstacles and allow full advantage of the enormous amounts of genomic information. This technology is the microarray, which allow simultaneous large-scale expression measurements on great numbers of genes using minute samples. Despite the current widespread use of microarrays, there is no single terminology on which science community agrees. Descriptive terms that have been used generally describe the technology itself. These include but are not limited to microarray, biochip, DNA chip, gene array, gene chips and more recently genomic chips (considered by some to include protein chips that mark protein expression using labeled antibodies). DNA microarrays, which are also, called DNA arrays or gene chips, are an example microarrays that use genome sequence information to analyse the structure and function of tens of thousands of genes at a time. Thus, biotechnology laboratories and pharmaceutical companies can identify the molecular targets with which drugs actually interact. Since it can also identify individuals with similar biological response patterns, microarray analysis can assist drug companies in choosing the most appropriate candidates for participating in clinical trials of new drugs. In an extension of these methods, in the future arrays will help medical professionals select the most effective drugs for individual patients.

Applications of Microarray Technology

As previously mentioned, gene chips facilitated a more comprehensive and inclusive experimental approach in which alterations in the state of entire genomes simultaneously can be assayed during health and disease or in response to a variety of stimuli. This ability to profile changes in gene-expression levels under different conditions makes microarrays the method of choice in many fields. Some of these fields are disease fingerprinting, drug targeting and evaluation, toxicity assessment, signal transduction, research and developmental cancer treatment. All of these potential uses increase the demand for microarrays in both academic and industrial settings.

Medical and Clinical Diagnosis

The ability of microarrays to provide genome-wide profiles of gene expression increases potential clinical application for this technology in disease diagnosis, prognosis and treatment. Particularly exciting is the potential of DNA array technology to provide individualized treatment for a wide variety of clinical conditions based on fingerprints for each of these conditions. Moreover, combining microarray expression profiling with other new array techniques such as tissue arrays (tiny biopsies that are robotically placed on the slide and then washed by the complementary test molecule) provides tissue mapping for the disease-specific candidate genes. The combination of robotically executed tissue arrays and with microarrays is an excellent candidate for many clinical applications, especially those that require mass testing, quick diagnosis and general diagnostics. As one of these applications is blood screening in which gene chips provided for the first time the prospect of an integrated platform for comprehensive donor and recipient testing, replacing conventional multiple individual assays. In addition, using microarrays to explore gene interactions facilitates testing many medical hypotheses concerning diverse human health-related issues such as alcoholism and drug abuse. It is also possible to use gene chips to identify the changes in specific gene expression that are associated with drug abuse and alcoholism. This allows establishing a cause-effect relationship between certain genes and drug abuse. Besides being able to track down the influence of administration of drugs or alcohol it is also possible to identify individual DNA sequence differences. The latter is important to the future growth of medical genetics. This has become increasingly apparent as epidemiologists, pathologists, and clinical geneticists focus more attention on the molecular basis of complex multi-factorial diseases. Accordingly, it is predicted that over the next decade microarrays combined with sophisticated informatics tools and genomic databases will provide molecular fingerprints of disease processes and prognoses. This should allow clinical laboratories to track down in depth the changes that accompany such diseases.

Pharmacological Industry and Microarrays

There are three major tasks with which the pharmaceutical industry deals on a regular basis: (1) to discover a drug for an already defined target; (2) to assess drug toxicity; and (3) to monitor drug safety and effectiveness. As well, determining the potential problems of a compound early in the drug discovery process will save time and money by focusing resources on compounds that are more likely to succeed. Additionally, the success of drug screening depends both on the speed or turnaround-time of evaluation and on the reliability of results. More to the point, for a long time one of the most pressing issues facing the pharmaceutical and biotechnology industry was the high dropout rate of lead drug candidates. Late-stage dropouts are both financially costly and undermine confidence in the company. Relief was found with the increase in the knowledge of total genomic sequences of many prokaryotes and eukaryotes. These represent gold mines for drug targeting and discovery. Accordingly, there was an urgent need to develop high throughput monitoring technologies that are meant to identify targets and provide lead candidate optimization. Accordingly, there was a call for developing a technology that can help in basic research, drug discovery and evaluation. Gene expression microarray allows the simultaneous monitoring of the expression levels of thousands of genes, this is an advantage that makes microarrays suitable for measuring the expression patterns of thousands of genes in parallel and generating clues to gene function that can help to identify appropriate targets for therapeutic intervention. Microarray also can be used to monitor changes in gene expression in response to drug treatments. This makes gene chips the current frontrunner technology in the drug industry. Moreover, this also provided prospective markets for this infant technology. With their ability to integrate genomics, pharmacology and molecular medicine, gene chips promise to revolutionize these fields as they reveal clusters of genes and gene regulation events involved in disease progression and to pinpoint potential drug targets and diagnostics. By comparing the ways in which genes are expressed in a normal and a diseased heart, for example, it is possible to identify the genes and hence the associated proteins that are part of the disease process. Researchers could then use that information to synthesize drugs that interact with proteins, thus reducing diseases effect on the body. Moreover, since that many drug targets are components of complex signaling pathways, and in order to understand the true biological consequences of modulating these targets it is necessary to understand the biology of the system in great detail. Since activation of signaling pathways leads to mRNA expression, microarray technology can be used to provide a detailed quantitative assessment of the consequences of this activation, providing a new biological perspective on well-established cellular systems. Because pharmacogenomics is the hybridization of functional genomics and molecular pharmacology it also aims to find correlation between therapeutic responses to drugs and the genetic profiles of patients as described above. These efforts will allow answers to questions about the effects of certain drugs on certain individuals, or ethnic groups, and identify appropriate targets for therapeutic intervention. Determining the potential toxicity of compounds early in the drug discovery process can be extremely beneficial in terms of both time and money saved. Moreover, to accurately evaluate the large number of compounds being produced, toxicology assays must have both high fidelity and high-throughput capabilities. Assays must be performed using limited amounts of compound though. Conventionally with traditional techniques these were very hard tasks. Currently, using microarrays to fulfill these goals and obtain detailed information concerning the molecular mechanisms behind toxic effects is becoming routine procedures in pharmaceutical companies replacing many conventional techniques. Besides being able to provide information on thousands of gene in one experiment which saves time it is rapid, cost effective, and relatively low-priced.

Microarrays Applications in Toxicogenomics

Toxicogenomics is the hybridization of functional genomics and molecular toxicology. The goal of toxicogenomics is to find correlation between toxic responses to toxicants and changes in the genetic profiles of the objects exposed to such toxicants. Currently, microarray technology is gaining popularity in predictive toxicology. The role of gene chips is to allow monitoring changes in gene expression, where change in mRNA or protein levels enables investigators to reject or embrace hypotheses about pharmacology or toxicology. This facilities testing drug toxicity, safety and effectiveness which are very critical where, determining the potential problems of a compounds early can save time and money by focusing on compounds that are more likely to succeed. Moreover, the use of animals in such studies is minimized because of the percussion and conservative nature of the testing. On the other hand, the possibility of utilizing microarrays in exploring genes from other organisms will allow practical studies of questions about basic cellular functions, including cell cycle and DNA repair mechanisms, which are the major issues in cancer and cytotoxicity research. Since major signaling pathways and cellular processes involved in cellular response to cytotoxic agents are conserved between organisms, as in yeast and mammals, these simple eukaryotic systems could be excellent models for testing sensitivity. e.g., anti-tumour drugs. Utilizing microarrays helped significantly to overcome many long-standing challenges when drug testing was required for organs with complex expression patters or in certain illnesses. As one of these organs is the human brain, for which measurement of gene expression with microarrays has potential applications in the study of many diseases, especially cognitive disorders for which animal models are typically not available.

Microarray-Based Oncology

Cancer is a genetic disease and as such, our understanding of the pathobiology of tumours derives from analyses of the genes whose mutations are responsible for those tumours. The cancer phenotype, however, likely reflects the changes in the expression patterns of hundreds or even thousands of genes that survived the primary mutation of an oncogene or a tumour suppressor gene. Recently developed functional genomic approaches, such as DNA microarrays have enabled researchers to determine the expression level of every gene in a given cell or cancer population, which represents that cell population's entire transcriptome. The application of microarray and related technologies to identify specific targets of defined genes that have clearly been implicated in cancer progression requires different experimental approaches than are conventionally used. The objective is to define changes in transcriptional profiles that occur in response to modulating the expression level of the gene to be studied. The resulting altered expression profile can then be viewed as a blueprint by which that gene affects its cellular function. In this approach the investigator has greater control over the question being asked as the experimental variables are artificially controlled. Therefore multiple repetitions of the experiment are not required and in general the experiment carried out in duplicate is sufficient to generate reliable data. Currently cancer research coupled with diagnostic DNA microarrays is playing a dominant role compared to the other developing technologies since they are relatively easy to make and use and are applicable to numerous scientific inquiries. In addition, the ability of microarrays to simultaneously analyse several thousands of genes in samples from sick or healthy tissues will provide information that is expected to improve our understanding of the complex molecular interaction networks of healthy cells and tissues. Such characterization of the molecular mechanisms involved in pathology should result in the identification of new therapeutic targets and the development of new medicines. The genetic profiles thus obtained should also permit the definition of new pathologic subclasses not recognizable by traditional clinical factors, as well as new markers for susceptibility to certain illnesses, and new prognostic markers or methods of predicting responses to treatment. The ability of microarrays to provide a systematic method to identify key markers for prognosis and treatment response by profiling thousands of genes expressed in a single cancer is significant when dealing with certain cancer types. Two of these are breast cancer, which is the most frequent and deadly cancer in women and prostate cancer in men. The great heterogeneity of breast cancer makes prognosis and response to current treatments highly variable and difficult to predict. The use of expression profiling with cDNA array techniques in mammary tumour cell lines and breast tumours will help classify controversial tumours and provide new prognostic tools and potential therapeutic targets. They will also boost our knowledge of the molecular events responsible for the development and progression of cancer. Moreover, this will also allow us to define specific molecular pathways of tumoral progression and to define markers of prognostic and diagnostic relevance, which will reflect significantly on cancer diagnostics and consequently its management.

Challenges

Microarray technology was initially limited to biotechnology companies and high-budget research laboratories. Gene-chip companies have developed marketing arrangements and collaborations to bring microarray chips to academic research laboratories as well as commercial users. However, microarrays continue to have limitations in addition to their technical difficulty, specificity, and reliability: Microarrays are victims of their own success since the large data sets generated by the chips add new statistical and informatics-related challenges and complexity. There is still no standard protocol for microarray data analysis. There is no uniform system for generally managing and provide a disbursement point for microarray data and for the products of various data-mining techniques and their applications. Today, while microarray experiments will undoubtedly enhance our knowledge about expression and regulation of genes they still raise many technical and informatics challenges. Novel statistical methods must be evolved for the analysis of these large and diverse datasets. Microarray Data Handling and Management Microarray technology has resulted in the generation of large complex datasets. This situation shifted the bottleneck in biological investigation from data generation to data analysis. This data handling and analysis has become one of the major bottlenecks in the utilization of array technology. The amounts of data obtained from each experiment can require powerful computing techniques for identifying clusters of genes and interpreting overall patterns of expression. At present it seems that the greatest challenges in microarray research are not the arrays itself but the way by which resulting data matrices are handed and analysed. Without the ability to handle the data reliably and expeditiously the development of knowledge about the underlying biological processes, etc., will not be practical.

Lack of a Standardized Protocol for Data Handling

Once the hybridized chip is scanned, the data flow encounters the following impediments: Data are collected and saved as both image-and text files. It is of critical importance that databases and tracking files are maintained by precise registration of the spot configuration of all genes on the chip. Then the saved files can safely be imported to software programs for performance of image analysis and statistical analysis functions without concern over possible overlap, etc. The data are mined for induced or repressed genes, patterns of gene expression, and temporal relationships of expression under different experimental conditions. A significant challenge exists in making sense of the vast quantity of data generated by microarray experiments. There is no single tool that meets all of the needs of the microarray researcher. Mixtures of software programs presently are used to perform a multitude of tasks, including data tracking, image analysis, database storage, data queries, statistical analysis, multidimensional visualization, and interaction with public databases on the Internet. Basic spreadsheet programs have been adapted to answer questions regarding the magnitude of change in gene expression. However, limitations often arise as a result of mismatched programs and inadequate memory capacity for managing the enormous datasets. It is only recently that more sophisticated and comprehensive analytical tools for cluster analysis, self-organizing maps, and principle component analysis have been applied to biologic data to extract higher-order relationships embedded in expression patterns.

Informatics Challenges

One of the challenges facing current microarray technology is the absence of many gene annotations and the inability to find with ease literature references to relevant genes. This extends to difficulties in linking observations with the actions of the products of the genes being studied, including the possibility of gene clusters. Future challenges for microarray research will therefore include development of databases and algorithms to manage and analyse vast scale genomic datasets. Some pertinent examples of available databases are shown in. Finally, the lack of a standard format for microarray data interferes with the creation of such a resource. This problem was recognized by the scientists at the European Bioinformatics Institute when they proposed that standards should be based on XML, a computer markup language that combines data and formatting in a single file for distribution over the World–Wide Web.

Statistical Challenges

Array technology is not old wine in new bottles and the same must be true for the statistical methods that deal with array data. Despite the previously mentioned problems, obtaining data with gene chips has already become routine in many laboratories. However, large-scale array data meets with the resistance or inappropriateness of conventional methods of data reduction and analysis: interpreting the data is a major difficulty. Firstly, there are serious difficulties in registering and solving microarray images. New types of software and statistical techniques are required for this. The U.S. government represented by the National Institute of Health (NIH) has made programs to support such developments (http://research.nhgri.nih.gov/microarray/ index.html and http://grants1.nih.gov/grants/guide/rfa-files/RFA-NS-02-001.html). Next, the usual paradigm in bio-statistics, few variables with many cases, is turned topsy-turvy in microarray work. In the latter, the question is which are the variables that are significant. Arrays can examine thousands of genes while virtually nothing is known about their functions and expressions. The problem of multidimensional applications of the data can be solved by building an understanding of what is (are) the function(s)of the genes that are the subject of analysis and is there any discernable logic to their expression. This problem is likely to be resolved by the development of matrices from biological and computational studies; it is a byproduct of the solutions. At present, it is more critical for microarrays to be sufficiently uniform that the data from one study can be displayed and solved along with other studies. In general it can be said that genes that are co-expressed are highly likely to participate in similar physiological programs in physical processes. This understanding will facilitate the design of novel, more accurate and intuitive statistical methods. This will not be solved by computation alone; the statistical problem of going backward from thousands or tens of thousands of measurements to fundamental biological processes is not a soluble mathematical problem. But, known interactions can be mapped onto the data that is being generated from these very large-scale experiments. On the other hand, because microarray experiments result in such large amounts of data, false-positive results are likely to be obtained, where, analysing multiple independent experiments may eliminate spurious results.

As a possible solution, a transition was made from experiments involving a small number of conditions (with an emphasis on the specific genes induced or repressed) to experiments involving hundreds of conditions. In these experiments patterns of global gene expressions are used to classify disease specimens and discover gene functions and drug targets. Currently, many groups are developing software that will help to interpret microarray data using non-conventional statistical techniques. The present methods of managing data are not sufficiently precise to allow proper statistical analysis. The use of fold difference (the ratio of the expression in cells of interest versus the control cells) as a quantitative measure of the differential expression is an example of this problem. Genes expressed at low levels require higher fold differences in order to rise above the level of the noise in the systems and furnish proper signal to noise ratio calculations. The present method of resolving this problem is performing replicates to minimize the effect of the noise in the system. But, this does not deal with the fundamental problem in the signal to noise ratio. There are other methods for assigning confidence for differentially expressed genes. But the noise in the data can still confound analyses and rank data are often more robust than absolute measured values. All of this is due to the relatively high noise to signal ratio in microarray techniques at this time. Either the systems must be made more precise or new statistical methods must be developed to correct for the noise without affecting the impact of the signal. There is also a problem stemming from gaps in the data flow from microarray assays. Methods have emerged for imputing incomplete data, which may mislead the outcome of the analysis. Finally, two generic problems remain major concerns for (statistical) analysis of array data, just as they are problems for any gene sampling exercise. First is aneuploidy (the number of copies of a gene in the effective genome), which has been shown to effect the measured expression level of a gene. This in turn either confounds the analysis or inhibits insight into mechanisms of abnormal biology. Second, is the occurrences of false positives that rise sharply when thousands of genes in a microarray dataset are evaluated simultaneously by fold changes and significance tests. Although beyond the scope of this review, these are serious, organic problems for gene analysis and a system must be devised to flag the data they produce and cleanse it from the datasets prior to final analysis. Microarray technology is becoming a corner stone in the development and integration of knowledge obtained from molecular targets. There are many microarray techniques with clinical applications. By facilitating exploration of the relationship of genes to disease, and then to the results of the application of this information great inroads will be made to the treatment of patients as well as the accumulation for its own sake. Arrays have added an important dimension to the breath and depth of knowledge that is obtained. The future of microarrays seems even more promising to the life sciences than PCR technology. Its potential is only limited by the practical constraints that we have mentioned briefly in this review. This technology will allow the application of new insights, some of which are self-generated. At present the use of microarrays studying variations at the single nucleotide levels (SNPs) for individuals for gene profiling has allowed the characterization of profiles that are consistent with the development or presence of cancer. That they will produce the ability to rule out the likelihood of cancer remains to be proven. Coupled with advanced informatics, microarray diagnostics will furnish individualized investigation of the physiology and disease patterns of individuals and may be able to dispense the need for large, costly and often difficult-to-interpret clinical trials. They may lead to true individualized prevention and treatment of diseases. The integration of genes and environmental factors now possible is thought to be critical not only for cancer, but for many other diseases, such as diabetes, heart disease, asthma, and some neurological disorders. In addition, this will radically improve the identification of both genetic and molecular causes of susceptibility to certain diseases. Array technology has already revolutionized fields such as pharmacogenomics. There was a clear progress in finding drug targets and hence drug discovery, drug development and safety and molecular diagnostics were very clear. Despite these phenomenal accomplishments and further potential there was a major drawback to defeat; microarrays are entirely dependent on the state of knowledge of the genome under investigation. But this is fading as a principal problem due to the solution of the human genome and the speed at which various other genomes have been or are in the process of being sequenced. Another criticism that has yet to be resolved is the concern that the expression levels achieved in these artificial systems are not physiologically relevant. This is undeniably of concern and the onus is on investigators to devise additional experiments to confirm that genes identified in this manner represent physiologically relevant targets. We have also mentioned several methodological challenges to the practicality of microarrays; but there seems little concern that these problems will yield in the foreseeable future. Looking at the larger picture it is clear that analysis of gene expression is only in its infancy. Even the rather obvious approaches, such as cluster analysis and finding differentially expressed genes, have been used only rather crudely. It is expected that gene-expression data analysis methods will develop similarly as sequence analysis methods have developed over the past decades. The amounts of gene expression data will continue growing and the data will become more systematic. Currently gene expression profiling is similar to gene sequencing before the era of genome sequencing: the measurements are carried out to attack particular questions or sometimes just to demonstrate the concept. Establishing a public repository for gene expression data would facilitate systematic construction of gene expression matrices for various organisms. Like genome sequencing, the systematic gene expression profile is not an end in itself. It is a long way from having detailed gene expression profiles to the real and practically usable understanding of underlying cellular processes. Bioinformatics methods and tools will be needed to cope with the huge amounts of data, but they will not bring any deep understanding by themselves. On the other hand, the traditional gene by gene methods will not be sufficient to understand increasingly complex gene regulatory networks consisting of thousands or tens of thousands of genes. Microarray technologies are improving to the point when an entire genome will be monitored with very high resolution. With such potentials, researchers can explore thousands of genes faster and more cheaply. With all these arguments a key question remains to be answered, what do we need to do? The most pressing need is for new methods of bioinformatics, especially data accession, reduction and analysis, which must be developed to handle the flood of new information. Secondly, as far as we deal with biological processes through the genes that are involved in these processes we must use all means available to understand these processes and as a result to be able to interpret the data obtained by microarrays. Microarray data analysis is hence an important area that would be crucial in determining the success of this technology.

Data Analysis

Microarray technology is rapidly becoming a standard technique used in research laboratories all across the world. In essence, all the variants of the technology allow simultaneous profiling of the expression levels of tens of thousands of genes, potentially whole genomes in a single experiment. This unique power provides scientists with an opportunity to look at the transcriptional profile of biologic systems, processes, and diseases in an unbiased fashion. The relative ease (despite the prohibitive cost) of performing microarray experiments in molecular laboratory settings, combined with the potential power of the technology, have captured the imagination of scientists in academic and industry research institutes. This combination of ease of use with unforeseen power also appealed to administrators in the same institutes and in funding agencies, thus leading to rapid spread of the use of the technology. Many research groups in the academy and industry have implemented microarrays in multiple experimental settings with varying degrees of success. Microarrays have been successfully applied to almost every aspect of biomedical research. However, many more experiments remain uncompleted, or even worse, unavailable to the scientific community. Interestingly, instead of excitement at the results of experiments that utilize microarrays, scientists often express some confusion, a tendency to focus on what is already known, and a sense of weariness. These feelings arise from the tension between the relative ease of producing the results and the objective difficulty of dealing with the results. Provided that one has the funding and setting, it should not take one trained laboratory assistant more than a week to run 810 samples (from total RNA extraction to microarray hybridization). A month's worth of results from such a laboratory assistant will leave the investigator with the task of dealing with half a million or more data points. The objective difficulty resulting from the large quantities of information generated by the experiments is further complicated by the lack of simple and accepted approaches to analyzing large-scale gene expression data (or even visualizing the data). Additionally, the diversity of experimental designs and schemes adds to the confusion. Microarrays are used to classify diseases, to identify the effects of a stimulus in vivo or in vitro, to single out genes that may play a role in a specific disease or a specific biologic process, and to distinguish transcriptional programs that underlie such a process. The size of experimental groups and the design of the experiments also vary widely, from timecourse experiments to cross-sectional studies, from single observations with no repeats to analysis of hundreds of samples. Naturally, these diverse experimental schemes pose diverse computational requirements the analysis of an experiment designed to discover a new class of a disease which is different from an experiment designed to test the immediate targets of a known transcriptional activator. The wealth and complexity of information that characterizes results of microarray experiments has led to the suggestion that there may not be a single best analytic approach and that indeed the application of several analytical and computational approaches to a dataset may aid in the exposure of different and complementary aspects of the data.

The Gene Funnel

Generally, one can describe the process of analysing gene expression data as a funnel-shaped process. You start out with many genes, and by applying filters this number is gradually reduced. The first step in the analysis involves looking at the experiments and detecting outliers and technically inferior experiments. It is noticed that because of the wealth of information it is often very hard to identify a bad experiment if it managed to escape this preliminary screen. Once convinced that all arrays lack defects and are relatively comparable, one can define a set of genes that are termed legal genes. These are genes that pass a certain threshold of expression in at least one of the experiments. The parameters used are the signal (the gene expression level) and the detection call (present or absent call). One can usually eliminate genes that do not have a present call in a fraction of the experiments (5% or 1 if less then 20 microarrays). In addition one can also set an expression level threshold that a gene needs to pass in a fraction of the experiments (5%) to be included in the dataset. To determine this threshold the scientist can hybridize the same sample on two microarrays and compare the expression levels. As a general rule, the consistency between values obtained from running the same sample on two microarrays is intensity-dependent, that is, the lower the intensity is, the lower the agreement between the two microarrays. Often, a discernible threshold can be observed. Over this threshold the consistency is indeed impressive and lies within the 2-fold range, a phenomenon not observed when two different samples are compared. This filtration process usually reduces the number of the genes in the dataset by a third to a half (depending on the sample number, the type of tissues being investigated, etc.). The next step is to define a set of active genes. These are genes in which something happened to their expression level. There are many computational and statistical ways to define an active gene. One can take a simple and straightforward approach by converting the data to be in terms of expression ratios. This is the natural representation in two dye-competitive hybridization systems but not in Affymetrix microarrays, where the readings are absolute gene expression levels.

What is done in such cases is termed virtual two dye. Basically, one creates a set of ratios by dividing the values of every gene by a number. This number may be the geometric mean of the controls if you are dealing with distinct groups, the geometric mean of all the experiments, or of any other group of experiments. This transformation allows us to query the genes for their activity. Relatively weak filters are used and genes that changed at least 1.5-fold in every direction in at least 5% of the experiments (or 1 if there are fewer than 20 experiments) are considered. Genes that did not vary are excluded from the active genes dataset, but may serve later for controls in verification experiments. This process usually greatly reduces the number of genes in the dataset. It also allows for more stringent and specific queries, such as genes that changed in a certain direction in a certain subset of the experiments. The active genes dataset will be the dataset that we will use for all other analyses.

Clustering Methods

The most commonly used analysis tools are clustering methods. Clustering methods attempt to identify genes that behave similarly across a range of conditions or samples. The motivation to find such genes is driven by the assumption that genes that demonstrate similar patterns of expression share common characteristics, such as common regulatory elements, common functions, or (in the case of mixed tissue studies) common cellular origin. This assumption is supported by the successful application of clustering algorithms to the analysis of basic mechanisms of cellular function in yeast and mammalian systems. The main advantage in clustering tools is that using them mitigates the inherent difficulty in becoming familiar with the results. By grouping the genes into clusters that behave similarly, these methods allow the investigator to browse the data in a less intimidating and chaotic atmosphere. Additionally, many of the methods are relatively easy to visualize, thus improving the accessibility of the biologically meaningful information that is in the data. Several clustering methods have been applied to gene expression data, hierarchical clustering being the most popular, but others include k means, deterministic annealing, self-organizing maps, combinatorial methods and graph theoretical approaches, and super-paramagnetic clustering. The results of clustering algorithms are highly dependent on the input, i.e., on the data that is used for the analysis. Methods for gene filtering are commonly applied. It is important to remember and note that various schemes for selecting genes that are legal (with expression levels that pass a certain level) and active (that change compared with a certain threshold) may affect the results of the analysis, as will inclusion of multiple and diverse experiments. In general, we can apply more than one clustering method to a dataset. One usually compares the results of several methods and of several gene-filtering schemes before deciding what is the true signal in the dataset. Many times the decision will lie on the reproducibility of the cluster using various methods. This process, although tedious, allows us to gain confidence that the patterns we observe represent true biologic phenomena that are independent of the analysis method and, as previously stated, to become familiar with the data and with the true signal that characterizes it.

Hierarchical Clustering

This is probably the most popular clustering approach to large-scale gene expression data. The principles behind this approach are reasonably easy to understand and very intuitive to visualize. Tools for hierarchical clustering (Cluster) and visualization (Treeview) designed by Michael Eisen were freely available to the research community from the very early stages of the introduction of microarray technology, making this clustering into a standard in the field. Basically, this is an agglomerative process in which single member clusters are fused to bigger and bigger clusters. In somewhat more detail, the procedure starts by computing a pairwise distance matrix between all the genes, the distance matrix is explored for the nearest genes, and they are defined as a cluster. After a new cluster is formed by agglomeration of two clusters, the distance matrix is updated to reflect its distance from all other clusters. Then, the procedure searches for the nearest pair of clusters to agglomerate, and so on. This procedure leads to a hierarchical dendogram in which multiple clusters are fused in nodes according to their similarity, finally resulting in a single hierarchical tree. There are several hierarchical clustering algorithms that differ in the way the distances are calculated. As mentioned earlier, Cluster and Treeview can be obtained from Michael Eisens lab at http://rana.lbl.gov/EisenSoftware.htm.

k-Means Clustering

This is an iterative procedure that searches for clusters that are defined in terms of their centre points or means. Once a set of cluster centres is defined, each gene is assigned to the cluster it is closest to. The clustering algorithm then adjusts the centre of each cluster of genes to minimize the sum of distances of genes in each cluster to the centre. This results in a new choice of cluster centres, and so we can reassign genes to clusters and repeat the process. These iterations are applied until convergence. This method has more of a global character than hierarchical approaches. It does not generate a hierarchical tree, but rather a predetermined number of clusters. The k-means clustering is especially useful in cases in which one knows how many distinct gene expression patterns to expect. The site previously mentioned also provides tools for k-means calculations.

Self-Organizing Maps

Self-organizing maps (SOMs) were introduced to the analysis of microarray data by Tamayo and coworkers (20). As in the k-means procedure, one assigns the data into a predetermined set of clusters. However, unlike k-means, what follows is an iterative process in which gene expression vectors in each cluster are trained to find the best distinctions between the different clusters. In other words, a partial structure is imposed on the data and then this structure is iteratively modified according to the data. SOM is superior when dealing with messy data that contains outliers and irrelevant parameters. We frequently apply SOM analysis to new datasets, as a means to browse the trends in the data and to detect outliers. Although many software packages contain SOM feature, we use GeneCluster that can be obtained from Whitehead Institute/MIT Center for Genome Research website at http://www-genome.wi.mit.edu/cancer/software/software.html.

Graph-Theoretic Clustering

Graph-theoretic clustering (CLICK) is an innovative clustering method that utilizes graph-theoretic and statistical techniques to identify tight groups of highly similar elements (kernels), which are likely to belong to the same true cluster. Several heuristic procedures are then used to expand the kernels into the full clustering. CLICK and a visualization tool (Expander) are available at http://www.math.tau.ac.il/~roded/click.html. Most often, one can apply more than one clustering method on a dataset and we feel more confident if clusters are reproduced using different methods.

Finding Relevant Genes

Relevant genes are defined as genes in which the expression level characterizes a specific experimental condition. Usually, these are genes in which the expression levels differ significantly between different experimental conditions. The simplest way is to compare the expression level of a gene in the different conditions and look at the relative expression fold change of the gene: the ratio of its expression level in one set of samples to its expression in another set. This can even be done between two samples. Naturally, the question immediately arises: what is the threshold of relative expression at which a gene is considered changed? Many published studies use a threshold of 2-fold change that is based on the earliest microarray works. However, the ratio itself is not a good predictor of the relevance of a gene. In an ongoing study of lung tumours in our lab, the fold ratio of 2 and more in a single tumour compared with its matching control did not predict the fold ratio of this same gene calculated from a comparison of 20 lung tumours to 7 normal tissues. Impressively, there were genes with fold ratios of more than 8 in the single sample comparison that had changed in the opposite direction when averages were calculated. This observation was also true when we compared fold ratios of a single pair of samples (tumour/normal) to fold ratios of a pool of tumour RNAs and a pool of normal RNAs, and in primary cell culture experiments. Furthermore, the genes that had the highest fold ratios in the single-sample comparisons were not necessarily those that had significant P values in the multiple replicate analysis. When one compared fold ratios calculated from the pools comparison with those obtained from the average comparison, one did observe that genes substantially changed in pooled samples comparison seemed to better overlap with genes with significant P values (Figure 2D) than in the other comparisons. This analysis suggests (as one would intuitively presume) that basing an observation on fold ratio may easily lead to spurious results, especially when a small amount of samples is analysed. The major problem with relying on fold changes for finding relevant genes is that they do not provide any means to measure the statistical relevance of the results. Scoring Methods Statistical considerations are essential when trying to find genes whose expression characterizes a specific experimental condition. The available methods can be divided into parametric and nonparametric methods.

Parametric Methods

These approaches model expression profiles within a parametric representation, and ask how different the parameters of the experimental groups are. A simple example is the classic t test. Other examples of parametric approaches are the separation score and the Bayesian t test. The significance analysis of microarrays (SAM) method developed at Stanford University Statistics and Biochemistry Labs is a method for identifying genes with statistically significant changes in gene expression. It deals with the specific issues of multiple testing (see STATISTICAL SIGNIFICANCE) and is easily installed as an Excel add-in. SAM is freely available from http://www-stat.stanford.edu/~tibs/SAM/ and we frequently use it. Nonparametric Methods No a priori assumptions are made about the distribution of expression profiles in the data. Instead, one attempts to directly examine the degree to which the two groups of expression measurements are distinguished. The methods that are used were introduced by Ben-Dor and coworkers and applied successfully to the study of breast cancer, melanoma, and recently to idiopathic pulmonary fibrosis.

Threshold Number of Misclassifications

The threshold number of misclassifications (TNOM) measures how successful we are in separating the two groups of samples by a simple threshold over the expression values. That is, we search for the threshold value of the genes expression that will distinguish the experimental conditions. A gene is scored by the number of misclassifications made by the best threshold that we can find for it. If the expression value of the gene allows us perfectly separate the groups, the gene has a TNOM score _ 0. On the other hand, if the two groups are interspersed, the gene has a score that may be close to the size of the smallest group of samples. INFO score is a refined version of TNOM that measures the misclassifications made by a simple threshold in terms of the information lost (or entropy) of the labels of samples in each side of the threshold. There are several other scores that we often use including the Gaussian score and Ds score (n-1) (23, 28). Score-Genea package that calculates the scores for a given data set is available on http://fgusheba.cs.huji.ac.il/software.htm. The main weakness of parametric approaches is the assumptions that they make about the data. For example, outliers can significantly skew the t-test score by changing the variance estimated in the sample. Similarly, scale transformation (e.g., working in logarithmic scale) can have drastic impact on the scores. Nonparametric approaches are more robust to this type of phenomenon. This comes at a price, as nonparametric approaches are less sensitive to the actual expression values.

Loading the Results with Biologic Meaning

One of the most painful stages in the analysis of the results of microarray experiments is the postcomputational stage. The long-awaited and dreaded computational analysis has finally been completed, and now the investigator has to examine the lists of genes he or she so eagerly anticipated and to generate new hypotheses. In the easiest cases, more than half of the genes in these lists will be familiar to the investigator; however, it will still be difficult to put them in a meaningful biologic framework without focusing mainly on what was previously known. To fully realize the potential of microarray experiments as hypothesis-generating tools, it is essential to have tools that will facilitate the process of looking for biologic meaning in the data.

Probabilistic Relational Models

Probabilistic relational models (PRM) allow the inclusion of multiple types of information in the computational process itself. For example, experiments can be annotated by their gene expression patterns, experimental or clinical data, the cell type or strain used in the experiment, the cellular phenotype triggered by each condition, etc. The genes can be annotated by the gene expression data, sequence elements present in the gene promoters, functional information, gene ontology definitions, protein motifs, and more. The analysis then creates context-specific groupings (clusters) of genes and experiments that are enriched with biologic information, thus facilitating the hypothesis-hypothesis-generating process. PRMs represent the first tool that allows statistically sound, unbiased integration of biologic information and microarray results. GenMAPP GenMAPP represents a completely different approach, which is to analyse gene expression data in the context of known biologic pathways. It allows the uploading of gene expression data unto known pathways and gene families using complex selection criteria. Furthermore, it allows investigators to author their own pathways using a simple and intuitive tool. GenMAPP, which is available at http://www.genmapp.org/. Pubgene Available at http://www.pubgene.uio.no/cgi-bin/PubGene.cgi, allows the user to upload results of microarray experiments and to search for interesting literature clusters. The literature clusters are then ranked according to the data analysis parameters and the user-defined number of the highest scoring clusters are displayed as a result. It allows the investigator to rapidly get a feeling about the relationship of the results of his microarray experiments to published literature. Naturally, this does not replace the actual work of looking for and reading the literature, but it can quickly and easily highlight a specific literature cluster that might have been otherwise overlooked.

CONCLUSION

As opposed to the situation a few years ago, today there is a wide range of analysis tools for examining microarray experiments. There is no single best method for analysis, but there are multiple tools that can be applied to the data and allow querying it from different angles. Clustering methods may help in looking at gene expression patterns, scoring methods to identify genes that are statistically associated with the process studied, and the advanced tools to facilitate the generation of new hypotheses.

REFERENCES

1. Dunham, I. *et al.* The DNA sequence of human chromosome 22. *Nature* **402:** 489–495 (1999).
2. Hattori, M. *et al.* The DNA sequence of human chromosome 21. *Nature* **405:** 311–319 (2000).
3. Lennon, G. G. & Lehrach, H. Hybridization analyses of arrayed cDNA libraries. *Trends Genet.* **7:** 314–317 (1991).
4. Kafatos, F. C., Jones, C. W. & Efstratiadis, A. Determination of nucleic acid sequence homologies and relative concentrations by a dot hybridization procedure. *Nucleic Acids Res.* **7:** 1541–1552 (1979).
5. Gillespie, D. & Spiegelman, S. A quantitative assay for DNA-RNA hybrids with DNA immobilized on a membrane. *J. Mol. Biol.* **12:** 829–842 (1965).
6. Southern, E. M. *et al.* Arrays of complementary oligonucleotides for analysing the hybridisation behaviour of nucleic acids. *Nucleic Acids Res.* **22:** 1368–1373 (1994).
7. Zhao, N., Hashida, H., Takahashi, N., Misumi, Y. & Sakaki, Y. High-density cDNA filter analysis: a novel approach for large-scale, quantitative analysis of gene expression. *Gene* **156:** 207–213 (1995).
8. Nguyen, C. *et al.* Differential gene expression in the murine thymus assayed by quantitative hybridization of arrayed cDNA clones. *Genomics* **29:** 207–216 (1995).
9. Fodor, S. P. A. *et al.* Light-directed, spatially addressable parallel chemical synthesis. *Science* **251:** 767–773 (1991).
10. Fodor, S. P. *et al.* Multiplexed biochemical assays with biological chips. *Nature* **364:** 555–556 (1993).
11. Pease, A. C. *et al.* Light-generated oligonucleotide arrays for rapid DNA sequence analysis. *Proc. Natl Acad. Sci. USA* **91:** 5022–5026 (1994).
12. Schena, M., Shalon, D., Davis, R. W. & Brown, P. O. Quantitative monitoring of gene expression patterns with a complementary DNA microarray. *Science* **270:** 467–470 (1995).
13. Shalon, D., Smith, S. J. & Brown, P. O. A DNA microarray system for analyzing complex DNA samples using two-color fluorescent probe hybridization. *Genome Res.* **6:** 639–645 (1996).
14. DeRisi, J. L., Iyer, V. R. & Brown, P. O. Exploring the metabolic and genetic control of gene expression on a genomic scale. *Science* **278:** 680–686 (1997).
15. Lipshutz, R. J., Fodor, S. P., Gingeras, T. R. & Lockhart, D. J. High density synthetic oligonucleotide arrays. *Nature Genet.* **21:** 20–24 (1999).
16. Bowtell, D. D. Options available—from start to finish—for obtaining expression data by microarray. *Nature Genet.* **21:** 25–32 (1999).
17. Edman, C. F. *et al.* Electric field directed nucleic acid hybridization on microchips. *Nucleic Acids Res.* **25:** 4907–4914 (1997).

18. Sosnowski, R. G., Tu, E., Butler, W. F., O'Connell, J. P. & Heller, M. J. Rapid determination of single base mismatch mutations in DNA hybrids by direct electric field control. *Proc. Natl Acad. Sci. USA* **94:** 1119–1123 (1997).
19. Gray, D. E., Case-Green, S. C., Fell, T. S., Dobson, P. J. & Southern, E. M. Ellipsometric and interferometric characterization of DNA probes immobilised on a combinatorial array. *Langmuir* **13:** 2833–2842 (1997).
20. Walt, D. R. Bead-based fiber-optic arrays. *Science* **287:** 451 (2000).
21. Michael, K. L., Taylor, L. C., Schultz, S. L. & Walt, D. R. Randomly ordered addressable high-density optical sensor arrays. *Anal. Chem.* **70:** 1242–1248 (1998).
22. Ferguson, J. A., Boles, T. C., Adams, C. P. & Walt, D. R. A fiber-optic DNA biosensor microarray for the analysis of gene expression. *Nature Biotechnol.* **14:** 1681–1684 (1996).
23. Spellman, P. T. *et al.* Comprehensive identification of cell cycle-regulated genes of the yeast *Saccharomyces cerevisiae* by microarray hybridization. *Mol. Biol. Cell* **9:** 3273–3297 (1998).
24. Cho, R. J. *et al.* A genome-wide transcriptional analysis of the mitotic cell cycle. *Mol. Cell* **2:** 65–73 (1998).
25. The Chipping forecast. *Nature Genet.* **21:** (Suppl.), 1–60 (1999).
26. Wodicka, L., Dong, H., Mittmann, M., Ho, M.-H. & Lockhart, D. J. Genome-wide expression monitoring in *Saccharomyces cerevisiae. Nature Biotechnol.* **15:** 1359–1367 (1997).
27. White, K. P., Rifkin, S. A., Hurban, P. & Hogness, D. S. Microarray analysis of *Drosophila* development during metamorphosis. *Science* **286:** 2179–2184 (1999).
28. Chambers, J. *et al.* DNA microarrays of the complex human cytomegalovirus genome: profiling kinetic class with drug sensitivity of viral gene expression. *J. Virol.* **73:** 5757–5766 (1999).
29. Gingeras, T. R. *et al.* Simultaneous genotyping and species identification using hybridization pattern recognition analysis of generic mycobacterium DNA arrays. *Genome Res.* **8:** 435–448 (1998).
30. Lockhart, D. J. *et al.* Expression monitoring by hybridization to high-density oligonucleotide arrays. *Nature Biotechnol.* **14:** 1675–1680 (1996).
31. Liang, P. & Pardee, A. B. Differential display of eukaryotic messenger RNA by means of the polymerase chain reaction. *Science* **257:** 967–971 (1992).
32. Shimkets, R. A. *et al.* Gene expression analysis by transcript profiling coupled to a gene database query. *Nature Biotechnol.* **17:** 798–803 (1999).
33. Ivanova, N. B. & Belyavsky, A. V. Identification of differentially expressed genes by restriction endonuclease-based gene expression fingerprinting. *Nucleic Acids Res.* **23:** 2954–2958 (1995).
34. Kato, K. Description of the entire mRNA population by a 3¢ end cDNA fragment generated by class IIS restriction enzymes. *Nucleic Acids Res.* **23:** 3685–3690 (1995).

35. Bachem, C. W. *et al.* Visualization of differential gene expression using a novel method of RNA fingerprinting based on AFLP: analysis of gene expression during potato tuber development. *Plant J.* **9:** 745–753 (1996).
36. Velculescu, V. E., Zhang, L., Vogelstein, B. & Kinzler, K. W. Serial analysis of gene expression. *Science* **270:** 484–487 (1995).
37. Boucherie, H. *et al.* Two-dimensional protein map of *Saccharomyces cerevisiae*: construction of a gene-protein index. *Yeast* **11:** 601–613 (1995).
38. Gygi, S. P. *et al.* Quantitative analysis of complex protein mixtures using isotope-coded affinity tags. *Nature Biotechnol.* **17:** 994–999 (1999).
39. Mann, M. Quantitative proteomics? *Nature Biotechnol.* **17:** 954–955 (1999).
40. Oda, Y., Huang, K., Cross, F. R., Cowburn, D. & Chait, B. T. Accurate quantitation of protein expression and site-specific phosphorylation. *Proc. Natl Acad. Sci. USA* **96:** 6591–6596 (1999).
41. Burns, N. *et al.* Large-scale analysis of gene expression, protein localization, and gene disruption in Saccharomyces cerevisiae. *Genes Dev.* **8:** 1087–1105 (1994).
42. Ross-Macdonald, P., Sheehan, A., Roeder, G. S. & Snyder, M. A multipurpose transposon system for analyzing protein production, localization, and function in *Saccharomyces cerevisiae. Proc. Natl Acad. Sci. USA* **94:** 190–195 (1997).
43. Ross-Macdonald, P. *et al.* Large-scale analysis of the yeast genome by transposon tagging and gene disruption. *Nature* **402:** 413–418 (1999).
44. Niedenthal, R. K., Riles, L., Johnston, M. & Hegemann, J. H. Green fluorescent protein as a marker for gene expression and subcellular localization in budding yeast. *Yeast* **12:** 773–786 (1996).
45. Zong, Q., Schummer, M., Hood, L. & Morris, D. R. Messenger RNA translation state: the second dimension of high-throughput expression screening. *Proc. Natl Acad. Sci. USA* **96:** 10632–10636 (1999).
46. Johannes, G., Carter, M. S., Eisen, M. B., Brown, P. O. & Sarnow, P. Identification of eukaryotic mRNAs that are translated at reduced cap binding complex eIF4F concentrations using a cDNA microarray. *Proc. Natl Acad. Sci. USA* **96:** 13118–13123 (1999).
47. Diehn, M., Eisen, M. B., Botstein, D. & Brown, P. O. Large-scale identification of secreted and membrane-associated gene products using DNA microarrays. *Nature Genet.* **25:** 58–62 (2000).
48. Weinstein, J. N. Fishing expeditions. *Science* **282:** 628–629 (1998).
49. Golub, T. R. *et al.* Molecular classification of cancer: class discovery and class prediction by gene expression monitoring. *Science* **286:** 531–537 (1999).
50. Alizadeh, A. A. *et al.* Distinct types of diffuse large B-cell lymphoma identified by gene expression profiling. *Nature* **403:** 503–510 (2000).
51. Mack, D. H. *et al.* in *Deciphering Molecular Circuitry Using High-Density DNA Arrays* (eds Hihich, E. & Croce, E.) 85–108 (Plenum, New York, 1998).
52. Alon, U. *et al.* Broad patterns of gene expression revealed by clustering analysis of tumor and normal colon tissues probed by oligonucleotide arrays. *Proc. Natl Acad. Sci. USA* **96:** 6745–6750 (1999).

53. Perou, C. M. *et al.* Distinctive gene expression patterns in human mammary epithelial cells and breast cancers. *Proc. Natl Acad. Sci. USA* **96:** 9212–9217 (1999).
54. Ross, D. T. *et al.* Systematic variation in gene expression patterns in human cancer cell lines. *Nature Genet.* **24:** 227–235 (2000).
55. Scherf, U. *et al.* A gene expression database for the molecular pharmacology of cancer. *Nature Genet.* **24:** 236–244 (2000).
56. Fambrough, D., McClure, K., Kazlauskas, A. & Lander, E. S. Diverse signaling pathways activated by growth factor receptors induce broadly overlapping, rather than independent, sets of genes. *Cell* **97:** 727–741 (1999).
57. Lee, S. B. *et al.* The Wilms tumor suppressor WT1 encodes a transcriptional activator of amphiregulin. *Cell* **98:** 663–673 (1999).
58. Harkin, D. P. *et al.* Induction of GADD45 and JNK/SAPK-dependent apoptosis following inducible expression of BRCA1. *Cell* **97:** 575–586 (1999).
59. Mei, R. *et al.* Genome-wide detection of allelic imbalance using human SNPs and high density DNA arrays. *Genome Res.* (in the press).
60. Pollack, J. R. *et al.* Genome-wide analysis of DNA copy-number changes using cDNA microarrays. *Nature Genet.* **23:** 41–46 (1999).
61. Pinkel, D. *et al.* High resolution analysis of DNA copy number variation using comparative genomic hybridization to microarrays. *Nature Genet.* **20:** 207–211 (1998).
62. Holstege, F. C. *et al.* Dissecting the regulatory circuitry of a eukaryotic genome. *Cell* **95:** 717–728 (1998).
63. Wyrick, J. J. *et al.* Chromosomal landscape of nucleosome-dependent gene expression and silencing in yeast. *Nature* **402:** 418–421 (1999).
64. Tamayo, P. *et al.* Interpreting patterns of gene expression with self-organizing maps: methods and application to hematopoietic differentiation. *Proc. Natl Acad. Sci. USA* **96:** 2907–2912 (1999).
65. Eisen, M. B., Spellman, P. T., Brown, P. O. & Botstein, D. Cluster analysis and display of genomewide expression patterns. *Proc. Natl Acad. Sci. USA* **95:** 14863–14868 (1998).
66. Wen, X. *et al.* Large-scale temporal gene expression mapping of central nervous system development. *Proc. Natl Acad. Sci. USA* **95:** 334–339 (1998).
67. Chu, S. *et al.* The transcriptional program of sporulation in budding yeast. *Science* **282:** 699–705 (1998).
68. Tavazoie, S., Hughes, J. D., Campbell, M. J., Cho, R. J. & Church, G. M. Systematic determination of genetic network architecture. *Nature Genet.* **22:** 281–285 (1999).
69. Wolfsberg, T. G. *et al.* Candidate regulatory sequence elements for cell cycle-dependent transcription in *Saccharomyces cerevisiae*. *Genome Res.* **9:** 775–792 (1999).
70. Marton, M. J. *et al.* Drug target validation and identification of secondary drug target effects using DNA microarrays. *Nature Med.* **4:** 1293–1301 (1998).

71. Gray, N. S. *et al.* Exploiting chemical libraries, structure, and genomics in the search for kinase inhibitors. *Science* **281:** 533–538 (1998).
72. Rosania, G. R. *et al.* Myoseverin: a microtubule binding molecule with novel cellular effects. *Nature Biotechnol.* **18:** 304–308 (2000).
73. Emmert-Buck, M. R. *et al.* Laser capture microdissection. *Science* **274:** 998–1001 (1996).
74. Bonner, R. F. *et al.* Laser capture microdissection: molecular analysis of tissue. *Science* **278:** 1481–1483 (1997).
75. Simone, N. L., Bonner, R. F., Gillespie, J. W., Emmert-Buck, M. R. & Liotta, L. A. Laser-capture microdissection: opening the microscopic frontier to molecular analysis. *Trends Genet.* **14:** 272–276 (1998).
76. Wang, A. M., Doyle, M. V. & Mark, D. F. Quantitation of mRNA by the polymerase chain reaction. *Proc. Natl Acad. Sci. USA* **86:** 9717–9721 (1989).
77. Dulac, C. Cloning of genes from single neurons. *Curr. Top. Dev. Biol.* **36:** 245–258 (1998).
78. Jena, P. K., Liu, A. H., Smith, D. S. & Wysocki, L. J. Amplification of genes, single transcripts and cDNA libraries from one cell and direct sequence analysis of amplified products derived from one molecule. *J. Immunol. Methods* **190:** 199–213 (1996).
79. Kwoh, D. Y. *et al.* Transcription-based amplification system and detection of amplified human immunodeficiency virus type 1 with a bead-based sandwich hybridization format. *Proc. Natl Acad. Sci. USA* **86:** 1173–1177 (1989).
80. Guatelli, J. C. *et al.* Isothermal, in vitro amplification of nucleic acids by a multienzyme reaction modeled after retroviral replication. *Proc. Natl Acad. Sci. USA* **87:** 7797 (1990).
81. Eberwine, J. *et al.* Analysis of gene expression in single live neurons. *Proc. Natl Acad. Sci. USA* **89:** 3010–3014 (1992).
82. Luo, L. *et al.* Gene expression profiles of laser-captured adjacent neuronal subtypes. *Nature Med.* **5:** 117–122 (1999).
83. Cho, R. J. *et al.* Parallel analysis of genetic selections using whole genome oligonucleotide arrays. *Proc. Natl Acad. Sci. USA* **95:** 3752–3757 (1998).
84. Bulyk, M. L., Gentalen, E., Lockhart, D. J. & Church, G. M. Quantifying DNA-protein interactions by double-stranded DNA arrays. *Nature Biotechnol.* **17:** 573–577 (1999).
85. Brent, R. Genomic biology. *Cell* **100:** 169–183 (2000).
86. McAdams, H. H. & Shapiro, L. Circuit simulation of genetic networks. *Science* **269:** 650–656 (1995).
87. McAdams, H. H. & Arkin, A. It's a noisy business! Genetic regulation at the nanomolar scale. *Trends Genet.* **15:** 65–69 (1999).
88. Bhalla, U. S. & Iyengar, R. Emergent properties of networks of biological signaling pathways. *Science* **283:** 381–387 (1999).
89. Weng, G., Bhalla, U. S. & Iyengar, R. Complexity in biological signaling systems. *Science* **284:** 92–96 (1999).

90. Arkin, A., Ross, J. & McAdams, H. H. Stochastic kinetic analysis of developmental pathway bifurcation in phage lambda-infected *Escherichia coli* cells. *Genetics* **149:** 1633–1648 (1998).
91. Winzeler, E. *et al.* Functional characterization of the *Saccharomyces cerevisiae* genome by precise deletion and parallel analysis. *Science* **285:** 901–906 (1999).
92. Fields, S. & Song, O. A novel genetic system to detect protein–protein interactions. *Nature* **340:** 245–246 (1989).
93. Roberts, C. J. *et al.* Signaling and circuitry of multiple MAPK pathways revealed by a matrix of global gene expression profiles. *Science* **287:** 873–880 (2000).
94. Brown, P. O. & Botstein, D. Exploring the new world of the genome with DNA microarrays. *Nature Genet.* **21:** 33–37 (1999).
95. Ly, D., Lockhart, D. J., Lerner, R. & Schultz, P. G. Mitotic misregulation and human aging. *Science* **287:** 2486–22492 (2000).
96. Cherry, J. M. *et al.* Genetic and physical maps of *Saccharomyces cerevisiae*. *Nature* **387:** 67–73 (1997).
97. Ball, C. A. *et al.* Integrating functional genomic information into the *Saccharomyces* Genome Database. *Nucleic Acids Res.* **28:** 77–80 (2000).
98. Mewes, H. W. *et al.* MIPS: a database for genomes and protein sequences. *Nucleic Acids Res.* **28:** 37–40 (2000).
99. Walsh, S., Anderson, M. & Cartinhour, S. W. ACEDB: a database for genome information. *Methods Biochem. Anal.* **39:** 299–318 (1998).
100. Kanehisa, M. & Goto, S. KEGG: Kyoto Encyclopedia of Genes and Genomes. *Nucleic Acids Res.* **28:** 27–30 (2000).
101. The FlyBase Consortium. The FlyBase Database of the Drosophila Genome Projects and community literature. *Nucleic Acids Res.* **27:** 85–88 (1999).
102. Karp, P. D., Riley, M., Paley, S. M., Pellegrini-Toole, A. & Krummenacker, M. EcoCyc: encyclopedia of *Escherichia coli* genes and metabolism. *Nucleic Acids Res.* **27:** 50–53 (1999).
103. Iyer, V. & Struhl, K. Absolute mRNA levels and transcriptional initiation rates in *Saccharomyces cerevisiae*. *Proc. Natl Acad. Sci. USA* **93:** 5208–5212 (1996).
104. Lee, C. K., Klopp, R. G., Weindruch, R. & Prolla, T. A. Gene expression profile of aging and its retardation by caloric restriction. *Science* **285:** 1390–1393 (1999).
105. Fan, J. B. *et al.* Parallel genotyping of human SNPs using generic oligonucleotide tag arrays. *Genome Res.* (in the press).
106. Lashkari, D. A. *et al.* Yeast microarrays for genome wide parallel genetic and gene expression analysis. *Proc. Natl Acad. Sci. USA* **94:** 13057–13062 (1997).
107. Winzeler, E., Lee, B., McCusker, J. & Davis, R. Whole genome genetic typing using high-density oligonucleotide arrays. *Parasitology* **118:** S73–S80 (1999).
108. Winzeler, E. A. *et al.* Direct allelic variation scanning of the yeast genome. *Science* **281:** 1194–1197 (1998).
109. Troesch, A. *et al.* Mycobacterium species identification and rifampin resistance testing with highdensity DNA probe arrays. *J. Clin. Microbiol.* **37:** 49–55 (1999).

Glossary

A

ACNFP Advisory Committee on Novel Foods and Processes.

ACRE Advisory Committee on Releases into the Environment.

Adaptive mutation Mutations in microorganisms those are most suitable for survival under altered conditions.

Adenosine deaminase deficiency Deficiency of an enzyme called adenosine deaminase. The patients have an extremely weak immune system and die unless kept in germ-free sterile environments.

Adenoviruses A commonly occurring class of viruses. The most common being the one that causes common cold.

Allele An alternative form of the same gene.

Allergen Any substance that induces a hypersensitive (allergic) reaction in the body.

Amino acid The basic unit that builds a protein. There are 20 basic types of amino acids that are found in nature.

Anther The male part of the flower that bears pollen.

Attenuated vectors These are vectors in which the infective component is removed, rendering it harmless. These are also called avirulent vectors.

Autosome One of the 22 matched pairs of chromosomes in which each pair is inherited from the mother and the other from the father.

B

Bacteriophage These are a class of viruses that infect bacteria, also known as *phage.*

Base These are the nitrogen-containing components of the DNA molecule whose sequence determines the genetic code.

Biodiversity The diversity of variety, species, families, genera, etc., of living organisms in a particular place.

Bioremediation Cleaning up waste or polluted land, waters, etc., using living organisms.

C

Cell The smallest structural unit of all living organisms that is able to grow and reproduce independently; a cell is formed of a mass of living material surrounded by a membrane and contains a nucleus. Cells can be classified as germ-line (sperm, eggs) or somatic (rest of body tissues).

Cell line A culture of cells, which can be kept alive indefinitely through the appropriate supply of nutrients.

Chromosome A microscopic particle containing thousands of genes (DNA) found in the nucleus of the cell. The human being has 23 pairs of chromosomes in each somatic body cell. Of them, 22 are matching (homologous) pairs called autosomes. The two remaining are sex chromosomes: in females two X chromosomes, in males an X and a Y chromosome. One chromosome of each pair plus one sex chromosome come from the mother, the other 23 from the father. Each chromosome is a double strand of DNA packaged in protein.

Chymosin One of the enzymes which causes milk to clot and turn into curds and whey during cheese making.

Clone

1. A group of cells that have descended from one parent cell.
2. Genetically identical organisms that have descended from one parent.
3. Identical sequences of DNA (genes) that have been produced by bacteria or other organisms.

Cloning The process of creating a clone.

Code The sequence of DNA bases which forms the instructions for a given characteristic or trait.

Conjugation The 'mating' process in bacteria which requires cell-to-cell contact being established, and in which genes are transferred between cells.

Cultivar A crop plant.

Cytoplasm The part of the cell outside the nucleus.

D

Deoxyribose A constituent of DNA.

Differentiation A process of development whereby properties of a cell or tissue become specialized for its particular task.

Diploid The presence of two chromosomes of a pair in a cell.

Directed Mutation See **Adaptive Mutation**

DNA Deoxyribonucleic acid, a large molecule that contains all genetic information in the cell.

DNAse Enzyme that breaks down DNA.

DNA polymerase An enzyme that makes DNA in the presence of a **DNA primer** and a mixture of the four **nucleotides**.

DNA primer A short stretch of DNA needed to start the DNA polymerase reaction.

Dominant gene or allele Most genes in the human body come in pairs. The pair's members can have codes that match each other or be different. If they are different, one is usually dominant, the other is described as 'recessive'. This means that it is this gene that is expressed in the phenotype (e.g., the gene implicated in Huntington's disease).

Double-helix The physical structure of DNA, comprising two parallel strands of DNA coiled helically.

E

Embryo Early developmental stage of a plant or animal after fertilization.

Entelechy That which makes something, what it is. The essence of a thing, organism.

Endogenous Inside, internal, within.

Enzyme A protein that speeds up the rate of a chemical reaction or helps a reaction that would not take place at the given conditions to take place. It is itself not permanently changed by taking part in the reaction. (Often described as a biological catalyst)

Epigenetic factors Factors having an influence on the form, function or behaviour of an organism but originating outside the **genome**.

Epistasis Interaction between genes.

Eugenics Deliberate manipulation of the genetic make-up of human populations, traditionally by selective birth control, infanticide, mass murder, genocide. Genetic engineering offers new opportunities for the eugenicist including genetic screening, in vitro fertilization plus preimplantation screening, germ line genetic modification, etc. Eugenic ideas can be traced to the ancient Greek philosopher Plato, but are more recently associated with Nazi Germany and Hitler.

Eukaryote The major class of living things including all multicellular, higher organisms and some single-celled organisms that have a **nucleus** in their cells containing the **chromosomes**.

Exogenous Outside, external, without.

Expression (of a gene) The central dogma of molecular biology says that genes (DNA) are transcribed into a message (RNA) that is translated into a polypeptide or protein, which in turn gives structural and functional features to a cell.

F

Foetus Fully developed embryo in womb.

F1 hybrid First generation after a hybrid has been formed.

G

Gamete A reproductive or 'sex' cell: sperm in males, ova in females. It carries inherited characteristics from parent to offspring.

Gel electrophoresis Analytical tool to separate substances, e.g., DNA fragments, from each other by having them move through a gel driven by an electric field.

Gene The so-called biological unit of inheritance; a segment of DNA that provides the genetic information necessary to make one protein or polypeptide.

Gene amplification The process whereby genes or a sequence of DNA in the genome is greatly increased in number of copies.

Gene cloning The technique of making many copies of a gene.

Gene expression The production by a cell of the protein or polypeptide for which the specific gene codes.

Gene pair Corresponding genes in each of a pair of matching (homologous) chromosomes.

Gene pool A term used to describe all the genes at a given locus in the population. The size of the pool is related to the genetic diversity in the population.

Genera Plural of genus, a group of similar species.

Gene silencing The process(es) whereby certain genes in the genome are prevented from being expressed by chemical modifications and other means.

Gene stacking Accumulation of several advantageous traits in an organism, usually considered in the context of herbicide tolerance genes in plants (weeds).

Gene therapy Correcting the genetic make-up of a patient by adding the correct genes in a form, which will be expressed, either to the somatic tissues that are affected (cystic fibrosis) or to the germ line cells (egg or sperm, currently illegal).

Genetic code The code establishing the correspondence between the sequence of bases in nucleic acids (DNA and the complementary RNA) and the sequence of amino acids in proteins.

Genetic determinism Determinism is the doctrine that all acts, choices and events are the inevitable consequence of antecedent sufficient causes. Genetic determinism is the doctrine that the organism is the inevitable consequence of its genetic make-up, or the sum of its genes.

Genetic engineering The manipulating of genetic material in the laboratory. It includes isolating, copying and multiplying genes, recombining genes or DNA from different species and transferring genes from one species to another, bypassing the reproductive process.

Genetic disability, disorder or disease Afflictions that can be associated with what some categorise as defects in a person's genetic inheritance. They may involve changes (mutations) in single genes or in whole chromosomes, parts of which may be lost, duplicated or placed in the wrong position; or from the interaction of multiple genes and external factors in fetal development. Later in life such interactions seem to be the basis of many of the common serious disorders, such as heart disease, diabetes and cancer, although these are not usually thought of as gene disorders.

Genetic fingerprinting A technique, which enables genetic relationships between close relatives, or the identity of individuals to be established—usually beyond reasonable doubt.

Genetic map The body of information on the relative positions of genes on chromosomes. Much of the effort of the Human Genome Project is directed to mapping chromosomes.

Genetic marker A harmless variable inherited change in DNA or protein that can be used to locate a diseased gene on a particular chromosome. Any segment of DNA that can be identified, or whose chromosomal location is known, so that it can be used as a reference point to map or locate other genes. Any gene that has an identifiable phenotype that can be used to track the presence or absence of other genes on the same piece of DNA transferred into a cell.

Gene therapy The genetic modification of body cells of an individual patient, directed at alleviating disease in that patient.

Genetic inheritance The transfer of genetic instructions for different characteristics, such as the colour and size of the cell or organism, to successive generations, via reproduction.

Genetic modification A technique where individual genes can be copied and transferred to another living organism to alter its genetic make-up and thus incorporate or delete specific characteristics into or from the organism. The technology is also referred to as genetic engineering, genetic manipulation and gene technology.

Gene A biological unit of inheritance; a segment of DNA that provides the genetic information necessary to make one protein.

Genome The totality of genes in an organism

Genotype A characteristic of an organism is said to have two aspects: genotype and phenotype. Genotype is the make-up of the gene itself—its code. The precise variant(s) of the gene(s) carried by an individual.

Genus A group of similar species.

Germ line Sperm and egg cells (ova) and their precursors.

Germ line gene therapy Used to describe genetic modification or 'therapy' of the sex cells (sperm, ova and their precursors). It is banned in the UK.

GMO (Genetically Modified Organism) An organism changed by genetic engineering. Often now called 'a transgenic'.

Growth hormone A protein, produced by the pituitary gland that promotes growth of the whole body.

GTAC Gene therapy advisory committee.

H

Herbicide A chemical compound used to kill weeds.

Haemoglobin The red oxygen carrying pigment of the blood. Haemoglobin incorporates two pairs of polypeptides called globins. Various gene mutations that affect the kinds of globins made, or their amounts, can result in diseases called the haemoglobinopathies, notably thalassaemias and sickle cell disease.

Haploid Used to describe a cell having half the usual number of chromosomes, for example, sperm and ova.

Herman the Bull A Dutch bull engineered to contain the human lactoferrin gene that codes for the protein lactoferrin used in the treatment of sepsis. Human lactoferrin is produced in the milk of cows bred from this bull. (Said to be named after Herman de Boer, the managing director of Gene Pharming, Holland)

Heterozygous An organism is described as heterozygous when it has inherited unlike alleles of a specific gene pair from its parents.

Heterozygote An individual who has two different alleles of a gene.

HFEA UK Human Fertilisation and Embryology Authority.

Homozygous An organism is described as homozygous when it has inherited alleles of a specific gene pair that are the same.

Homozygote An individual who has two identical allells of the gene.

Horizontal gene transfer Transfer of genes from one individual to another, of the same or different species, usually by means other than crossbreeding. It is said to have happened naturally in evolution and can now be done artificially.

Hormone A chemical messenger that circulates in the blood, e.g., insulin, sex hormones.

HUGO Human Genome Organisation, responsible for the Human Genome Project.

Huntington's disease A single gene disorder which affects about 1 in 2,000 people and in which, during adult life, there is worsening involuntary movement, accompanied by progressive dementia. It is inherited as a dominant characteristic.

Hybrid The progeny of a cross between parents of different genetic types or different species.

I

Insulin A protein hormone secreted into the blood from the pancreas and which is involved in controlling, amongst other things, the amount of sugar in the blood.

Interrupted genes Genes whose coding sequence is interrupted at intervals by long stretches of non-coding sequences. The coding regions came to be known as *exons* and the non-coding regions as *introns.* This structure is now found to be characteristic of most eukaryotic genes. The number and size of introns vary greatly, and they are often much longer than the coding sequences. After transcription, the intron regions are removed, or *spliced* out form the RNA transcript before it is translated into protein.

Intracytoplasmic Into the cytoplasm, the part of the cell outside the nucleus.

In vitro Used to describe a biological process conducted outside the body of a living organism. Literally 'in the glass', as in a test-tube.

In vivo Describes biological processes happening in a living organism.

IVF In Vitro Fertilization.

J

Jumping genes More correctly called transposons or mobile genetic elements that can move round genomes and, assisted by suitable carriers, transfer between organisms.

L

Liposome Microscopic particle of lipid (fat or oil).

Locus In genetics the place on a chromosome which is occupied by a gene. Plural: loci.

M

Meiosis The form of cell division (splitting) by which gametes (sperm, ova) are formed. It is described as a reduction division, the gametes each receiving only one of each pair of chromosomes, the haploid state.

Messenger RNA RNA intermediate in protein synthesis containing a transcribed copy of the gene sequence that specifies the amino acid sequence of the polypeptide it encodes.

Metabolism The sum total of the chemical processes that take place in living organisms resulting in growth, development and all other forms of energy transformation.

Metabolite One particular chemical intermediate generated in metabolism.

Microorganism A microscopic organism.

Mitochondria Small structures in living cells which are found outside the nucleus and which contain a small number of the cell's genes. Mitochondria are often described as the 'power house of the cell'. They carry their own complement of DNA and are replicated independently so that when the cell divides, each daughter cell will receive half of the mitochondria.

Mitosis Cell division of somatic cells. Each daughter cell receives both chromosomes of each pair, the diploid state.

Mobile Genetic Element Also called **Transposon** or **Transposable Genetic Element**: A sequence of DNA that can move (transpose) from one place to another in the genome of a cell.

Molecule The smallest unit of any chemical substance that can have an independent separate existence and still retain the properties of that substance; it is composed of atoms bonded together.

Molecular biology The study of proteins and nucleic acids, substances that make up the living world, their structures and their relationships to biochemical activity; and the substances that are the repositories of genetic information and the agencies for its communication from one generation to the next.

Monoculture Growing the same crop over a large area year after year.

Multigene families Genes that exist in multiple copies in the genome form several copies to many thousands or hundreds of thousands of copies.

Mutagen A substance or agent that causes genetic mutations, or chemical alteration of the genetic material, DNA.

Mutation A change in the genetic code (DNA) in a particular gene. A gene that has undergone a mutation is called a mutant; so also is an organism in which the mutant gene is expressed.

N

Nucleotide A building block of DNA containing a base.

Nucleus The place in the cell containing the chromosomes. A structure in the eukaryote cell bounded by a membrane that contains the genetic material, in the form of DNA organized into chromosomes.

O

Oncogenes Genes associated with cancer.

Ooplasmic transplantation Adding some of the cytoplasm of an ovum from a donor to the ovum of another woman so that the recipient will receive healthy mitochondria. As the mitochondria contain some 30 or so genes the child resulting from the fertilized treated ovum would strictly have 3 genetic parents. By May 2001 some 30 children were reported to have been born by this IVF variant.

Organism An individual plant, animal or microorganism that can independently carry out all life functions.

Ovulation Release of egg cell (ovum) from the ovary of an animal.

P

Patent Intellectual property protection which gives the owner exclusive right to exploit an invention for a fixed period (e.g., 17–20 years) in exchange for full disclosure of how the invention is made.

Pathogen Any agent that can cause disease.

Phage A virus that infects bacteria.

Phenotype How a gene manifests itself as an observable phenomenon (e.g., a visible characteristic of an organism). The expressed characteristics, or an expressed character of an organism due to its genotype.

Phytoremediation Cleaning up waste or pollution using plants.

Placenta The organ that links a developing fetus to the wall of the womb and the maternal blood supply.

Plasmid A piece of parasitic genetic material found in a cell that can propagate itself using the cell's energetic resources.

Polypeptide An amino acid chain, one or more of which are needed to make up a protein.

Polygenes The (hypothetical) many genes affecting a character, each having a small, additive effect on the character.

Polygenic Controlled by or associated with more than one gene.

Polymer A molecular chain, ring or web formed of units of smaller molecules, often repeating in an ordered fashion.

Polymerase An enzyme that links together the building blocks of a polymer to form the polymer.

Polymerase chain reaction (PCR) A laboratory process in which a specific DNA sequence is amplified many millions of times in only a few hours

Polyploidy Refers to the number of chromosomes in a cell. Most cells are normally diploid, i.e., have two sets of chromosomes. When the number is increased the cell shows polyploidy.

Positional cloning Isolation of a gene through knowledge of its specific location on a particular chromosome.

Primer See DNA primer.

Progeny Offspring.

Prokaryote The class of living things, including all bacteria that do not have a nucleus in their cell.

Promoter The region of a gene at which the RNA polymerase binds to start transcription. Most promoters are located upstream of the gene, except some eukayotic genes which have promoters internal to the gene.

Protein A chemical substance made up of the elements carbon, nitrogen, oxygen and hydrogen which either makes up the structure of living tissues (e.g., muscle) or helps chemical reactions in living tissues to take place (e.g., enzymes). Egg white is the purest example of a protein in everyday life. Made of one or more polypeptides. Enzymes and some hormones are proteins.

Proto-oncogenes Cellular genes which when mutated, or over-expressed become oncogenes.

Provirus A virus that has inserted its genome or a complementary copy of its genome into the host cell genome.

R

Recessive gene or allele A gene that exercises little or no outward effect unless it is present in both of a pair of chromosomes, and therefore has been inherited from both parents, is said to be recessive.

Recombination The formation of new combinations of alleles or new genes that occur when two homologous DNA or chromosomes break and exchange parts.

Recombinant DNA technology The procedure for 'cutting' and 'splicing' DNA to make new combinations of genes.

Reductionism The doctrine that complex systems can be completely understood in terms of its simplest parts. For example, an organism is to be completely understood in terms of its genes, a society in terms of its individuals, and so on.

Refuge A stand of non-genetically modified crop in a GM crop designed to minimize the build up of pesticide resistance genes in the pest population by interbreeding of the two pest populations, i.e., exposed and unexposed to the pesticide concerned, thereby diluting the resistance genes.

Replication The process whereby DNA makes copies of itself when a cell divides. The two strands of the DNA molecule unwind and each strand directs the synthesis of a new strand complementary to itself (and identical to its former partner).

Retinoblastoma Eye cancer.

Retrotransposon A mobile genetic element that depends on a reverse transcription step to move and to duplicate.

Reverse Transcription The reverse of transcription—making a copy of complementaly DNA (eDNA) from an RNA sequence—catalyzed by the enzyme, reverse franscriptase.

Ribose A constituent of RNA.

Ribosome An organelle in the cell required for protein synthesis.

Ribosomal RNA RNA molecules that make up the ribosome.

RNA Ribonucleic acid. Forms of nucleic acid which are concerned with the synthesis of a protein. Messenger RNA is the means whereby the recipe for a polypeptide encoded in DNA that is contained in the nucleus of a cell, is transcribed and carried to the place in the cell where proteins are synthesized, namely the cytoplasm. The transcribed recipe then guides the assembly of the corresponding polypeptide, a process called translation.

RNA editing The process in which the base sequence of the RNA transcript is changed by addition of bases to the RNA molecule or by chemical transformation of one base to another. This subverts the genetic information carried in the genes.

RNA polymerase An enzyme that makes RNA.

S

Sex chromosomes The X and Y chromosomes in human beings that determine the sex of an individual. Females have two X chromosomes in most body cells; males have an X and a Y chromosome.

Sickle cell disease A single gene disorder caused by an altered polypeptide in the globin of haemoglobin. It was the first disorder in which a mutation in a structural gene was shown to produce an altered amino acid sequence. Inheritance is autosomal recessive. In homozygotes there is severe chronic anaemia, episodes of painful infarction (tissue death due to obstruction of the blood supply) especially of lungs, spleen and bones, and proneness to infection. It is connected with a resistance to malaria.

Somatic Used to describe body tissues which do not include the sex cells (sperm & ova and their precursors)

Somaclonal variation Genetic variations of plant cells arising in cell culture, due to enhanced genetic instability.

Species Smallest unit of classification of living organisms commonly used; the group whose members have greatest mutual resemblance; a group of individuals able to breed amongst themselves but—disregarding geographical constraints—generally not with organisms of other groups. A reproductively isolated group whose genes do not usually combine with outsiders.

Species barrier A hypothetical boundary believed to exist between one species and another (**not a scientific concept**!).

Stem cell A cell that throughout life is able to produce all the cells within an organ. A change whether accidental or engineered, in the genetic complement of a stem cell will be passed to its progeny, by mitosis, and may be expressed in them. Appropriate stem cells are therefore an obvious target for somatic cell gene therapy.

Substantial equivalence A non-scientific term to describe novel food crops produced by biotechnology which have passed all compulsory tests for stability, nutritional properties and safety.

Substrate A chemical substance that takes part in a chemical reaction catalyzed by an enzyme.

Superweeds Weeds that are tolerant to one or more herbicides.

Sustainable agriculture Agriculture which minimizes inputs in the form of fuel and chemicals (pesticides, pertilisers) and outputs in the form of air, groundwater and soil pollutants whilst producing soils that retain fertility for future generations (e.g., biodynamic, organic and permaculture systems).

T

Terminator technology A biotechnological 'technology protection system' that renders sterile the seeds which are saved the first generation after the first sowing.

Terminator genes Genes exploited in terminator technology.

Thalassaemias A group of disorders in which there is reduced production of one or more of the globin chains which constitute normal haemoglobin. This imbalance of globin chains results in chronic anaemia, which is a common cause of premature death and much suffering. Inheritance of each of the many genes which may be responsible is recessive.

Tissue A collection of cells specialized to perform a particular function. Tissues are usually composed of several cell types. Organs, e.g., heart, are made up of such tissues.

Toxicity Poisonousness.

Toxin A poisonous substance.

tPA (Tissue-type plasminogen activator) A protein in the blood that has an enzyme-like function in helping the enzyme plasminogen to dissolve blood clots.

Trait A character difference associated with a particular gene.

Transcription The process of making a complementary sequence of the gene sequence in the genome, which is either used directly, as in case of Ribosomal RNAs (rRNAs) and Transfer RNAs (tRNAs), or is further processed into the messenger RNA, which is translated into protein. The enzyme known as DNA-dependent RNA polymerase catalyzes the process.

Transcription factors Proteins in eukaryotes that regulate the transcription of other genes by binding to regulatory sequences of the gene, interaction with one another and with the RNA polymerase.

Transduction The transfer of genes by viruses from one organism to another.

Transfer RNA RNA molecules that transfer specific amino acids to the messenger RNA so that the polypeptide it encodes can be synthesized.

Transformation In genetics, the transfer of genes by one organism taking up DNA belonging to another organism of the same or different species.

Transgenic Describes an organism that has had DNA (genes) foreign to that organism put in it by genetic engineering **(Transgenesis)** in the laboratory. Related words: transgene, transgenesis.

Translation The step in protein synthesis in which the messengerRNA directs the synthesis of a polypeptide of a particular amino-acid sequence by 'decoding' the genetic code.

Translocation A rearrangement of chromosomal material between different chromosomes, not of the same pair.

Trisomy The existence of three chromosomes instead of the normal two, e.g., trisomy 21 involves three copies of human chromosome 21.

Transposon See **Mobile genetic element.**

U

Umbilicus Organ connecting fetus to placenta.

V

Virus A tiny infectious parasitic organism, too small to reproduce outside a host cell. Viruses carry either DNA or RNA, surrounded by a protein coat. Some cause diseases, e.g., chicken pox, influenza. Some viruses, however, suitably modified, can be used in research as a means of delivering a gene into cells. Here it is their ability to infect cells that is exploited. Its genetic material can become integrated into the cell's genome to form provirus.

Virus vector A virus used as a carrier for a gene while making a GMO.

Virulence Ability (of pathogens) to infect organisms and cause disease.

Vector Carrier.

X

X-linked Describes the form of inheritance of diseases for which the gene responsible is carried on the X chromosome.

Xenotransplantation Animal to human organ transplants or xenografts. Their association with gene technology is because the animals will have to be genetically modified so that their organs when removed will 'look' sufficiently like human organs to the recipient's immune system that they will not be rejected.

Z

Zygote A fertilized ovum.

Selected Bibliography

1. Molecular Cell Biology
   Fifth Edition
   Harvey Lodish (Massachusetts Institute of Technology)
   Arnold Berk (U. of California, Los Angeles)
   Paul Matsudaira (Massachusetts Institute of Technology)
   Chris A. Kaiser (Massachusetts Institute of Technology)
   Monty Krieger (Massachusetts Institute of Technology)
   Matthew P. Scott (Stanford U.)
   Lawrence Zipursky (U. of California, Los Angeles)
   James Darnell (Rockefeller U.)
2. Principles of Gene Manipulation: An Introduction to Genetic Engineering (Studies in Microbiology) by R.W. Old, S.B. Primrose
3. Genes VII by Benjamin Lewin
4. Molecular Biology of the Cell, Fourth Edition by *Bruce Alberts*
5. Molecular Biology of the Gene, Fifth Edition by *James D. Watson*
6. Molecular Cell Biology, Fifth Edition by *Matthew P Scott*
7. Gene Cloning and DNA Analysis *by* T. A. Brown
8. Genomes 2 T.A. Brown
9. An Introduction to Environmental Biotechnology, Kluwer Academic Publishers, Milton Wainwright (1999)
10. A textbook of Biotechnology, S. Chand & Company Ltd., R.C. Dubey (1993)
11. USGS (U.S. Geological Survey), URL:http://water.usgs.gov/wid/html/bioremed.html
12. www.geobiotics.com
13. Alexander, M. 1994. Biodegradation and Bioremediation. Academic Press, San Diego.
14. Introduction to Bioinformatics T.K.Attwood & D.J.Parry-Smith
15. Bioinformatics Edited by Andreas D. Baxevanis & B.F. Francis Oullette
16. Textbook of Molecular Biology by Sivarama Sastry, G. Padmanabhan & C. Subramanyam
17. Basic Biotechnology Edited by Colin ratledge and Bjorn Kristiansen
18. Animal Tissue Culture by Freshney
19. Plant Tissue Culture by Razdan.
20. "Lingo,Lingo,Lingo" by Ravi Ramesh Pathak in COMBAT LAW Vol 4 issue 2. Pg.86
21. "Getting in Tune" by Vivek Sharma, H.S. Oberoi & N. Raghuram in COMBAT LAW Vol 4 issue 2. Pg.44

Websites

www.accessexcellence.org
www.mun.ca/. . . /images/
encarta.msn.com
www.emc.maricopa.edu
www.mun.ca
www.biochem.unizh.ch.
www.bio.miami.edu/dana
www.basic.northwestern
smccd.net/accounts
Adapted from www.cambiaip.org
biochemie.web.med.uni-muenchen.de
www.ncbi.nih.gov/index.html

Index